**PROPERTY OF
BAKER COLLEGE OF AUBURN HILLS
LIBRARY**

Punched-Card Systems
and the Early Information Explosion
1880–1945

Studies in Industry and Society
Philip B. Scranton, Series Editor
Published with the assistance of the Hagley Museum and Library

Punched-Card Systems and the Early Information Explosion 1880–1945

LARS HEIDE

The Johns Hopkins University Press

Baltimore

This book has been brought to publication with the generous assistance
of the Robert L. Warren Endowment of the Johns Hopkins University Press.

© 2009 The Johns Hopkins University Press
All rights reserved. Published 2009
Printed in the United States of America on acid-free paper
9 8 7 6 5 4 3 2 1

The Johns Hopkins University Press
2715 North Charles Street
Baltimore, Maryland 21218-4363
www.press.jhu.edu

Library of Congress Cataloging-in-Publication Data

Heide, Lars, 1950–
    Punched-card systems and the early information explosion, 1880–1945 / Lars Heide.
       p. cm.
  ISBN-13: 978-0-8018-9143-4 (hardcover : alk. paper)
  ISBN-10: 0-8018-9143-4 (pbk. : alk. paper)
  1. Punched card systems—United States. 2. Information technology—United States.
  3. Punched card systems—Europe. 4. Information technology—Europe. I. Title.
  HF5548.H387 2009
  004'.9—dc22   2008028731

A catalog record for this book is available from the British Library.

Special discounts are available for bulk purchases of this book. For more information, please contact Special Sales at 410-516-6936 or specialsales@press.jhu.edu.

The Johns Hopkins University Press uses environmentally friendly book materials, including recycled text paper that is composed of at least 30 percent post-consumer waste, whenever possible. All of our book papers are acid-free, and our jackets and covers are printed on paper with recycled content.

Contents

Introduction   1

1 Punched Cards and the 1890 United States Census   15

2 New Users, New Machines   38

3 U.S. Challengers to Hollerith   68

4 The Rise of International Business Machines   105

5 Decline of Punched Cards for European Census Processing   128

6 Punched Cards for General Statistics in Europe   138

7 Different Roads to European Punched-Card Bookkeeping   164

8 Keeping Tabs on Society with Punched Cards   211

Conclusion   252

Acknowledgments   269
Appendix: Financial Information: Tables and Figures   271
Notes   279
Essay on Sources   351
Index   361

Introduction

In the 1930s and 1940s, three large-scale registers of citizens that relied on punched cards were initiated in the United States, France, and Germany, demonstrating that industrial nations—whether democracies, autocratic states, or dictatorships—found use for and began to establish huge administrative systems from the 1930s onward. Punched cards, also known as punch cards, were the first technology to facilitate large, machine-readable registers that improved the abilities of the nation states to locate and control their individual inhabitants, for better and for worse.

In the United States, the Great Depression had caused severe social problems. Twelve million Americans had lost their basis for existence, as their jobs had vanished or their farmsteads had been ruined.[1] "Social justice through social action" was one of Franklin D. Roosevelt's presidential campaign promises in 1932. A major component of Roosevelt's policy of social action was the Social Security Act of 1935. This provided income for the elderly in the form of a pension, a program for unemployment compensation funds, and federally funded relief to the blind and to dependent children. As of 1937, twenty-one million citizens were entitled to an old age pension that was financed through compulsory payments from their employers. The salaries and wages paid were recorded under each employee's name by the Social Security administration in Baltimore, Maryland,[2] so that their pensions could be calculated, an operation that was depicted by a contemporary newspaper as "the world's biggest bookkeeping job."[3] This massive assignment was accomplished by the use of enormous punched-card registers, processed on machines operated by large numbers of government employees.

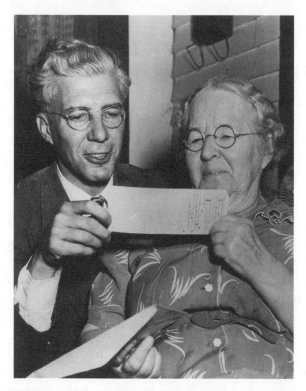

Ida Fuller, of Ludlow, Vermont, receives a Social Security benefit check in 1950 in the shape of a punched card. The message is that the Social Security cares for the individual and the punched card's role is underlined by turning its printed side toward us. The side facing the two individuals is blank. (Social Security Administration, Baltimore, Maryland)

Similarly, France used a military mobilization register made up of punched cards to bolster its government after capitulation to the invading Germans in the summer of 1940. The armistice had divided the country: three-fifths came under direct German military rule, while Philippe Pétain was to govern the remainder from the city of Vichy—the area known as Vichy France. Pétain was an aging hero from the First World War, and his autocratic regime was not content with the army of only a hundred thousand men permitted by the armistice, a figure determined by the Germans to minimize the threat of this potential opponent.

Shortly after the defeat, the Vichy regime quietly started to prepare for a mobilization of an additional two hundred thousand men. They established a register of punched cards listing every male fit for military service. To conceal this project, a national register of all inhabitants in Vichy France was established, ostensibly to create a permanent tool to avoid gathering census information every five or ten years. The mobilization itself was intended to be kept quiet to avoid German detection—no radio announcements, no public notices on the squares to be studied by agitated citizens—as punched cards would allow each man's order to be mailed to his current address. The military mobilization register was destroyed after the German occupation of Vichy France in 1942 rendered the plan impossible. The national register, however, was completed.

In Germany, starting in 1941, several registers using punched cards emerged from the attempts by the Armaments and Munitions Ministry to make the war effort more effective. Two years later, the ministry initiated the development of a German national register, but by the end of the war this had still not been implemented.

The punched cards applied in these administrative systems in the United States, France, and Germany in the early decades of the twentieth century found their origins in more modest ambitions half a century earlier in the 1880s. The first punched-card system had been built by the engineer Herman Hollerith to process the United States population census in 1890, a job requiring only counting, not calculation. He completed this assignment with great success, but the market for counting-based processing of census data was small. To reach a broader market, during the 1890s and 1900s, the original numeric punched-card system was developed for other kinds of statistics requiring addition, in private business and in public organizations. Hollerith incorporated his business as the Tabulating Machine Company, which later became International Business Machines.

In the 1910s, the first bookkeeping systems using punched cards were designed, punched-card machines were launched that printed both the calculations computed by and the information stored on punched cards, and challengers to the Tabulating Machine Company emerged in the United States. During the 1920s, the punched-card machines gained improved calculation capability and incorporated letters as well. Improved calculation was required to compose an invoice with the numbers of the various items acquired and their item prices or to write a pay slip based upon the number of work hours and the hourly rates. Punched cards and their corresponding machinery were improved to meet these growing needs.

Alongside these developments in the United States, punched cards spread to Europe. European companies adapted the original American technology and developed their own information systems and technologies; some of these were based on standards and basic patents from the United States, while others applied different ideas and principles. Though the European companies remained smaller than their American competitors or parent companies, they contributed significantly to the shaping of punched-card technology. The ways in which the punched-card business developed in the various European countries echoed differences in their technological cultures, business structures, and roles of government. The outcome was various nationally shaped machines with emphasis on different aspects of the technology.

In the 1930s, the scope of punched-card applications started to expand from business statistics and bookkeeping to include record management. Until then, punched cards had been a data-processing tool to be discarded once the process was completed. Punched cards became a storage medium. Several insurance companies, public utilities, and other businesses introduced registers of customers and wage earners in their punched-card-based bookkeeping systems. Various national governments adopted and developed this concept to make record-keeping the core of the system.

The ability to mobilize people proved essential in the interwar years. Roosevelt, Hitler and, later, Churchill accomplished this through their radio broadcasts. Building up large punched-card registers provided a supplemental, more tangible way to mobilize people. Punched cards were less charismatic than the nations' leaders, but, for the first time, they offered the nation an efficient technology, allowing direct access to the individual citizen. The major register projects in the United States, France, and Germany exemplified this. Through these and other large registers, punched cards came to contribute significantly to the shaping of modern societies. The press release photograph of an elderly lady receiving her pension check from the Social Security Administration in 1950 illustrated how the state cared for the well-being of each citizen. But these registers could also represent a threat, as with the possibility that the national register in Vichy France could have been exploited to locate Jews for deportation.

Thus punched cards were developed from an ad hoc technology by the end of the nineteenth century into a pivotal technology for managing advanced industrial nations in the 1930s and in the Second World War. In this book, I first analyze the invention of the original punched-card technology to process the 1890 U.S. census. Then I explore the reshaping of the original technology and its manufacturing, which grew into a large industry. This process encompassed both the innovation and production sections of industry, as the users and customers who bought equipment and services contributed suggestions and ideas for inventions and improvements. In Western societies, the shaping and reshaping of punched cards shared the same general characteristics, though the process varied in the United States, Great Britain, Germany, and France.

This complex story aims at enhancing our understanding of technological and business developments in information management in these four countries, encompassing the shaping of the technology, the related dynamics of business, and the interaction of businesses among nations.

Analytic Framework

The punched card was the basis for the most advanced information technology from the 1920s to the Second World War. Punched cards facilitated storing information through combinations of holes in individual cards that various machines processed. Each job required the punched cards to be handled in a predetermined order. For example, in the 1930s, one card was needed for every entry in issuing invoices. The cards were punched on a key punch and the perforations verified by use of a separate device. Afterward, a sorter arranged the cards in a specific order, before their subsequent tabulation. The tabulator was a combined calculating machine and printer that performed the additions—and, in advanced versions, the subtractions—needed to figure the total amount due before printing the invoice.

Punched-card technology distinguished itself from the competing information technologies by facilitating more complex tasks, like producing statistics and printing invoices, with little human interference after the initial setup of the machines. Further, machine processing of punched cards made it feasible to tackle large projects. In contrast, processing the same assignment by competing technologies, that is by hand or by the use of standard office machines, usually meant that the project had to be divided up into several individual tasks, which complicated management and was more labor intensive. From the 1920s, the advantages of punched-card machines increased as they acquired printing capability and gradually improved calculation capacity. However, it was not easy to introduce punched cards for a task. Their use demanded a high degree of standardization and formalization of the tasks to be processed, which, in turn, made greater demands on the user organization than did competing technologies.

Accomplishments and failures in applying technology in many installations ultimately were the basis for punched cards' success, as interaction with users enabled producers to develop and improve the equipment. This user-oriented approach makes the history of punched cards more complicated than would be an exclusively producer-centered approach. Therefore, theoretical and empirical studies are used in this book to select an appropriate set of concepts for its analytic framework.

The sociology of technology has produced essential concepts to describe a technology and its setting but, at the same time, provides only little guidance for analyzing why a particular path of development was chosen—and in many cases pursued for many years. In contrast, empirically based studies in the history of technology have focused on the dynamics of technological development and settings, but these studies write at a modest theoretical level that has curtailed their analytic contributions. A key problem in technological development is determining why one of several technical options is chosen in a development process and why a certain path of development is selected, changed, or terminated. These choices, which could be intentional or incidental, are made in both private companies and public organizations, and this makes understanding the interplay of technology and organizations essential.

Traditionally, sociologists viewed technology as an external factor in their studies of the nature and development of society and social behavior, and they focused on technology's social implications. In the 1980s, several sociologists broadened their studies to encompass the shaping of technology, and two approaches emerged: the *actor-network theory* and the *social construction of technology*. Originally, both approaches strongly emphasized social factors over technical in the shaping of technology, but they have since reduced this emphasis.[4]

The actor-network theory was developed by sociologists Bruno Latour and Michel Callon in the 1980s. They approached technology as a generalized network of the relations between the nodes constituted by essential individuals and technical components. In this way, they provided a basic symmetry to analyze both human and nonhuman components of a technology.[5] Latour and Callon's approach opens up for an extremely transparent analysis. However, it is based on an assumption of perpetual change that fits well with epistemological reflection but is less suited to empirically based studies of the development of a technology like punched cards, which became stabilized for long periods.

The development of punched-card technology was characterized by a combination of stable standards over long periods combined with infrequent reshaping of the technology and ceaseless smaller changes. Social construction of technology theory is related to a perception of technology as being reasonably stable for an extended time and supports distinguishing between minor changes and infrequent basic reshapings. Engineer Wiebe E. Bijker and sociologist Trevor Pinch developed this theory in the 1980s and 1990s.[6] They adopted the term *construction* in their social studies of the development of technology based on results from sociology of science. However, the meaning of "construction" seems somewhat obscure. In contrast to the laws of natural science, all technologies are undisputed constructs. They are made by people for people, which makes "social shaping" a preferable term.[7]

Wiebe E. Bijker has been the major exponent of the social construction of technology approach. Based on his study of the development of bicycles, Bakelite, and fluorescent lighting, Bijker classified the description of technological development into three phases.[8] The first phase is *genesis,* characterized by the interpretative flexibility of the nascent technology. Different "relevant social groups" or interested parties are formed through their diverse and hence flexible interpretations of this technology, which may develop in several directions or along varied paths.

The second phase is *closure and stabilization.* During this phase, the spectrum of possible interpretations narrows, and a "lock-in" takes place on a specific technological interpretation, with alternatives being scrapped. The term *stabilization* implies that the chosen interpretation or embodiment of the technology subsequently remains unchanged for a period. Originally, the third phase was described rather vaguely as the "wider context," while Bijker subsequently used it to explain the closure and stabilization process.

The social construction of technology approach was a tool to describe technological development. This approach provided a method to identify the people and motives shaping a technology but offered no appraisal or critique of the development process.[9] The basic problem was the absence of explicit concepts to handle the various ways that power is exerted—ways that are crucial to understanding the dynamics of the important closure and stabilization processes. Bijker introduced the terms *micropolitical* and *symbolic power* as a means for this understating, but he did not develop a theory of how they affected these processes. So, Bijker's approach needs to be supplemented with concepts and arguments to facilitate the understanding of the dynamics of the shaping process. Inspiration for this can be found in empirical studies of various technologies and business history.

Many scholars in the history of technology tradition have worked to gain new insights into the dynamics of the shaping of technology and the impact of technology on society. Thomas P. Hughes compared the histories of electrification in the United States, Great Britain, and Germany, which created one of the most important technological structures in the modern world.[10] As his unit of analysis, he introduced the concept of *technological system*, which addressed his technological and organizational aspects as a connected whole, composed of interacting components—technical, economic, and social. Hughes used this term in a pragmatic and fruitful way, but he did not offer an explicit definition, which limited its analytic power. Through his analysis, Hughes demonstrated the great advantages of the systems concept, which enabled him to move the research focus from individuals to organizations, while still leaving space for the individual person. "System builders" replaced lone inventors, and the coworkers of the system builders were appraised.

To analyze the dynamics of a technological system, Hughes borrowed the concept of *momentum* from physics, where momentum is determined by mass, velocity, and direction. In technological systems, mass consisted of several components, such as machines, devices, and other physical artifacts requiring considerable capital investment. Momentum also arose from the people and organizations involved. Organizations that shaped and were shaped by the technology could be business concerns, government agencies, professional societies, and educational institutions.

Hughes' momentum concept further implied that social development shaped technology and was shaped by it, yet the interactions of technological systems and society were not symmetrical. Systems in creation were the most malleable. There could be several alternative solutions, both technically and organizationally. Social factors had a high degree of shaping impact but as the technological systems grew larger and more complex, gathering momentum, the systems became less shaped by and more the shapers of their environment. This suggested that it was easier to shape a system before it acquired momentum. However, a system with great technological momentum could be made to change direction or speed when its components were subject to significant forces of change.[11]

Hughes applied the notion of momentum to several growing technological systems. A basic question is what made momentum change. Hughes indicated that the cause was forces outside the system.[12] One of his examples was railroads in the United States, for they lost momentum as the competing automobile system acquired momentum in the early twentieth century.[13] But why did the automobile system gain momentum? Was it due to factors within that system, for example, the user's individual freedom? Or was it due to limitations within the railroad system? Railroads were able to carry from one railway station to another, but passengers and freight had to travel to the station of departure and from that of arrival. Two analogous examples are Atlantic liners and transatlantic flights and punched-card systems and computers. Did the old technology disappear, to be replaced by the new one? This appears to be the case for the punched card and computer example, but it was not the case for the two prior examples. Though not comprehensively, the momentum concept can facilitate complex analysis with human, organizational, and technical components, like the dynamics of Bijker's closure and stabilization, and it can help to explain why a stabilized technology can become subject to reassessment.

Business history provides a different perspective on industrial development. Business historian Alfred D. Chandler Jr. has made a fundamental contribution to understanding the crucial role of organizations in the development of industrial capitalism between the 1850s and the 1930s.

Since the 1950s, he has studied the strategies and organizational forms of big companies in the United States.[14] He showed that scale and speed were winning weapons when managed through efficient organizational hierarchies. Then in *Scale and Scope* (1990), he extended his analysis to encompass the largest industrial enterprises in Great Britain and Germany.[15] In this work, Chandler applied the economics of scope as the additional critical element in the evolution of the largest enterprises (economies of scope referring to the economies of joint production or distribution). Chandler offered an extended, structured comparison of the big industrial firms in Great Britain, Germany, and the United States.

Chandler considered his scholarship to be a contribution to the understanding of the emergence of modern industrial society. From this perspective, two limitations of his work are that he studied marketing, not users or consumers, and that he reduced the essential role of government to antitrust regulation and education. The history of the punched-card industry exemplifies these limitations and shows how productive it can be to go beyond them. The punched-card industry demonstrates the importance of moving beyond an aggregate approach to marketing and markets to one of looking at the users and their problems. This is needed to appreciate the shaping of the technology itself and its industry. Government was a formative, major purveyor that both nourished the emergence of both the first interpretation of punched-card technology and the establishment of the interpretation of the technology for managing large registers.

On a more general level, Chandler's scholarship focuses on documenting the history of business and not on analyzing its dynamics. His dynamic arguments were based on transaction economics, supplemented by the economics of scale and scope.

Economist Ronald H. Coase introduced the term *transaction costs* to understand the price mechanism, dismissing the classic economists' claim of free transactions.[16] Transaction costs became the theoretical stepping stone for Douglass C. North's development of an analytic framework to explain how the performance of economies was affected by institutions and institutional change in Western history since the sixteenth century. North analyzed the role of transaction costs in institutional development. Such expenses included the costs of paying for the exchange of goods, of measuring the valuable attributes of what is exchanged, and of protecting rights and policing and enforcing agreements. He argues that transaction costs were the sources of social, political, and economic institutions. Classical economists extolled the benefits of specialization and the division of labor. They argued that output could be increased without increasing the number of producers, simply by reallocating production to those producers with the lowest opportunity costs. However, such a reallocation raised the number of transactions and thus increased total transaction costs.[17]

Extending the scope of analysis from classical economics to transaction costs provided an important new theoretical insight. Furthermore, for analyses based on empirical studies of companies and public organizations, the extension from simple costs to transactions is important to understand the dynamics of a prolific industry operating across four countries and two continents.

Analytic Strategy

I have selected Wiebe Bijker's construction of technology approach as the basis for this study of the punched-card industry, as it facilitates analysis of reasonably stable standards combined with infrequent reshaping of the technology, observed from the empirical material. However, "shaping" technology is preferred to "constructing" technology, as outlined above.

When using Bijker's approach to the shaping of technology, it is essential to enhance his concepts for understanding how one among several technological alternatives was chosen for the closure and stabilization phase of development. The empirical evidence shows that each alternative appears to have been related to a social group of users or customers. Even in some cases, one user seems to have been the exclusive focus of the producer, for example, the Census Office in the United States in 1890 for Herman Hollerith's development of the first punched-card and the French army's administration of conscription for the Bull company's format for bookkeeping in the mid-1930s. These were "prime users," while the concept of *prime application* is applied in cases in which the selection of alternatives focused on an application at several companies or public institutions, for example, operational statistics for the shaping of punched-card technology at the British Tabulating Machine Company and Deutsche Hollerith Maschinen Gesellschaft in the 1920s.

Further, understanding the dynamics of a technology-based industry requires both the engineering activities of innovation and production—and the business contributions of management, marketing, and sales. Results from studies of business organizations facilitate the analysis of these aspects of the shaping of technology and industry.

The additional dynamic problem in the shaping of technology is explaining why and when an established technology closure was renegotiated, modified, or completely blown open for reshaping. For this end, the technology in a punched-card closure is perceived as technological system, as this was conceptualized by Thomas P. Hughes.

Analyzing the shaping of punched-card systems and Western society using this framework requires access to both corporate and engineering materials on the punched-card producers and information on the development of users. For most of the bigger punched-card producers, board-level material is preserved and provides the needed information on corporate decisions.[18] However, engineers below the executive level made most decisions on the shaping of technology. Only fragments of contemporary material on this kind of decision making have been preserved in any of the companies.

Unfortunately, very little material still exists from the small producers. This problem of the lack of contemporary empirical material has been alleviated by including sources from punched-card users and patents. A patent is informative because it contains a detailed description of the device and its expected applications. Patents proved particularly valuable for challengers to the punched-card trade's major producers, as they showed the facilities for and development of the challengers' equipment and provided essential information about related organizations, particularly in the United States, where the patent office recorded transfer of patent rights. Also, patents offered fruitful access to related information on the processing of the original application, litigation, and the variations of patent laws in different countries, which had considerable implications for the industry.

This book's objectives cover three levels of analysis with diverse dynamics: Individual actions involving the design and manufacture of the technology, company-level decisions and strategies, and conditions on national levels. The shaping of punched-card technology is the focal point. Individuals shaped technology—on the first analytic level—and they worked either alone or in organizations, companies, or the machine shop of the United

States' Bureau of the Census, which from this perspective resembled a company. For the individual inventor or innovator, the analysis focuses his acts, which correspond to the location of the shaping process.[19]

The situation on the second analytic level is more complex for individuals employed in a company, because the technology was also shaped by individuals whose acts were facilitated or curtailed by the company. A company managed established strategies that its inventors and innovators had to pursue, for example, the choice of primary users or applications for development. Moreover, management decided which new equipment was to be produced. Therefore, the ideal research strategy for developing and producing technology in a company would be to study the actions of every relevant individual. However, the preserved material does not, generally speaking, facilitate this research strategy. Consequently, the choice has been made to use company-level information as the main focus of analysis for the shaping of technology. This reduces detail, but the outcome still is a robust analysis of the main features of the development of punched cards, and it facilitates distinguishing the technology's main features.

The third level of analysis—the national level—enables a multi-society perspective and provides insight into aspects of its diversity. This book covers four major industrial societies—the United States, Great Britain, Germany, and France—and four successive punched-card closures. Two concerns contributed to the organization of the analysis. First, the narrative facilitates comparisons between the development in the four countries that would provide important insights into industrial and societal variations. To achieve this goal, the analyses of the various development paths in the four countries are kept separate. Second, the study presents a history of a technology in which the industry in the United States was dominant and the first of its kind, and yet where the shaping and production of punched-card equipment in Great Britain, Germany, and France made significant contributions to the overall development.

Topics Covered in the Book

First I investigate the dynamics of census processing in the United States, including Herman Hollerith's inventions and innovations for the first punched-card system for the 1890 census. I show how the original punched-card system grew out of the organizational shortcomings caused by the absence of a permanent census office in the United States.

Next I analyze America's reshaping of the first punched-card system into a standardized, general statistics-processing system between 1894 and 1907. The application field was extended from the original processing of population data to encompass the compilation of general statistics, notably operational statistics. This opened up an extensive expansion of the trade.

Then I look at how the shaping of punched cards for bookkeeping tasks in the United States from 1907 to about 1933. Once more, this reworking of punched cards facilitated another substantial expansion, and several U.S. competitors emerged to challenge Hollerith's company.

Then the scene shifts the Europe. Already in the 1890s, Hollerith attempted to introduce and spread the use of the punched card for census processing across the Atlantic but had limited success. A discussion of the reasons for the cool reception in Europe compared with its success in the United States is presented followed by various transitions to punched-card-based bookkeeping in Great Britain, Germany, and France from 1920 to 1939. Causes and implications are analyzed for discernible national and technological distinctions. Relations are explored within the national structures and cultures, as these countries became mass societies. All three countries experienced opposition to the American version of the technology and the development of distinct national forms shaped by currency restrictions and growing national sentiments in the 1930s.

Punched-card use was expanded greatly to build large registers of people in the late 1930s and during the Second World War. Such registers' advantages for combating crises in diverse societies justified their design, but their implementation proved difficult and contained hazards.

I conclude my analysis of the invention and reshaping of punched cards with a review of the shaping of punched-card technology and business in the four countries. This complex story illustrates the scope of concepts needed for a comprehensive analysis of the interactive development of business organizations and their technologies. Further, it reflects basic changes in Western industrial society between the late nineteenth century and the Second World War.

ONE

Punched Cards and the 1890 United States Census

The first major application of punched cards—processing information from the 1890 United States census—was hailed on the front page of *Scientific American* and praised in other contemporary science and technology publications as a great advance over earlier, manual methods of processing and as a manifestation of American efficiency and technical ingenuity.[1] This view that the invention of punched cards was a testament to America's being in the forefront of technological inventions was also expressed in later, technical books on census processing and the punched-card industry.[2] However, these later authors were operating with hindsight, as they had experienced the subsequent success of punched cards, in various applications, using substantially improved equipment.

By contrast, social historian Margo J. Anderson has shown that the introduction of punched cards to process census returns grew out of public demand for more and better census data, combined with Congress's hesitation to establish a permanent census office. The latter decision resulted in inadequate and sporadic funding, and a frequent turnover in administrative management.[3] This situation triggered the development of the first punched-card system to process population census returns.

This chapter extends Anderson's insights by examining how returns in the United States were processed in successive Census Offices. These offices provided the institutional setting for the invention and development of the original punched-card system to process the census in 1890.

The United States decennial census was one of the compromises resulting from the Constitution of 1787. The census was to create a basis for apportioning seats among the states in the federal House of Representatives and for sharing the burden of paying direct taxes to the federal government. The

THE NEW CENSUS OF THE UNITED STATES—THE ELECTRICAL ENUMERATING MECHANISM.—[See page 1?

The front page of *Scientific American* on 30 August 1890, depicting the technical success of Herman Hollerith's punched-card processing of the U.S. census. (*Scientific American* 30 August 1890: 127) fathers of the Constitution did not invent the concept of a census, but the United States was the first country to introduce regular population counts.[4]

The Constitution prescribed that all people in the Union be counted every ten years, and the first census took place in 1790. The Constitution required an enumeration of the number of free persons and the number of slaves.[5] From the first census, however, more than the constitutionally minimum information was collected, and the objective of the censuses became to obtain accurate data on the military and industrial strength of the country. After the Civil War (1861–1865), the Constitution's racial distinction was amended so that all people had to be counted on an equal footing.[6] In practice, however, the racial distinction was upheld through elaborate statistical reports in the census publications.

During the nation's first century, the task of processing the censuses grew in scale and particularly in scope. The increasing population yielded a larger scale effort, as Americans grew from 3.9 million people in 1790 to 76 million in 1900 (a factor of 19). This expansion was surpassed by the simultaneous extension of the scope of the endeavor, created by the burgeoning ambitions of politicians, bureaucrats, and statisticians. The outcome was a dramatic increase in the size of the censuses' statistical reports, which grew from fifty-six pages in 1790 to 26,408 pages in 1890 (a factor of 472).[7]

This growth in scale and scope was a product of the absence of a permanent census organization. A new office was established for each census, only to be closed down when it had completed its assignments or ran out of funding. All these factors caused great variations from one census to the next, which can be substantiated through the lack of continuity in the position as head of the successive Census Offices and the substantial variation census publications. During the first twelve censuses, from 1790 to 1900, only twice did same person head two census operations.[8]

In the same period, U.S. census publications varied much more than in comparable countries with permanent census organizations. However, as the United States had no permanent census office, many statisticians were recruited for each census, even though their main affiliation remained elsewhere. Therefore a large part of the statistics community in the United States came to work on the censuses, which promoted a network of statisticians.

In the nineteenth century, no leader of the census operations, who considered himself a statistician, had any academic training in statistics. Even in the 1880s, few American statisticians had formal instruction before they entered the field. In 1887, statistics was only taught at three universities in the United States: Johns Hopkins (in Baltimore), Massachusetts Institute of Technology (in Cambridge, Massachusetts), and Columbia College (now Columbia University in New York). Moreover, no federal or state statistical institution offered systematic instruction.[9] Professional statisticians only joined the census office in the 1930s.[10] Therefore, the collection and processing of information for earlier censuses lacked both the organized transfer of experience from one census to the next and trained personnel.

From the 1840s onward, interest groups tried to establish a permanent census office. The most vigorous argument against a permanent office was that it would require eternal federal appropriations. However, permanent statistical departments existed in many other governmental bodies and in several states in the late nineteenth century, and every nation state in Europe had a permanent organization responsible for census taking. Within the U.S. federal government, the Department of the Treasury compiled foreign trade statistics as early as 1789, and in 1866 a permanent office was established for this purpose. From 1862, the Department of Agriculture compiled and tabulated agricultural statistics. From 1884, the Bureau of Labor in the Department of the Interior kept labor statistics. From 1891, the Federal Bureau of Immigration in the treasury compiled immigration statistics.[11]

All these federal institutions published statistics regularly—either yearly or more frequently. Further, several states of the United States established their own statistics offices, such as Massachusetts (1869), Pennsylvania (1872), Connecticut (1873), and Ohio (1877).[12] In addition, the American Statistical Association had been founded in 1839 as a professional body. Its activities included publications, a library, and seminars.[13] European states had established central statistical offices responsible for census operations much earlier, for example, France (1799), Prussia (1810), England (1837), and Austria-Hungary (1863).[14]

Processing Census Schedules

Two basic changes in producing U.S. census statistics took place in 1850: the individual became the analytic unit and processing returns became centralized at Washington, D.C. From 1790 to 1840, the household had been the key unit and the individual districts had tabulated the results. The 1850 changes can be traced to criticism about inaccuracies in the 1840 census.[15] Introducing the individual as the census unit allowed the generation of more detailed statistics. Centralized processing could provide more uniform and, therefore, more exact outcomes than did processing in the hundreds of enumeration districts. These two changes reflected a growing general interest in detailed and reliable statistics, and the same changes took place in Great Britain, France, and Germany between 1840 and 1900.

In the United States, these changes in the way the census was handled caused the number of units to be processed to rise by a factor of 7.5 compared with the previous count. In 1840, data on 3.1 million families had been tabulated, whereas information on 23 million individuals was processed in 1850.[16] This made processing a critical problem that became conspicuous as a result of the simultaneous centralization of processing in Washington. Several alternatives existed that could solve this problem: reducing the task to the minimal constitutional requirement (a headcount, including slaves), employing additional clerks, or relying on technical aids.

Reducing the scope was never tried; on the contrary, the census continued to grow. From 1850 to 1890, the number of inhabitants rose by almost threefold, from 23 million to 63 million. Yet at the same time, the administrators' ambitions relating to the statistics grew, as expressed in the size of the double-entry tables published. This can be observed from the tables with race and sex as one entry and age as the other. In 1850, this table had sixty entries. In 1890, there were 1,696 entries, a massive increase.[17] (In 1850, slaves were recorded separately.)

A growing number of clerks were employed. However, this did not solve the problem because of the sporadic nature of the census operations caused by the congressional requirement of separate appropriations for census operations each year. By the end of the nineteenth century, T. C. Martin reported claims that census returns were about to reach a number that would prevent their processing by traditional means—before the following census, ten years later.[18] Regardless, any census can be processed manually. It is only a question of obtaining adequate appropriations and creating an appropriate-size organization.

Therefore, the main reason for introducing technical aids was the lack of a permanent census office combined with the growing desire for statistics. In contrast, the British and German statistical offices processed their censuses by hand until 1910. In 1872 and 1890, the census offices introduced mechanical aids for processing the United States census, first the Seaton device and then Hollerith's first punched-card system.

Census statistics were compiled from returned, completed census forms. Until 1840, enumerators recorded one row for each family, with columns for entering the number of persons of specified categories, such as white male from thirty to thirty-nine years of age. The totals for one schedule page could be obtained by adding the columns and writing the result on the form. Clerks transferred the totals to summary or consolidation sheets. Further additions gave totals for each census district, which were then transmitted to the census office to be aggregated and published.[19]

The 1850 census form had one row for each individual in a family. Each row had a number of columns indicating the various items of information wanted. Some questions, such as name or occupation, were answered by a name or a noun. The age called for a figure. The remainder, for example, questions on civil conditions and education, could be answered by making a checkmark in the proper position.

As each census form held information on several persons, there were two methods of tabulation. The first method was to have each table summarize the information in one row on a new form, a tallying list, repeating the process until the desired aggregation was achieved. This was a simple, tested method that already had functioned well in processing many censuses. It had the disadvantage, however, that tabulating each table required a separate handling of all the census forms. And, as the number of table entries grew, either the tallying sheets became unwieldy or compiling the table had to be split into sections; both results required separate handling of the census forms.

The second method was to begin processing by copying the information on each individual to a card or slip and compiling all the tables by sorting and counting the cards.[20] Here, the initial copying of the schedules gave no immediate benefit, but it eased the subsequent compilation of tables very much. The tallying method was applied in the censuses from 1850 to 1880, and the card method was used from 1890.

Although a substantial leap in the number of statistical units took

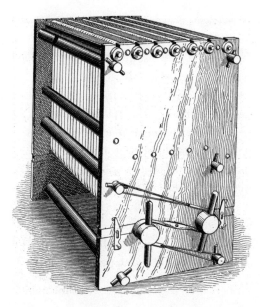

Charles W. Seaton's device for tabulation. This model is from 1880. The drawing is from 1903, when this copy was owed by Herman Hollerith. (W. R. Merriam, "The Evolution of American Census-Taking," *Century Magazine* 65 [1903], 836)

place at the 1850 census, tabulating the returns required no basic change in the method or technology applied. The same held for processing the 1860 census. The first attempt to introduce new technology in processing the information was made in 1870 by Charles W. Seaton, a manager in the 1870 Census Office. His idea was to improve the way a clerk organized the tallying lists on his or her table. For this purpose, Seaton built a relatively simple mechanical device out of wood. It had parallel rollers by which a roll of paper was unwound so that a number of tallying columns were placed side by side.[21] It seems that it enabled tabulation of 160 entries at a time for one or more tables.

The main difference between hand tallying and the Seaton device was that the latter allowed the simultaneous tabulation of up to eight tables or table sections.[22] The 1870 census office assessed the Seaton device as a significant improvement over older methods. On the basis of this appraisal, Congress awarded Seaton a sum of money corresponding to about twenty-nine years of a clerk's salary, which was a substantial congressional reward for solving problems of census processing in the United States.[23]

The Seaton device was, however, a mere extension of existing manual methods and had limited potential. It reduced the size of tallying sheets that a clerk managed, but a major source of error remained, as the operator still had to oversee the same number of entries.

Several of these devices were used during the compilation of the 1880 census.[24] The number of entries in the table having race and sex as one entry and age as the other was comparable to the number a decade earlier, so were the advantages of using the Seaton roller. For the 1890 census, a doubling of entries was planned. This would have been difficult to organize by use of either manual methods or the Seaton device. However, the decision to double the entries for the 1890 census was based on knowledge of the Massachusetts state census' successful application of single handwritten cards in 1885, and of Herman Hollerith's successful testing of his first punched-card system on a grand scale.

Systems for Processing the 1890 Census

The Hollerith punched-card system introduced for processing the 1890 census represented a momentous change. While Seaton's machine had organized the tallying lists, the punched-card system broke the tallying lists into single cards and comprised complex mechanical artifacts. These could not yet be described as machines, as they were exclusively operated by hand and had no moving parts. At the 1890 census, punched cards became the cornerstone of a technological system consisting of the producer (Hollerith), the machine shops building his hardware, the printing shops supplying the cardboard cards, and the Census Office as the only user of the technology.

This new technological system derived from three basic principles: representation of the information of each census unit on a single card or slip, mechanical tabulation, and electric card reading. John S. Billings conceived the first two in 1880 and was an experienced manager and statistician in the 1880 Census Office.[25] He mentioned the idea to Herman Hollerith, also a staff member, who developed this idea during the remainder of the 1880s.

Billings had introduced the representation of a census unit on a card for processing the 26,000 deaths reported in an 1880 investigation.[26] The

1885 Massachusetts state census also used single cards.[27] Further, mechanizing production processes was a principal feature of that age. One historian has drawn attention to the fact that, on several notable occasions, academics stimulated the problem choice of independent inventors.[28] This is another example of that stimulation, although Billings was trained as a medical doctor and had received no formal education in statistics. His ideas were based on his experience with library card files and fatality statistics, not on his formal education. Herman Hollerith adopted the electrical card reading in the innovation process.

The superintendent of the 1890 census chose Hollerith's first punched-card system for processing the census after a competition in September 1889. Three systems sought the contract: Hollerith's and two others presented by the Massachusetts statisticians Charles F. Pidgin and William C. Hunt. The task was to transcribe accurately and tabulate as quickly as possible the data on 10,491 people collected in the preceding census. The major argument for introducing Hollerith's punched-card system was its speed of processing and the consequent savings in labor.[29] Punched cards also offered more possibilities in terms of tallying, but a written card of a similar size could hold more information. For example, a figure of several digits could be used to represent a criterion. Representing multiple digit figures as holes required more space than writing by hand.

The competition and the three systems presented indicate the scope of census-processing improvements within statistical thinking in the United States in the late 1880s. Pidgin and Hunt designed their systems based on their experiences in the Massachusetts state census in 1885, where the returns were processed by use of one card on each inhabitant. Both Pidgin and Hunt used cards sorted by hand, and they only differed in the way the census data were transcribed. Pidgin used specially designed cards, printed in different colors, to facilitate transcribing, sorting, and counting. Hunt transcribed the data onto slips of paper by use of various colored inks.[30] Really, the three systems offered were very similar. All were based on the unit card concept. The difference lay in the way the data were represented on the cards—and in Hollerith's mechanized counting. Furthermore, the systems' originators, Billings, Hollerith, Hunt, and Pidgin, were all members of the statistics community and connected to the federal and the Massachusetts state censuses.

In the competition only speed and accuracy mattered, while the costs

1	2	3	4	CM	UM	Jp	Ch	Oc	In	20	50	80	Dv	Un	3	4	3	4	A	E	L	a	g
5	6	7	8	CL	UL	O	Mu	Qd	Mo	25	55	85	Wd	CY	1	2	1	2	B	F	M	b	h
1	2	3	4	CS	US	Mb	B	M	O	30	60	0	2	Mr	0	15	0	15	C	G	N	c	i
5	6	7	8	No	Hd	Wf	W	F	5	35	65	1	3	Sg	5	10	5	10	D	H	O	d	k
1	2	3	4	Fh	Ff	Fm	7	1	10	40	70	90	4	0	1	3	0	2	St	I	P	e	l
5	6	7	8	Hh	Hf	Hm	8	2	15	45	75	95	100	Un	2	4	1	3	4	K	Un	f	m
1	2	3	4	X	Un	Ft	9	3	i	c	X	R	L	E	A	6	0	US	Ir	Sc	US	Ir	Sc
5	6	7	8	Ot	En	Mt	10	4	k	d	Y	S	M	F	B	10	1	Gr	En	Wa	Gr	En	Wa
1	2	3	4	W	R	OK	11	5	l	e	Z	T	N	G	C	15	2	Sw	FC	EC	Sw	FC	EC
5	6	7	8	7	4	1	12	6	m	f	NG	U	O	H	D	Un	3	Nw	Bo	Hu	Nw	Bo	Hu
1	2	3	4	8	5	2	Oc	O	n	g	a	V	P	I	Al	Na	4	Dk	Fr	It	Dk	Fr	It
5	6	7	8	9	6	3	O	p	o	h	b	W	Q	K	Un	Pa	5	Ru	Ot	Un	Ru	Ot	Un

The layout of the punched card for the United States census in 1890 with the abbreviations in each position. Columns 1–4 were used to indicate the number of the enumerations district. The remaining positions were used for coding the information of an individual. (Leon E. Truesdell, *The Development of Punched Card Tabulation in the Bureau of the Census 1890–1940*, Washington, DC: Government Printing Office, 1965, 47)

of acquiring the method or system chosen did not count. Five hundred eighty thousand dollars in wages was estimated to be saved using punched cards.[31] The costs of using the proposed systems or methods were only calculated for Hollerith's punched-card system, subsequent to the test on speed and accuracy. Hollerith was paid $230,390, or about 40 percent of the estimated the wage savings.[32] The total cost of the 1890 census was $5.8 million, 143 percent more than the 1880 census in real dollars. This was far more than the population growth during the decade of 25.5 percent, which leaves the increasing ambition regarding statistical details as the main reason for the higher costs.[33]

The 1890 punched-card system consisted of three components: punched card, punch, and tabulator.[34] The card, made of thin cardboard measuring 6⅝ by 3¼ inches (16.8 × 8.3 cm), featured twenty-four columns having twelve punching positions each. These positions could record all the information returned in the census on one individual, organized in a compact layout specially designed for the information collected in the 1890 census.

Gathering census information on the population involved completing one form per household, with one row of data for each member, compiling age, sex, marital status, place of birth, occupation, and so on. First, the clerk punched the information on each individual onto a card by use of a pantograph punch (described below), and, at the same time, handwrote a serial number on the card, enabling verification of the punched information against the census list.

A pantograph was a device invented in the seventeenth century for replicating drawings. The pantograph punch used a similar method to replicate items punched on a perforated board, which contained abbreviations noting the positions for entering each item of data, onto cardboard cards. Over the punch board swung an arm, rather like that of a record player, with a needlelike "finger" at the end. In the manner of a pantograph, this finger's movement was followed along at the rear of the machine by a punch device. Hence, when the finger was pressed down into a depression in the plate in front indicating a punching position, the punch at the back created a hole in the cardboard card. The arm the finger was attached to worked as a lever and, thus, lessened the amount of pressure required for each punch operation.

The tabulator, with its electric card reading capabilities, was Herman Hollerith's main contribution to the punched-card machine. It had three main parts: the press, the counters, and the sorting box. The manually operated electric circuit closing press had a hard rubber plate containing 288 holes, or pockets, each corresponding to the full set of the card's punching positions. Each pocket was partly filled with mercury. The card to be read was placed between a series of spring-loaded pins and the rubber plate with the mercury cups. In each position containing a hole, the corresponding pin made contact with the mercury and closed an electric circuit that included the counter. Activated by an electromagnet, the counter moved ahead one unit. It displayed a hundred divisions and had two hands, like a watch. The big hand counted units, the small hand hundreds, allowing counting up to 9,999. Before reaching that figure in a tabulation, the operator read the counters, noted the figure on a sheet, and reset the device.[35] An operator could read up to about forty cards a minute.[36]

The third part of the tabulator was a sorting box, easing manual sorting of cards for a subsequent tabulation. The sorting box was divided into compartments, each closed by a hinged lid. The lid, which would be opened by a spring, was normally kept closed by a pin. The pin could be released

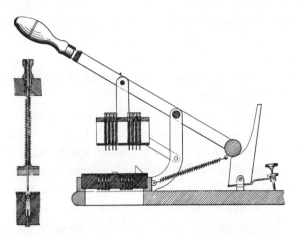

Cross-section of the tabulator's circuit-closing press, which was Hollerith's first card-reading device and was built for processing the U.S. census in 1890. Herman Hollerith, "An Electric Tabulating System," *School of Mines Quarterly* 10 [1889]: 302)

by an electromagnet connected by a wire to the press. For example, in a classification by sex, race, and marital status, it was possible, at the same time, to sort the cards in age groups in preparation for a following count of this distribution. When the operator closed the press, the lid of the compartment containing the relevant age group opened. The operator placed the card in the open bin, closed the lid, and was ready for the next card. If there was no age indication on the card, no lid opened and an error was indicated. In this way, sorting became a part of the counting process. When sorting could not be combined with a counting operation, it was easier to sort by hand than by the sorting box.

The total population figure was tabulated in two separate ways. First, clerks took a gross count based on the census forms that each contained information on a household. A special keyboard facilitated this work; it had keys numbered from one to twenty, as well as twenty-one counters on the tabulator. The operator took a form and struck the key indicating the number of persons in the household. This actuated the counter of that number, and the twenty-first counter recorded the number of households. In this way, households with up to twenty members could be processed mechanically. A few exceeded this and were noted down by hand. Finding the total population number simply involved multiplying the number of households in each group by the number of persons per family. This quick gross population figure was produced to check the subsequent punched-card calculation. Once the accuracy of the subsequent punched-card processing had been established, this special keyboard was discarded.[37]

When the total population figure of nearly 63 million was announced, some people who believed the Union to have at least 75 million inhabitants criticized the punched-card method as being inaccurate.[38] As the calculation had been performed twice, first directly from the census forms and then using punched cards, this figure hardly contained significant inaccuracies. Punching was not verified, but clerks checked small samples to monitor the key punch operators.[39] In statistics processing, punched cards were hardly ever verified simply because of the cost; verification required nearly as many hours of work as the punching operation.

The individual, written identification number provided a safety precaution that enabled each punched card to be checked against the original entries on its census form. The 1900 census also used the individual identification number, but then it was dropped.[40] This indicates that, like the original gross count, the individual identification number had been a safety precaution designed to ease the acceptance of the punched cards. The adjustment of the tabulator for each count—"programming" in today's usage—was done by making wire connections using screws between the relevant contact points on the counters, circuit closing press, and sorting box.[41]

Herman Hollerith

Born in Buffalo, New York, in 1860, Herman Hollerith moved to New York with his family in 1869.[42] At that time, American industrialization was gaining force and speed, which meant opportunities for hundreds of thousands of workers. The industrial cities grew rapidly. Nine years before Hollerith moved to New York, the city had 1.17 million inhabitants. The year after his arrival, it held 1.48 million.[43] The city's noise and its immense construction work proved both noticeable and attractive to a teenage boy. Industrialization called for skills in devising, designing, and managing technologies.[44]

At the age of sixteen, Herman Hollerith began studying mechanical engineering at the School of Mines, in today's Columbia University, founded only in 1864. In the 1870s, the school offered three curricula: mining and mechanical and civil engineering.[45] Hollerith graduated at the age of nineteen, an average student who met the requirements solidly but without class honors. Education at Columbia laid great emphasis on school subjects, and less on workshop experience, a significant shaping factor in Hollerith's career.[46] Later, his inventions would apply electricity as a prominent feature, but he attended no classes in electricity. Electrical engineering was only born with Thomas A. Edison's invention of the incandescent lamp and the installation of the Pearl Street electric power system in New York in 1882. Electrical engineering at Columbia appeared in 1889, ten years after Hollerith's 1879 graduation.[47]

Professor William P. Trowbrige of the Columbia School of Mines, already a special agent at the Census Office, recruited Hollerith for the coming census. Hollerith joined another student and several instructors to perform this short-lived task. The professional census employees he encountered offered this young graduate the possibility to get acquainted with various members of the network who would shortly be dispersed across the country. Hollerith's assignment in the 1880 Census Office was to work on statistics of manufacturing.[48] By late 1881, the Census Office started to downsize because of a funding shortage.[49] In 1882, through the census network, Hollerith took a job as an instructor in mechanical engineering at the Massachusetts Institute of Technology (MIT).[50] Hollerith left the following year to become an assistant patent examiner in the United States Patent Office in Washington, D.C. This shift could have had several reasons behind it: Hollerith's choleric temperament was hardly compatible with teaching; he wanted to live near Washington, D.C., and he earned more in the Patent Office.[51] The shift could also be related to his activity as an inventor, a vocation he had pursued only in his spare time. By the early 1880s, his first statistics machine was approaching the patent phase. In 1884, Hollerith resigned from the Patent Office and six months later filed a patent application on his statistical machine, thereby embarking on his main career as an independent inventor and entrepreneur, which lasted until 1911. This profession lay outside the census network, which Hollerith left temporarily. A few years later Hollerith would return to the network to seek "employment" for his invention.

Thus Hollerith's choice to become an independent inventor and his subsequent accomplishments placed him in the company of the heroes of the U.S. age of invention. In 1876 Alexander Graham Bell invented his telephone and Thomas Alva Edison opened his Menlo Park laboratory. An average of 20,260 patents applications were filed a year in the 1870s. In the 1880s this figure rose to 32,277.[52] At that time, there was no graduate training in engineering but Hollerith's employment in the census of 1880 and at the Patent Office had fulfilled this role in his career. At the census he collected and processed statistics on power and machinery used in manufacture. In the Patent Office, he learned about all varieties of machinery components.

The basic problem in the 1880 census had been the gigantic task of tabulating the population. During the 1880s, Hollerith developed John S. Billing's original idea and several times discussed his suggestions and constructions with him. The census office management was also keen to devise a way to mechanize tabulating population statistics, but Hollerith received no financial support. However, he did learn the duties of a clerk during his working hours through a temporary reassignment, and he actually operated the Seaton device.[53]

Hollerith's original tabulation system used a continuous roll of paper, somewhat like Seaton's, and consisted of a punch and a reader. Two lines of holes across the width of the paper made by a primitive hand punch represented the record for each individual. Holes in successive lines were staggered not to weaken the paper. The blank paper was fed from one roll to another, running under a metal template with a hole in each punch position through which the hand punch made perforations to provide the record. On the reader, the perforated roll of paper ran between a metal cylinder and one pin or pointer for every punching position across the paper. A hole in the paper enabled a pin to touch the metallic cylinder and an electric circuit was closed that, by use of an electromagnet, moved a simple mechanical counter.[54]

Although Billings provided the ideas for punch representation and mechanization, Hollerith developed the actual embodiment of these ideas. Though he had received no college training in electricity, he chose basic electromechanical technology, which was well-publicized through books and periodicals.[55] In the machine's first version a paper strip replaced Billings' single cards. The electric telegraph transmitter, the Seaton device, or the Jacquard loom could all have been the inspiration, as all of these

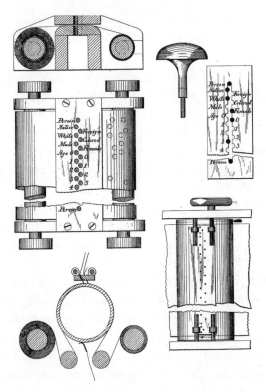

Herman Hollerith's first statistics processing system used punched paper tapes. The top drawing show details of his punch, and the two lower drawings depict his tabulator. (Herman Hollerith, U.S. Patent 395,783, filed and issued 1889. Reference numbers deleted.)

used strips. The French punch-strip-controlled loom was developed during the eighteenth century, and Joseph Jacquard built the definitive design in 1804. This loom was widely used in the United States in the 1880s.[56] Another consideration might have been that it was more difficult to read single cards than a strip, and that Hollerith's strip-reading construction allowed a mechanized feed.

The basic electromechanical technology provided simple construction as well as programming flexibility. It is often simpler to construct a prototype device using electromechanical technology than to use an exclusively mechanical solution. An example is the transmission of information from reading of a hole to the counter. In mechanical technology a physical connection would be established, for instance, by use of levers—as in traditional typewriters. Hollerith just used electric wires to establish the desired connections. If different connections between holes and counters were needed for different statistical compilations, it would have been easier to change wiring than to rearrange levers. The simple technical construction enabled Hollerith to bypass his limited workshop experience.

The price of programming flexibility was that an operator had to check all electrical connections for each program or machine setup. This problem continued to plague the electromechanical punched-card machines for as long as they were used. Initially, Hollerith solved this problem by doing all the programming himself.

Hollerith resigned his job as patent examiner in March 1884 to devote his time to the innovation of his machines for the projected mid-decade state-level censuses in June 1885. A federal project was mounted to establish a national census in 1885 based on censuses in the many states using the 1880 schedules. The states were to be persuaded through receiving a federal grant-in-aid if they deposited a copy of their return. Federal tabulations would be based on this material.[57] In addition, Hollerith aimed at applying his system to finish the tabulation of the 1880 census, which was still under way even if at a low key. In September 1884, Hollerith was ready to file a patent application. His brother-in-law loaned him the money to cover the expenses and to build a version of his census machine to try out on mass data.[58] Soon after receiving the money, he filed to patent his invention.[59] In any case, the plans for a mid-decade census vanished, and the office of the 1880 census was abolished in March 1885, yielding no income for a prospective supplier before 1890.[60]

The dwindling chances of census tabulations in 1885 caused Hollerith problems. Through a relative, he got a job as manager at a Missouri company producing railroad brakes, which he held until 1887. During those years, Hollerith also worked on a small keyboard-driven mechanical adding machine designed by Tolbert Lanston. From 1865 to 1887 Lanston had been a clerk in the Pension Office in Washington, D.C. Meanwhile, Lanston studied law and qualified as an attorney, while retaining his interest in mechanics and making several inventions in various fields.[61]

In the 1880s, adding machines were not common. The slide-based Thomas de Colmar calculators had been batch-produced since 1821, and about 1,500 examples had appeared by 1878.[62] The first keyboard-driven adders emerged in 1850, but no big producer had yet been established.

Burroughs only started production in 1885 and the next year followed Felt and Tarrant, who were known for their comptometers.[63] Most adding machines were related to bookkeeping, but the Lanston machine originated from statistics.

Charles W. Seaton of the 1880 Census Office sponsored Lanston's work on the adding machine in the early 1880s. The objective was to facilitate the compilation of statistics based on more than units, that is, on more than statistics requiring addition. This, and the simultaneous support for Hollerith's early punched-card work, shows the importance attached by the Census Office, especially by Charles Seaton, for further mechanization of statistical production than the Seaton device offered.

In 1882, Hollerith tried a working model of the Lanston adding machine in Charles Seaton's office. Shortly afterward Lanston lost interest in this calculator, as he became greatly occupied with machines for composing type and secured several patents in that field. In 1887, Lanston resigned from his clerical job and set up a company to develop and produce his machines for type composition. He then undertook the task of converting his patented idea into a practical machine for commercial use. For ten years he labored and, finally, in 1897 he launched a highly successful machine, the Lanston Monotype.

Hollerith interest in the Lanston adding machine grew. In 1884, while drawing up the first patent for his own census machine, he bought the rights to it. He went to the Pratt and Whitney Company of Hartford, Connecticut, to explore production options, but it proved too expensive.[64] The Lanston episode demonstrated that Hollerith was already concerned with mechanizing the compilation of statistics requiring addition at the time he filed his first patent in 1884. In the office of the census from 1880 to 1882, he worked on statistics requiring addition. Machines for addition are more complex than those for counting. None of Hollerith's inventions or innovations outside census processing was quickly commercialized. He returned to his census machine and, in 1887, moved back to New York.[65]

Even during his job at the railroad brake company, Hollerith tended his census machine, introducing two major changes. The first was to replace the paper strip with individual cards. This required a new card-reading device but enabled sorting. For this purpose, he added a sorting box. This transformation grew out of discussions with Billings. Major arguments in favor of the changes were the use of single cards in the Massachusetts State census of 1885, and the ability, in a simple way, to compile statistics on small social groups, like the 105,000 people of Chinese descent in the United States in 1880.[66]

The second innovation was the prototype of an adding tabulator.[67] In some statistics only units are counted. This is the case for the greater part of population census tabulations. The remaining statistics require the addition of larger numbers than one. Hollerith's first census system only could count, but he recognized the importance of machines that could add. Perhaps he devised the adding tabulator for the processing of the 1890 agricultural census, but the census first used adding tabulators ten years later.[68]

Hollerith's innovations of the statistics system were not confined to his drawing board and workshop. As early as 1885, he sought returned and completed census forms to use in his developmental experiments. Experience processing actual returns could prove useful in getting his system accepted. Hollerith tried, in vain, to obtain contracts to process the 1885 Massachusetts state and Baltimore city censuses.[69] In 1886, he used his system to tabulate mortality statistics for Baltimore and Jersey City.[70] During these trials, punching the cards proved a weak point. Hollerith perforated the cards with a conductor's punch and strained his arm. Further, the conductor's tool could reach only two rows of punching positions along the edges. Therefore, Hollerith designed the lever-based pantograph punch, which was easy on the arm and enabled the operator to perforate the full card.[71] Its design resembles an early kind of typewriter.[72]

In 1889, the Surgeon General of the Army applied Hollerith's system for statistics on illness and the city of New York used it for mortality statistics.[73] The army discontinued their application within a few years. With 27,000 military personnel on active duty in 1890, a system based on transcribing by hand the information from a report card for each incident to statistics sheets was entirely satisfactory. This system remained in place even as the number of personnel grew to 99,000 by 1914.[74] The Surgeon General then revived punched-card processing of illness statistics, as the army planned to introduce draft service and an army of a million men in the First World War. The commercial contracts with the Surgeon General and with the city of New York contained a provision of long-standing importance to the Hollerith company, and later to IBM, as the tabulators were rented out, not sold. IBM continued this practice until they were compelled by an antitrust case in 1956 to sell tabulators. Two reasons can explain this practice.

The first reason was technical: To Hollerith, leasing the tabulators gave the advantage of assuring their maintenance, which was crucial. There had to be mercury in the card reader's pockets, and all electrical circuits had to be wired correctly and to function properly.[75] Electricity provided flexibility and simple constructions, but it also required checks and maintenance. The first Eastman Kodak standardized camera provides a contemporary example in which the company also took care of a crucial process: loading and taking out the film. Kodak's camera was sold loaded with film for a hundred exposures. Once the owner took these pictures, he returned the camera to the Eastman factory, which developed the film, printed the pictures, and reloaded the camera.[76]

The other reason was financial: Renting the tabulators saved customers the full expense of buying the machines. This was important in the ad hoc Census Office, which only used these machines for a couple of years, as well as for the first application in the Office of the Surgeon General of the Army and in New York. Further, saving money was a major argument in the official report on the three systems or methods proposed to process the 1890 census.[77] Cost considerations were not uncommon when new machines were introduced in the nineteenth century. For example, George Corliss accepted a percentage of the fuel saved as rent for his steam engines, and telephone rental to the customers funded Bell Telephone's growth from the 1870s.[78] While the financial reason originally appears to have been the most important, the technical reason of securing the maintenance of the machines later became fundamental to the famous IBM sales organization, as it involved regular visits to the customers by IBM personnel.

The rent for a tabulator was $1,000 a year, which probably was reasonable, as an urban clerk or wage earner then earned about $550 a year.[79] At the same time, it was good business for Hollerith. Two decades later, when the tabulator was more highly developed and a separate mechanical sorter could be supplied, a tabulator plus a sorter cost about $1,200 to build, and the rent was $600 a year.[80] Further Hollerith's business and private accounts show that he had a substantial income during the processing of the 1890 census. Although the savings argument was important in 1889, Hollerith's total income from his tabulators became a major objection to using his machines once the Census Bureau became permanent in 1902.

In the spring and summer of 1889, Hollerith still had no major customer, and only in December of that year would he gain the contract to process the population census for 1890. Therefore, he worked to find other customers, but his efforts were confined to public statistics on individuals. In June 1889, he demonstrated his first punched-card system machine before a convention of United States Labor Commissioners.[81] Earlier that year, in April, he traveled to Europe and exhibited the machine in Berlin and Paris, which caught attention of census officials in Europe. In contrast, an approach from an insurance actuary in March 1889 led only to a demonstration the following year.

The Problem of U.S. Census Processing and the Shaping of the First Punched-Card System

The United States changed fundamentally during its first century. In 1776, there were thirteen states hugging the Atlantic coast and in 1790 a total population of 3.9 million. At the dawn of the twentieth century, the country stretched from the Atlantic to the Pacific and held 76 million inhabitants. The rising population and greater territory caused the scale of the censuses to grow and, simultaneously, their scope rose as well. The absence of a permanent census institution hindered the transmission of information from one census to the next, which in turn hampered controlling the censuses' growing scope and operations. Every decade, the management had to reinvent how to organize collecting, processing, and publishing the information. Each census operation in the United States had an individual history in the nineteenth century. France, Germany, and Great Britain had established permanent statistics institutions for processing census returns. For example, in England about 20 percent of the staff was re-employed from previous censuses, thus transferring prior knowledge.[82]

The problem of organizing processing the census returns was only slowly recognized in the United States. A historian has used the concept of a *reverse salient* to uncover the process of recognizing a problem. He borrowed the concept from military historians, who delineated those sections of an advancing line or front that had fallen back as reverse salients. Having identified them, the strategists analyzed them as critical problems. The important distinction is between the existence of a reverse salient and the identification of the critical problems.[83]

This distinction is useful for analyzing the history of the censuses in United States. From 1790 to 1840, the reverse salients in population statistics were the statistical unit applied and the decentralized processing. These limited the statistics' degree of detail and accuracy. Both introducing the inhabitant as the statistical unit and centralized processing in 1850 opened the way for more detailed statistics, but these steps made the processing the reverse salient. This problem might only have been identified in 1872, when the Seaton device was built and introduced. The Seaton device helped, but it was not a definitive answer. During the census operation in 1850, Charles W. Seaton looked for better aids, and he later encouraged Herman Hollerith's endeavor to develop John Shaw Billings' idea of mechanizing the tabulation based upon a single card representation of every individual's census information.

In planning for the 1890 census, Hollerith, Hunt, and Pidgin offered different processing methods. The census chose Hollerith exclusively on the consideration of minimizing processing time. The scope of these systems was remarkably narrow. They were all based on the representation of the information about every individual on a single card, which allowed sorting. All three eased the division of labor between the many operators and the few statisticians. The operators had limited training, and they compiled the tables from the census returns. The experienced—later trained—statisticians took on the complex assignments. They designed the census forms and the tables to be published, and they wrote the comments to the tables. The narrow scope of the methods offered can be explained by the fact that Hollerith, Hunt, and Pidgin were all members of the same networks of statisticians. The importance of such networks will fully emerge later.

This does not explain why other alternatives did not come into consideration. Why did the agenda not include a reduction of the scope of the census operation to something more manageable or the establishment of a permanent census institution? The elimination of the "relevant" alternatives is essential to understanding the shaping of this technology as is the actual choice of one of the relevant alternatives. That the United States eliminated alternatives can be explained through arguments based on institutional dynamics. Any reduction in the scope of the censuses was to be avoided, as the censuses had developed into a tool to demonstrate the capabilities of the United States as a great power. On the other side, Congress repeatedly resisted the establishment of organizations requiring permanent government funding. A permanent census institution would have had more focus on costs, as did the U.S. Bureau of the Census when it was established in 1902. The choice of Hollerith's method over those of Hunt and Pidgin was motivated by speed. A decision based on cost would have turned out differently.

TWO

New Users, New Machines

American business enterprises radically changed their scale and scope between the Civil War and the First World War. The small local firms that had previously dominated production were complemented by big corporations. Also, the new industrial society was based on large transportation, communication, and insurance companies that only had existed on a very small scale before 1830. Alfred D. Chandler and other business historians have described the organizational forms essential to operate the new complex companies. A key component of administrative coordination was establishing the managerial tool of upward communication. Upward communication was designed to provide top management with an accurate and current picture of the company's various operations and to ensure accountability along the line.

Various kinds of operational statistics were devised for collecting and analyzing data generated by the day-to-day operations in enterprises. Cost accounting was such a tool that railroads had been using since the 1870s, and similar instruments emerged in diverse industries by the 1880s.[1] Many companies enhanced their processing of business figures through the introduction of tabular report forms, and adding machines started to become available to mechanize processing collected data in the 1890s.[2]

Herman Hollerith's first punched-card system had been built and applied with great fanfare for processing the United States census in 1890. However, applying punched cards to compile operational statistics in businesses demanded a substantial improvement in Hollerith's equipment and an enlargement of his business approach. The first punched-card system had been shaped exclusively to count people for census data, and this embodiment had been stabilized to the degree required for building fifty-six tabulators.³ These tabulators were rented to the Census Office, but remained Hollerith's property and were returned when the census operation was terminated in 1894. To process operational statistics, this equipment needed to be improved to facilitate adding numbers punched on cards.

Hollerith's business had been based on the 1890 Census Office as his sole substantial customer, posing modest organizational demands. Extending the business to encompass many customers' processing different kinds of operational statistics would require establishing a more elaborately organized company.

The history of the punched-card application in census operations until 1905 and the emergence of business statistics as an application field illuminate how stabilized the first punched-card system actually was and provide insight into the dynamic of opening a stabilized technology to reshaping.

Hollerith's Innovations in the 1890s

The contract to supply punched cards for the 1890 census kept Hollerith busy for a while. However, he did not feel that his first punched-card system had reached a stable position. In 1892, he revived innovations aiming at significant changes to the system along three different paths: introducing pneumatics as a new basis for card reading, developing a keyboard-based adding machine, and constructing an adding tabulator. All three paths probably derived from experiences and discussions in the Census Office. Only in 1894 did Hollerith start to look outside the census network. Although he never implemented his ideas of a pneumatics-based punched-card system or the improvements to the Lanston adding machine, the adding tabulator became the cornerstone of his punched-card system.

Hollerith filed patents for a system using pneumatic technology in 1892. The major difference between this system and his first punched-card system for the 1890 census was the use of pneumatics instead of electricity for card reading.⁴ Hollerith's pneumatic innovations were probably triggered by operational problems with the mercury cups in the pin-box card reader. Mercury could accidentally be spilled from the cups, breaking the reader.⁵

Hollerith pursued two paths toward processing addition-based statistics from the mid-1880s on: a keyboard-based adding machine and an adding tabulator. The former was an alternative technology to punched cards, though Hollerith was not the only one to consider this solution. In the late 1880s, efficient adding machines started to emerge on the market. Hollerith's development of a keyboard adding machine was based on the Lanston adding machine that Hollerith had acquired in 1882. At that time, however, it had proved too expensive to produce. Now, he revived his interest and made additional innovations with Lanston. In 1897, the outcome was a strong and serviceable construct, which could be built cheaply and in viable quantities.[6] However, Hollerith was prevented from marketing the product by a problem with the patent. When the problem was finally resolved in 1899, Hollerith no longer had any need to pursue this path, as he had a well-functioning adding tabulator and customers interested in buying it.[7]

Building an adding tabulator was the alternative path to processing addition-based statistics. In the mid-1880s, Hollerith invented his first adding unit for a tabulator.[8] Basically, this was an electric version of the step wheel from Gottfried Wilhelm Leibnitz's mechanical calculating machine from the 1690s, of which an improved version was batch produced as the Thomas calculator in the nineteenth century. This machine was well-known in the United States by the 1880s.

In the early 1890s, Hollerith modified his adding tabulator from a spring-driven prototype to a more dependable version. This project was inspired by discussions at the 1890 census of the problems of tabulating the agricultural census. His development work took place in the Census Office, using agricultural census material, but punched cards were not applied to compile agricultural statistics in the 1890 census.[9]

As early as the mid-1880s, Hollerith was involved in processing both counting- and addition-based census statistics. His choice of the counting-based population census as his first application field provided a substantial job with limited technological complexity. It is far more complicated to build an adding unit than it is making a counter, as he discovered during his subsequent innovations between 1892 and 1894. His adding unit only became reliable in 1896 through trials on business statistics. However, his focus on the development of addition capability rather than process improvements indicates that he did not perceive speed as a *reverse salient* to his system using Thomas P. Hughes's concept. He only recognized this problem during the processing of the United States census in 1900.

Hollerith's business efforts were focused on governmental statistics from the outset. He found time to travel abroad to receive honors and cultivate potential customers, which indicates that he was never planning to exclusively be a supplier to the federal censuses in the United States. He also looked up state census jobs in the United States. He tried, in vain, to get the order to process the 1895 state censuses of New York and Massachusetts,[10] but these jobs were small compared with a federal census. From 1894 to 1895 he traveled four times to Europe. He won a big contract for processing the Russian census in 1897 and contracts for censuses in Canada, France, and Norway. However, he did not gain any further census contracts in Europe, which was disappointing because of the large number of national statistical organizations that would have been natural contacts for him.

Hollerith had attained inside knowledge of the United States census statistics community, and the European statistics communities appeared similar. They received him well and awarded honors in recognition of his work. He was elected member of the Royal Statistical Society in London in 1894, and in the following year his speech was well received at the Institut International de Statistique, the international statistical association, in Bern, Switzerland.[11] It was one thing to address learned audiences, but it was quite another to bring home orders. After all, he was the lone salesman for his system, and foreign governmental orders took time to materialize. He had not yet recognized the drawbacks to the efficiency of his first punched-card system, and its rejection by most European census offices could be attributed either to an opposition to progress or to the lack of funds. However, it was more serious that in 1900 France and Austria discontinued their use of punched cards after both countries had processed a census.

Hollerith struggled with the problems of supplying the nonpermanent census office in the United States and of getting other census offices in the United States and abroad to apply his first punched-card system after 1890. Therefore, he would probably have run into severe problems if he had decided to rely on census processing as the sole application of his first punched-card system. So, he chose to extend his business to other customers as well.

Early Challengers

Life insurance companies were the first businesses to emerge as commercial punched-card users. American life insurance companies had extended their strategy to address a wider public in the 1840s and the new policies, which were later called "ordinary insurance," were written for relatively large amounts and premiums were generally paid yearly. As the number of insurance companies rose quickly, so did their number of policies and their assets. In the 1870s, the American insurance companies introduced "industrial insurance," consisting of small policies on the lives of industrial workers and their families, with very small premiums collected weekly. Compared with the ordinary policies industrial insurance involved more transactions for smaller amounts, leading to bigger total transaction costs. Further, the extension of life insurance to encompass different social groups caused problems, as the insurance companies lacked a detailed tabulation of life expectancy as a function of various conditions, such as occupation and present age, that could be used to calculate premiums. An intercompany mortality study was suggested in 1889 at a meeting of the Institute of Actuaries of America. A collaborative study would be based on a much larger sample than could data in any single company.[12]

By that time, the life insurance companies had accumulated vast amounts of data from managing policies over several decades using index cards to process statistics. The information on each policy was copied to a written card, the cards were sorted by hand into piles according to the categories to be tabulated, and each pile was counted manually. This procedure had the advantage that each step could easily be checked. Further, card-based processing enabled cheap unskilled workers to do most of the processing, whereas earlier methods based on individual processing of every policy had relied heavily on actuaries and actuary assistants.[13]

In March 1889, the actuary of the Metropolitan Life Insurance Company in New York City approached Hollerith.[14] The Metropolitan was among the five biggest insurance companies in the United States, but Hollerith declined, as he was busy working to gain the contract to process the census in 1890, which he got nine months later. Further, during the summer of 1889, he found time to travel to Berlin and Paris to exhibit his machines. Eventually, thirteen months later, in April 1890, Hollerith demonstrated his first punched-card system to twenty-five members of the Institute of Actuaries of America.[15]

The Prudential Insurance Company was another of the top five insurance companies in the United States. Hollerith's demonstration in April 1890 prompted them to rent two tabulators, but no additional insurance orders appeared for several years. Prudential used punched cards to produce statistics to monitor their huge number of policies. First, the cards were used to compile statistics to monitor acquisitions, that is, the kind of policy and the age of the policy holder. Later, the cards were also used to produce statistics on which kinds of policies generated claims.[16]

There is no indication of a significant change in the tasks in the early 1890s that could have provided Prudential with a reason to introduce punched cards for processing their actuarial statistics. No new governmental regulations appeared, and most other life insurance companies kept using their existing methods to compile statistics for another decade. It might well have been the Prudential actuary, John K. Gore, who initiated the introduction of punched cards to the company, as he was active in the subsequent development of punched-card processing at Prudential.[17]

Gore made a mechanical sorter the core of the punched-card system he built subsequently, which indicates that he found Hollerith's simple sorting box inadequate. As the actuarial department was a permanent organization, they had no problem managing their statistics production, which distinguished them from the nonpermanent Census Office. Further, their data processing used written cards, and the actuarial departments carried out substantial sorting to reduce the labor of computation. Hollerith's punched-card system from 1890 included a manually operated sorter box coupled to the tabulator, but the sorter box was slow and could not be mechanized, so Hollerith had to start from scratch to build a mechanical sorter. Consequently, Gore and his brother-in-law started to build a punched-card system suited for life insurance statistics. They possessed the necessary mechanical skills to build their own punched-card sorter, which became the core of their system. Their system used a smaller card, only 57 percent the size of the card in Hollerith's first punched-card system. Further, Gore's card had only ninety punching positions, compared with the 288 positions on the Hollerith card.[18] In addition, the Gore system had a punch but not a tabulator.[19] Prudential's contract with Hollerith was probably terminated by 1895 as Gore started to build his system.

CLASS	AGE	DURATION
00 0	00 0	00 0 ●
10 1	10 1	● 1 D
20 2	20 2	20 ●
30 3	● 3	30 3
● 4	40 4	4
50 5	50 ●	● 5
60 ●	60 6	6
70 7	70 7	7
80 8	80 8	8
90 9	90 9	9

THE ACTUARIAL SOCIETY OF AMERICA.

The Gore punched card for the American Prudential Assurance Company. (David Parks Fackler, "Regarding the Mortality Investigation, Instituted by the Actuarial Society of America and Now in Progress," *Journal of the Institute of Actuaries* 37 [1902], 11)

Gore's sorter introduced mechanical sorting and was the first punched-card machine to be run by an electric motor. It was completed in 1895 and was used for several decades at Prudential.[20] Gore's sorter was based on a circular configuration, quite different from Hollerith's subsequent sorter design, and it enabled sorting by ten different punching positions scattered over the card—most of the later sorters could only sort one column at a time. Gore's sorter had a speed of only sixty-five cards per minute, but that was about twice the speed of an operator using Hollerith's sorting box.[21] The mechanical process had great advantages when a great number of cards were to be sorted.[22]

It is noteworthy that the Actuary Society of America chose the Gore system at Prudential for its cooperative mortality investigation of about 2.3 million life insurance policies issued between 1870 and 1900. In fact, the amount of data to be investigated was reduced so that it could be stored on the small Gore punched card.[23] The actuaries accepted the construction of Gore's punched-card system with a sorter but without a tabulator.

In contrast to Gore's approach to punched-card processing, Hollerith built a prototype adding tabulator, which he tested by using the returns of the agricultural 1890 census. While working on this in 1894, Hollerith was approached by J. Shirley Eaton. Eaton was a railroad accounting manager at the New York Central and Hudson River Railroad and suggested using punched cards for railroad audit and statistics. Hollerith liked the proposal and developed an application that he presented in late 1894.[24]

Timing was one reason that Hollerith preferred railroad audit and statistics to actuarial statistics. Hollerith's census contract would expire later in 1894, making him susceptible to a suggestion from a potential customer—especially as no new insurance order had emerged since the two tabulators were rented by Prudential in 1890. Further, railroad accounting required the adding tabulator that he was experimenting with, and not a mechanized sorter, which he would have had to build from scratch.

In the big railroad companies of the 1890s, freight accounting and statistics had become huge tasks for the accounting departments. A major problem was to audit the freight shipments to ensure that each package reached its destination and that the shipments were paid for. This had previously been accomplished by comparing the goods shipment reports from the forwarded and the received stations. The waybills formed the basis for this task, but they were awkward to handle and sort. After the waybills had been audited, the shipment reports were first used to compile the distribution of records of the freight incomes to the railroad's own various track sections and to foreign companies. Subsequently, the reports were used to compile statistics on the frequency and length of the freight trains on all the railroad's track sections.

Hollerith's presentation to New York Central and Hudson River Railroad persuaded the company, in 1895, to improve their waybill audit by using a punched-card system supplied by Hollerith. The Pennsylvania Railroad, by contrast, chose to audit their waybills using typewriters. This decision revealed the main task at hand, namely, to compile lists of the reports from the forwarded and from the received stations for comparison. Punched cards had the advantage that they could be rearranged by sorting and they could "automatically" be added on a tabulator. In practice, at this early stage of development, the advantages of using punched cards were small compared with manual methods, but they proved an appropriate tool for railroad freight auditing.

Railroads did not rush to adopt punched cards. Only in 1903—nine years after the initial test—did the Long Island Railroad Company become the second railroad to use punched cards for this purpose. This railroad was a part of the Pennsylvania Railroad Company, which followed in

1904.²⁵ By then Hollerith had improved his punched-card machines substantially. But at this point the adding tabulator worked well and a sorting machine was available.

The system that Hollerith developed for New York Central in 1894 was based on an improved version of his original spring-driven adding unit from 1887, now operated by use of an electric motor. For this application, he developed a new punched-card layout to accommodate the waybill information. The information was punched in columns with the numerals from zero at the top to nine at the bottom, which was an ingenious and enduring design.²⁶

Auditing railroad freight accounting began by punching the information in the waybill copies from the received stations on cards. Then, the cards were totaled using the adding tabulator and the result was checked against the reports from the dispatching stations. When that had been done, the cards were sorted according to forwarding station, checked against the reports from the stations, and added on the tabulator. This concluded the basic processing and the cards were then available for compiling various statistics on goods traffic.²⁷ This audit procedure ensured that the cards were accurately punched and, consequently, no separate verification was needed, in contrast to later bookkeeping applications.

Hollerith's system for railroad auditing from 1894 had two weak technical elements: the sorting box and the adding tabulator. As the cards were sorted without a simultaneous tabulation, it was easier to sort by hand. Thus, the company soon abandoned the sorting box. The real stumbling block proved to be the adding tabulator. After working for nearly a year to get the contract, Hollerith's system was tested at the railroad in 1895. However, after a few months, the railroad turned it down because the tabulator's adding unit was not reliable.²⁸

This left Hollerith in a critical position. The census contract had expired the previous year and he had relied on getting the New York Central and Hudson River Railroad as a customer. Unfortunately, he had lost this vital customer and was without income. Fortunately, within a few months, he managed to build a new adding unit for his tabulator that the railroad accepted for a second trial. The new adding tabulator proved to be a success, and the New York Central Railroad became a customer in 1897.²⁹ At first it appears impressive that Hollerith was able, in just a few months, to build a radically redesigned adding unit. But he was not without experience in this field. He had gained knowledge from his work on the Lanston adding machine, and the new adding unit was basically an electrical version of a mechanical adding unit.

Hollerith's new adding tabulator had four adding units, which resembled cash registers. In lieu of the keys, nine electromagnets were activated through the electric reading of numbers on a card, and the results were displayed through small glass windows. An important new feature was the introduction of plugboard programming, which facilitated the tabulator adjustments for a new application.[30] Until then, Hollerith had programmed his tabulators by screwing wires to contact points. Plugboards were used by operators to route telephone calls before automatic dialing systems came into use and worked by establishing electrical connections. Austrian Otto Schäffler had introduced the first plugboard to program a punched-card system in 1890.[31]

The features of the column-based punched card, addition, and plugboard programming became the basis for most subsequent punched-card machines in the company, which eventually became IBM. At the same time, the punched-card system for the New York Central Railroad highlighted the punching mechanism as a weakness in the apparatus. The pantograph punch was slow to operate, and its punching positions were imprecise and caused false readings. To alleviate this problem, Hollerith devised a new punch that operated like a one-handed typewriter and was radically different from his old pantograph punch. Using the new punch, the card was placed on a carrier and moved, column by column, under the punches as the keys were operated. Eleven keys were provided, ten for punching the numerals zero to nine and an additional key for skipping over and leaving a column blank. The new punch was more precise than the pantograph punch was, thus enabling a narrower column width, but it was harder to operate as there was no amplification of the punch operator's pressure. Simultaneously, Hollerith introduced a longer card than the card from the population census in 1890, $7\sfrac{3}{4}$ inches (19.7 cm) instead of $6\sfrac{5}{8}$ inches (16.8 cm).[32] The new card held thirty-six columns, instead of twenty-four, which later increased to forty-five columns in 1907 and to eighty columns in 1928. The last two formats became industry standards.

The new key punch was patented by Hollerith, but he developed it in cooperation with Eugene Amzi Ford from one of Hollerith's machine

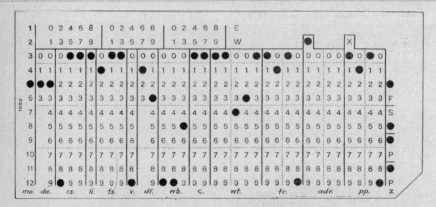

Fig. 2.—New York Central Traffic Card. (Three-fourths the actual size.)

NOTE.—The letters at the bottom do not appear on the card, but were added by the engraver. They mean, in their order, month, day, receiving station, line, forwarding station, via, date forwarded, way bill number, commodity, weight, freight, advanced charges, prepaid.

Hollerith's first column-based punched card designed for New York Central and Hudson River Railway. However, the transformation to column layout is not complete, as shown by the twelve-digit column to the left and the two top nonstandard columns. ("Recording Waybill Statistics by Machinery," *The Railroad Gazette* 34 [1903], 527)

producers, the Taft-Peirce Manufacturing Company of Woonsocket, Rhode Island. Ford had gained his experience from building typewriters in the 1880s. Now, he used his expertise with keyboard devices to innovate Hollerith's key punch.[33]

The outcome of Hollerith's collaboration with the New York Central Railroad in the 1890s was a punched-card system comprising a column punched card, a keyboard punch, and an adding tabulator. The main differences compared with Gore's punched-card system were that Gore had a sorter, while Hollerith had an adding tabulator. If they wanted to expand their businesses beyond their current applications and market segments, Hollerith needed a sorter and Gore lacked a tabulator. Hollerith did work on a sorter but considered it less important than an adding tabulator.[34] Gore never found a second customer, while Hollerith only got his second railroad customer six years after his first. The punched-card trade still moved very slowly, which made the contract to process a Russian census important for Hollerith.

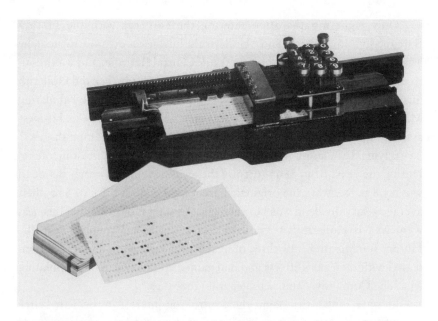

The Hollerith key punch from 1901 with perforated 45-column cards. This punch remained the standard for decades and shows the shift to a column perception of the punched card from the field-oriented perception expressed in the pantograph punch. (IBM Corporate Archives, Somers, New York)

In December 1896, Herman Hollerith was awarded the contract to supply punched-card machines to process the first census of all of Russia the following year, shortly after this his contact started with the New York Central and Hudson River Railroad. However, in the two years that followed the expiry of the contract to process the United States 1890 census, Hollerith had had no substantial income and he needed money to live on and to pay his basic business expenses. To raise money and secure his family, he considered selling his business to Western Electric Company, one of his suppliers.[35] In March 1896, he got a contract for two tabulators to process returns from the French census of 1895. He went to the Library Bureau in Boston and arranged that they would market his punched-card system abroad for ten years.[36] The Library Bureau had been founded in 1876 as an offshoot of the American Library Association and had built up a good business in supplying library equipment and supplies. By the mid-1890s it was a general office supplier and had established offices in London (1894) and Paris (1896).[37]

Only eight months after entering the contract with the Library Bureau, in November 1896, Hollerith's business fortunes turned; he attained the contract with the New York Central Railroad and was about to sign the contract to process the Russian census. Only then did Hollerith incorporate his business, and the new company was named the Tabulating Machine Company (or TMC). He assigned his patents, sold his business assets to the company, and got a controlling 50.2 percent of the shares.[38]

Buying Tabulating Machine Company shares proved most lucrative. During the years from 1897 to 1905 the average nominal yearly dividend was 13.8 percent. As the original shareholders paid a price of $70, their average yearly dividend was 19.7 percent. Another indication that business went well is Hollerith's termination, in 1899, of his contract with the Library Bureau after only three of its ten years' duration. He was discontented with their sales efforts and had gained confidence in the Tabulating Machine Company's ability to generate revenues.[39]

In 1901 Hollerith bought the controlling interest in the Taft-Peirce Manufacturing Company of Woonsocket, Rhode Island, that produced some of his machines. The Tabulating Machine Company was earning well, and the motive for the purchase may have been to implement a vertical integration of the development, innovations, production, marketing, and maintenance of his punched-card equipment. However, Taft-Peirce had financial problems and over the next few years it proved a liability to the Tabulating Machine Company, which invested a great deal of money in the company. In 1903 Taft-Peirce went into receivership, but within two years its acting receiver managed not only to pay off the firm's debts but also to repurchase the shares held by the Tabulating Machine Company. Taft-Peirce thus continued to manufacture Hollerith's punched-card machines.[40]

The 1900 Census and Hollerith's Break with the Census Bureau

The system to process the returns of the 1900 population census was chosen by a contest, as had happened ten years earlier. This time, Hollerith and Charles F. Pidgin were the contenders.

During the 1890s, Charles F. Pidgin, as chief statistician in Boston, was in charge of the Massachusetts state census in 1895 and numerous statistics assignments that demanded addition. Through this experience, he developed several systems for processing statistics. He had no formal training in mechanics, but his patents demonstrate electromechanical and mechanical knowledge and the ability to design devices of limited technical complexity.

For counting statistics, he had developed a single-card system, proposed for the federal census in 1890, using transcription by markings and card colors. During the 1890s, he added a sorting box, which enabled the tabulation of a table with up to 144 entries.[41] Pidgin also invented a new tabulation system using a large keyboard with up to 540 keys, one key for each entry in the table under compilation and a counter for each key to count the number of individual entries made in that entry.[42] Compared with punched cards, the large keyboard had the disadvantage that it was necessary to use the census forms to compile every table. Furthermore, it was not possible to verify the entered information—although this was of minor importance for census processing. Except for processing the United States census in 1890, the punched cards were rarely verified in statistics applications, as 100 percent accuracy was not required.

Pidgin also developed two systems for tabulating statistics that required addition. The first was a sorting arrangement to generalize his card system. The other was a new adding machine with slide entry, which the National Cash Register Company considered sufficiently important to acquire as late as in 1909. This gives one indication that Pidgin's role should not be neglected. Another indication is Hollerith's nervousness about Pidgin's inventions. To satisfy his curiosity, Hollerith even hired a private detective from Pinkerton's detective agency in New York to go up to Boston to carry out a little industrial espionage. But Hollerith did not learn much information from this mission.[43]

Both Pidgin and Hollerith were members of the United States census network. They shared a preoccupation with mechanizing the tabulation process. Both were full of ideas, but only Hollerith proved to have the technical and entrepreneurial skills needed to design, produce, and maintain systems for mechanized processing of statistics on millions of units.

At the contest for processing the 1900 census, Hollerith entered the same system that had secured him the contract for the 1890 census, and once again it was the only mechanized system that was presented. Pidgin offered two systems: an improved version of the card system offered ten years earlier and his large keyboard system. Not surprisingly, Hollerith's system was the fastest, and he was awarded the order.[44] Another mechanized system would have been required to beat Hollerith. The 1900 census returned a population of 76.2 million, a rise of 21 percent in ten years, and the tabulations were similar to those a decade earlier. The total cost of tabulating the population statistics rose by 6 percent, a fall in cost per capita of 13 percent.[45]

This contract was a great business opportunity for Hollerith, as thirty-five of the tabulators had already earned money from both the United States census in 1890 and the Russian census in 1897, and so maintenance was his only expense. A contest on processing speed alone offered no incentive to introduce the new adding tabulator, built from 1894 to 1896 for the New York Central Railroad. The main bottleneck in population census processing was the speed at which the information could be processed; addition offered no advantage. Both the counting tabulator from 1890 and the new adding tabulator were fed by hand, but the counting tabulator from 1890 held more counters and this meant faster processing.

As well as the systems proposed, two alternatives had emerged during the 1890s: John K. Gore's punched-card system in the Prudential Insurance Company and keyboard-based adding machines. Gore's card was limited in capacity, and he would have needed to build a system that used a bigger card to get the order for the 1900 census, including a tabulator to count the large number of cards. However, no evidence exists to suggest that Gore ever tried to get customers outside the Prudential Insurance Company.[46] The other was a system of keyboard-based adding machines. Since 1890, both Burroughs and Felt and Tarrant had marketed these machines, but adding machines as such offered few advantages when they were applied to process the population census, which only required counting.[47]

The competition in 1899 to be awarded a contract to process the upcoming United States census was based exclusively on processing population statistics. However, during the following year, punched-card processing of the agricultural census became an issue, probably inspired by the successful adding-based processing of information at New York Central Railroad. After a thorough investigation in the Census Office of the technological options available, the Hollerith system was chosen. This time the investigation included the Burroughs and Felt and Tarrant adding machines. The Hollerith system still had no mechanical sorter, but this problem was not raised by the Census Office.[48] It was easier to hand sort punched cards than paper forms, as would be the case if adding machines were chosen.

The contract for the agricultural census required Hollerith to supply two punched-card systems: one for farm statistics and another for crop statistics. The farm statistics were compiled from forms covering all information on each farm, using a farm card. The farm card system was the same as that used at the New York Central Railroad, 7¾ by 3¼ inches (19.7 × 8.3 cm) with thirty-six columns, which allowed more information to be stored on a card than on the twenty-four columns of the first punched-card system from 1890. Crop statistics covered all produce or kinds of animals on each farm, and one card was used for each crop or animal on every farm, giving an average of twenty cards per farm.

The card for processing crop statistics was smaller than the farm card, as fewer data were recorded. It was 5⅝ by 3¼ inches (14.3 × 8.3 cm) and had only twenty columns, that is, it was shorter than the farm card but had the same width. This enabled the new key punch developed for the New York Central Railroad to be used in both systems, but the tabulators differed. The tabulators for crop statistics had fewer sensing units and fewer adding units.[49] The two agricultural punched-card systems show that Hollerith still custom-built punched-card systems. Now he had three systems in use in the Census Office, all to be returned when the census work was finished. However, the innovations for the agricultural applications were limited to the adaptations necessary for the small crop card.

Sorting created a bottleneck, especially for crop statistics that were stored on 116 million cards. Therefore, Hollerith resumed his earlier sorter considerations and developed a horizontal sorter that introduced a mechanical feed in the Hollerith punched-card machines. Gore's sorter had had this facility since 1895, and it was the only way to get around the monotonous work of manual sorting.[50] Hollerith used this opportunity to get rid of the mercury cups in the sensing unit, which had probably caused problems in processing the 1890 census. In the new sensing unit, the cards were stopped momentarily for sensing and then ejected into the sorter chutes. The speed was about two hundred cards a minute, several times faster than manual sorting or than using the Gore sorter.[51] Twenty sorters were manufactured and sold to the Census Office on their order. This indicates that Hollerith did not believe that there would be a general demand for the sorter from punched-card users, but shortly after New York Central Railroad rented several sorters.

The sorter equipped Hollerith with a mechanical feed that could be a way to achieve a higher processing speed. It was obvious that a mechanical feed could relieve the operators from the monotonous card-reading chore. Hollerith soon adapted his mechanical feed from the sorter for the counting tabulator. The average speed was about 210 cards a minute, about five times the speed of a manual feed. As the cost for a tabulator with the mechanical feed was only 50 percent higher than for one with manual feed, the cost performance was improved by a factor of 3.3. In 1901, the Census Office rented several tabulators with a mechanical feed.[52]

Inventing a mechanical feed and constructing a sorter machine were crucial contributions of the 1900 census to the shaping of Hollerith's punched-card systems. By introducing these features, punched cards gained a significant advantage over the more manual-based systems for processing statistics compiled from the census. While Hollerith's first punched-card system from 1890 had enjoyed only limited success in Europe, the new punched-card systems introduced from the 1900s were highly successful. The reasons for this were the technical improvements and that Tabulating Machine Company established agencies in Europe. The mechanical feed and machine sorting became basic parts of Hollerith's standardized statistics-processing system, which he introduced in 1907.

By 1902, the ad hoc office of the census in 1900 had become the permanent Census Bureau by Congressional legislation. This changed its relationship to governmental bodies and to Hollerith as a supplier, which, eventually, led to the termination of Hollerith's census contract and a lawsuit brought by Hollerith against the Census Bureau from 1910 to 1912 for patent infringement.[53]

The simplest way to understand the conflict between the Census Bureau and Hollerith is as a personal feud between two individuals. This was the approach adopted in the court case, as well as in John H. Blodgett's analysis from 1968.[54] Geoffrey D. Austrian extended his analysis to include the influence of government in industry as the context, specifically the problem of how public activities were limited by patent law, but this is not substantiated in contemporary material.[55] However, an analysis of the organizational position of the Census Bureau provides a context essential to the understanding of the conflict between the Bureau and the Tabulating Machine Company. It also complements the understanding gained so far of the origin of the first punched-card system in the organizational limbo created by the absence of a permanent census organization prior to 1902.

The Census Bureau was created as a large and potentially powerful federal agency under the Commerce Department but had few responsibilities beyond providing data for apportionment. A quarter of a century of political controversies over its role and location followed. The Census Bureau tried, in vain, to become the central federal statistics bureau. Little departmental control had been exerted on the successive temporary census offices.[56]

The establishment of a permanent organization offered Hollerith improved business opportunities. But a permanent institution had to look more carefully at costs. While an ad hoc institution only had to raise money a few times to get equipment, a permanent institution had to do it every year. President Theodore Roosevelt's administration (1901–1909) was keen on statistics as a solid basis for social legislation, but he was also at the forefront of the struggle against trusts and monopolies. From the late 1880s, Hollerith had supplied punched-card systems and had also held a monopoly, as Gore and Pidgin never succeeded in becoming challengers to his business.

Statistician Simon Newton Dexter North became director of the Census Bureau in 1903. His position in relation to Hollerith was weak, as only the Tabulating Machines Company was able to supply reliable punched-card equipment, and by then a return to manual methods was inconceivable. The approaching expiry of Hollerith's early patents from 1906 would bring them into the public domain. This made a possible Census Bureau production of punched-card machines a bargaining position in relation to Hollerith. In addition, a machine production could strengthen the Census Bureau's position in the controversy over its role in the civil administration and the degree of control by the Commerce Department to which it should be subject. In the negotiations between Hollerith and North, a private company was up against a governmental body and two uncompromising men faced each other. Both stood by their rights, and both lost.

Until 1904 the Census Bureau continued to rent tabulators on Hollerith's conditions but at the yearly renewal of the contract that year, North demanded a lower rent. Simultaneously he approached the Secretary of Commerce and asked for funds for the Census Bureau to build punched-card machines and the reason given was to save money. The Tabulating Machine Company accepted a price reduction for the fiscal year 1904–1905. Hollerith was against this concession, but he was overruled by his board of directors.[57] In early 1905, North obtained an appropriation for the costs of experimental work to develop tabulating machinery. At the same time, he tried to extend his contract with the Tabulating Machine Company for further use of their tabulating machines for the fiscal year 1905–1906. Hollerith made the new contract conditional on the government's abstention from experimental work on tabulating machines. North could not accept this, as he was confident that the Census Bureau would succeed in building punched-card machines. For his part, Hollerith did not want to supply his machines to serve as models for Census Bureau machine constructions, and his machines were withdrawn.[58] North saw no conflict between building machines and renting machines from the Tabulating Machine Company. He would, of course, not circumvent any valid patents, and he believed that he would have no reason to do this, as Hollerith's early punched-card patents would expire shortly.

The break left the Census Bureau with only its purchased punches and sorters. The Bureau was not ready to return to manual methods, and the development of new machines in its machine shop would take time. Therefore, they contracted with Charles F. Pidgin for his manual card system with a sorting box. The Pidgin system was used to compile immigration statistics, but it caused great problems, and the contract was terminated within six months.[59]

The tacit cooperation between the federal government and Hollerith from 1880 to 1905 on census processing illuminates their options and limitations. The federal government had a basic and critical problem with census processing as long as the census offices remained temporary. An alliance centered on a network of leading personalities within the Census Office and the engineer Herman Hollerith was established. Within this network, which was based on personal relationships rather than contractual agreements, Hollerith invented and developed his punched-card system.

The basis of this tacit cooperation disappeared as Congress solved the fundamental census management problem by establishing the Census Bureau on 1902. However, the rift between the Bureau and the Tabulating

Machine Company only surfaced when the Bureau, the following year, got a new director, Simon N. D. North, who did not consider himself a member of this network. The Census Bureau became concerned with the cost of processing equipment. Further, a production of punched-card machines in the Census Bureau itself became a tool in the organization's struggle to gain a key role in the federal administration. The main reasons for the break with Hollerith were a change in the census institution's economics and the strife within the federal government, but the actual breakdown of the relationship was brought about by two unyielding personalities.

When the relationship broke down in 1905, Hollerith lost his most important customer. Even so, that year seemed to be exceptionally good for the Tabulating Machine Company shareholders, as they received dividends of 60 percent of the nominal share value. Although this was 7.5 times higher than the year before, no dividend was paid for the following three years.[60] So, the high dividend in 1905 might have been a deliberate action to boost confidence and to cover up the company's crisis.

For Hollerith, the main problem was the loss of his primary user since 1890. He was known in America and abroad for his census machines but no longer held a census contract. Even though he had already found new customers, this was a personal blow. More ominous problems, perhaps, were that he had been overruled by his board of directors and he was having difficulties running the Taft-Peirce Company. Was he losing the grip of his company? He had been its sole manager and was able to run it as long at his main activities were limited to a small number of customers and based on machines and punched cards produced outside his company. But his business had grown, and he had internalized machine production by acquiring Taft-Peirce. Hollerith seemed to be approaching the limits to his one-person management style. The drifting census contract exposed this problem for the first time, and expansions during the following years would provide further confirmation of its existence.

Business Market Breakthrough and Standardization

At a time when the market for keyboard office machines was becoming substantial, the loss of the census contract in 1905 compelled Hollerith to take the plunge into the private business market. Since 1901 he had been developing a more offensive business strategy with a wider range of customers. Previously he had used most of his efforts to respond to challenges defined by others, in the census offices and at New York and Hudson River Railroad. In 1900, he had been approached by two big insurance companies, but he turned them down as he then was busy on the contract to process the returns in the census in 1900.[61] He focused on building a mechanical feed and a sorting machine for that assignment. He got the order to process the census and soon his machine building projects moved to the workshops of Western Electric Company, his main machine producer. Through innovation, Western Electric was able to convert Hollerith's feed and sorter design into reliable machines.

Hollerith started to expand the scope of his applications to attract additional business customers in 1901. He used a lot of energy to attract private industry's interest in his punched cards, and the preserved correspondence paints a picture of Hollerith as his company's sole salesman, an impression substantiated by other sources. First, he responded to the insurance approach he had received the year before by circulating brochures to several life insurance companies.[62] In addition, he started to develop punched-card applications for three business applications: wage administration, sales analysis, and cost accounting. He found customers for punched-card-based cost accounting and sales analysis, and he used this customer base to attract additional business customers.[63]

When Hollerith's pain from the blow of losing the census contract in 1905 eased, the private business market proved attractive, fast growing, and able to sustain the Tabulating Machine Company. However, while his business strategy so far had been based primarily on the large-scale task of census processing, he would now have to rely on a large number of smaller customers. This larger number of customers compelled Hollerith to standardize his various ad hoc punched-card systems. The outcome was a second set of punched-card machines that embodied his second punched card closure and formed the basis for the Tabulating Machine Company's product range for two decades.

In 1905, Hollerith built, marketed, and maintained three different punched-card systems: The first punched-card system for the census office from 1890 with a 24-column punched card, the punched-card system originally built for the New York Central and Hudson River Railroad in 1894

Hollerith standard 45-column punched card from 1907. Hollerith based his second punched-card closure on this card, which stayed a standard within the Tabulating Machine Company and its successors until 1928. When the Powers and the Bull companies' punched-card machines were developed, they too used this card. (Robert Feindler, *Das Hollerith-Lochkarten-Verfahren für maschinelle Buchhaltung und Statistik*. Berlin: Verlag von Reimar Hobbing, 1929, 24)

with a 36-column punched card, and the system built for the crop statistics of the 1900 agricultural census with a 26-column punched card. In addition, he had planned a never-implemented wage system from 1901, which used a small card with twelve columns. The new applications between 1903 and 1906 were based on machines in production by 1902, several of which reused tabulators returned from processing the agricultural census in 1900.

Simultaneously, Hollerith developed a new series of punched-card machines, produced from 1907, which was his second punched-card closure with two new basic features: the standard 45-column punched-card and brush reading. The first closure was his punched-card system for the 1890 census, with a counting tabulator and entirely manual operation. Later Hollerith introduced several new features—a column-based card, a keyboard punch, an adding tabulator, a mechanized card reader, plugboard programming, and a sorter—into his subsequent ad hoc punched-card systems. By 1907, he united these features using a 45-column punched card and a new mechanism to read cards. The 45-column card was the same size as the card for the New York Central and Hudson River Railroad, but it had more columns.[64] The additional columns were squeezed

NEW USERS, NEW MACHINES 59

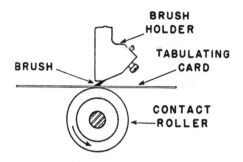

Hollerith introduced dynamic brush reading of cards in 1907. It remained the only card-reading method used in electromechanical punched-card machines as long as they were produced. (Wallace John Eckert, *Punched Card Methods in Scientific Computation*. New York: Columbia University, 1940, 4)

into the card by introducing a smaller tolerance, which was made possible by the introduction of the new dynamic brush reading.[65] Now, the mechanized reading took place while the card was moving.

In the old pin box, the reading took place while the card was halted and functioned as a stop to the collapsible reading pins, whereas in the new system the card acted as an electric insulator. Now the card passed between steel brushes and a row of brass rollers, one set for each column on the card. A hole in the card allowed an electric current to pass between the brush and the brass roller, thus closing the circuit. All parts of the machine were synchronized with the movement of the card and registered the digit value of a hole according to its row on the card. Dynamic card reading was more complicated to implement, but it enabled faster card handling, because the card no longer had to stop during the reading process.

The new tabulator was very different from the wood casing of its predecessors. Now black metal panels were used, mounted on steel frames, which lent the new tabulator a more functional appearance. New, more compact adding units were introduced, and they were incorporated into the base of the machine. Plugboard programming, originally invented by the Austrian Otto Schäffler in 1890, now became a standard feature. The new tabulator held up to five adding machines and ran at a speed of one hundred fifty cards per minute.[66]

The new sorter was vertical to save space in the office. The cards were fed from the top of the 5-feet-high (1.50-m-high) unit into thirteen chutes, one for each of the twelve positions in a column of the card and a reject chute for a blank column. The machine's speed was improved to process two hundred fifty cards per minute with the introduction of the new continuous card reading rather than the previous intermittent action. A drawback of the vertical design was the limited chute depth. A chute held up to two hundred cards and had to be emptied when it became full; otherwise the cards became severely crumpled.

Hollerith introduced his standard punched-card system from 1907 without fanfare, and there is no indication that it was the result of a grand plan. It was the outcome of the dynamic innovation of cards, machines, and production, based on the increase in users and steadily growing scope of primarily statistics applications. However, though the outcome proved to be sound from a technical and business perspective, the development, production, and sales of the new standardized punched-card system strained Hollerith's personal development approach, made him modify his pricing principles, caused production problems, and necessitated an increase in share capital.

Hollerith was basically a drawing board engineer with limited machine shop experience. However, Eugene A. Ford of the Taft-Peirce Company, an experienced shop engineer, played a crucial role in the development of his new series of machines. He was needed to make new machines that were simple to produce and easy to maintain. While Hollerith's first punched-card system from 1890 had no turning parts, his new sorter and tabulator operated by electromotors, which made design and production more demanding. Ford had played a critical role in designing Hollerith's keyboard punch. Hollerith acknowledged this by engaging Ford as an inventor in 1905, which did not, however, imply that Hollerith established a research and development department. In the years around 1900, the first American industrial research laboratories were established at the American Bell Telephone Company and the General Electric Company.[67] Hollerith did not follow their lead because the Tabulating Machine Company was small. Ford became a single inventor, as Hollerith was, and Ford's primary workplace was in Uxbridge, Massachusetts, far from the Tabulating Machine Company in Washington, D.C., and from the firm's machine production.[68]

The advent of Hollerith's new standardized punched-card system caused three sets of problems: pricing, supply, and financing. Originally, Hollerith's machine rental was based on the number of rented machines. For his new private business customers after 1901, he had changed this practice to price by the number of cards supplied, plus a fee per card for the machines. All contracts specified that cards were to be purchased solely from Hollerith. This was important for his business, as 48 percent of the company's total revenue from 1909 to 1913 originated from card sales.[69]

Hollerith's second problem with the production of his new standardized punched-card machines was establishing a production of reliable machines able to keep up with the orders in 1907 and 1908—a problem that was aggravated by his reluctance to produce more of the machines for his earlier punched-card formats. The outcome was an order backlog, which he spent two years filling. Some companies, like Union Pacific Railroad, waited for two years to receive their machines, but no one is recorded as having abandon punched cards because of the delay.[70] However, once the production of his new machines was on track, the success of Hollerith's new standardized set of punched-card machines was evident from the Tabulating Machine Company's soaring total revenues, which rose by 55 percent in 1909 and by 51 percent in 1910.[71]

This expansion caused financial constraints in the company due to the policy of renting out of their machines, as the investment was returned more slowly than in selling the machines. Most of the expansion was based on money earned by the company, and this remained the general strategy throughout the punched-card era, only deviated from a few times. However, the substantial expansion of the Tabulating Machine Company's business was a major reason that no dividend was paid to the shareholders from 1906 through 1908. The share capital was increased by $100,000 in 1908, which was the only external money injected into the company to accomplish this transition and subsequent expansion of business.[72] Dividend payment resumed in 1909.

Punched Cards and World War

The First World War provided growing revenues to the Tabulating Machine Company. The vast resources of modern industrial society enabled the operation of huge armies employing vast armament and munitions resources. Millions of soldiers were drafted or recruited, equipped, fed, housed, trained, equipped, transported, and clothed. Armaments and munitions were produced on a so-far-unseen scale and scope. This made the First World War a test of the capability of modern industrial societies at the expense of dead and wounded soldiers and of impoverished populations. The United States and Great Britain prevailed, so did Germany in spite of the defeat; France got by; Russia succumbed.

The task of mobilizing men and society for warfare was approached according to the experience of the various nations in government regulation and in organizing business. To a large extent, this task was accomplished by blowing up to a national scale known ways and tools of organizing big production and distribution. A key tool was operational statistics processed by using punched cards.

The First World War started in August 1914, but the United States only entered in April 1917. Though the United States only waged war for nineteen months, by the end of hostilities in November 1918, it had managed to send one million soldiers to France in addition to providing a vast amount of weaponry. This was accomplished through a controlled expansion of the American army and by an introduction of a government command economy. The bureaucratic means applied reveal the methods and ambitions in the United States at that time.

The United States armed forces were based on voluntary service and some 180,000 individuals were serving in 1916. Then conscription, which had not been known since the Civil War, was introduced. Registration of young men commenced, and about 24 million men were registered. In 1917 there were 640,000 people under arms, 2.9 million in 1918.[73]

Scaling up the army when the United States entered the war stressed the manual system for processing medical and casualty statistics. Therefore, punched cards were reintroduced to compile medical and casualty statistics. Originally, Hollerith's first punched-card system had been tested for this task from 1889 to 1890, when the army had 27,000 people on active duty. At that time the army had preferred its well-established manual method. Then the basis for this choice had changed in two ways: Punched-card technology had been substantially improved in the intervening quarter of a century, and the number of people on active duty had grown substantially. Managing a vast army dispersed over several continents required quick and reliable statistics. A special medical record section was organized in the Surgeon General's office in October 1917, and this introduced punched cards to process the vital statistics from the United States' entrance into the war.[74] The cards held a printed "Man Number" identifying the soldier. This enabled checking the data on the card against that in

Card for processing illness statistics in the U.S. Army during the First World War, using a 45-column system. The limited capacity of a card made it necessary to split the data onto two cards: one for medical cases and another for surgical cases. This is an early example of problems caused by limited data capability of the 45-column card. Note: In the illustration the numbers are typewritten. In practice, they were penciled. (Albert G. Love, "Medical and Casualty Statistics," in *The Medical Department of the United States Army in the World War.* Washington, DC: Government Printing Office, 1925, vol. 14, 1221)

the personnel file, but the card held no perforations to facilitate mechanized location of the individual, for example, through his "Man Number." As in the census in 1890, the cards were designed exclusively for statistics processing and were not used to locate individuals.

The national economy was mobilized through the introduction of the command economy.[75] To control the economy, the government in the summer of 1916—before entering the war—carried out a census of the production capability of about 80,000 industrial establishments. This census was managed by a special administrative body, the Industrial Preparedness Committee of the Naval Consulting Board, and was processed using punched cards.[76] Notably, it was not carried out by the Census Bureau. Later several special administrative bodies pushed to control the economy, such as the Food Administration, the Fuel Administration, the Railroads Administration, and the War Industries Board.

The War Industries Board controlled the production and distribution of virtually all goods and services. Hundreds of subordinate departments, boards, and committees collected and processed data for the War Industries Board. To sustain such extensive control of the material economy, the government applied Hollerith punched-card equipment.[77] The Railroad Administration controlled the railways that still were operated by their prewar managements. A basic tool for this control was extensive standardized operational statistics. Punched-card processing within the various railways eased this task, which further encouraged the diffusion of punched cards.[78]

The First World War in the United States became a showcase for the application of punched cards for various operational statistics. Punched-card processing was quick and exact, but neither the army nor the government applications could be distinguished from the statistics processed at the Census Bureau and the operational statistics produced by private companies. Over the next two decades, improved punched-card systems would mirror changing bookkeeping practices. Also, they would provide a way to reach a large group of individuals, as applied in the Social Security pension program. During the Second World War punched cards would be used in the United States, France, Germany, and Great Britain to improve their warfare capabilities.

Reshaping Punched Cards

The first closure of punched-card technology was challenged, ameliorated, and the second closure was shaped between 1890 and 1907. Hollerith's limited commitment to the first closure is remarkable. Though he had built and owned more than fifty tabulators that embodied this version, he proceeded until about 1905 to develop machines that significantly extended the capabilities of punched cards. However, subsequent history shows that Hollerith saw the punched-card system for the census in 1890 as one system, the system for New York Central Railroad as another, the system to process farm statistics in the census in 1900 as a third, and the system to process crop statistics in the census in 1900 as a fourth. Until around 1905, his perception of technology was closely related to its application. This explains his stepwise improvement strategy in this period. He built an adding tabulator, designed the column-based punched card, built an improved and mechanized card reading tool, and constructed a machine sorter.

Each of these improvements was based on endeavors to attract and improve concrete punched-card applications, all of them statistics tasks, except the auditing at the New York Central Railroad. The limited scope of each improvement project and its intimate relation to a concrete application explains how business opportunities of limited scale came to contribute to the form the technology took and how business and technology were linked. This was a business strategy based on direct and close contact with his relatively few customers, with the United States census operation as his prime user.

This strategy was not challenged by Hollerith's success at the New York Central Railroad and other early private business assignments. Private businesses did not have sufficient volume to change his business strategy fundamentally until his census contract was terminated in 1905 and private business market increased considerably. The outcome was the standardized punched-card system for statistics processing, the technology's second closure. The machines had improved processing performance, durability, and maintenance, but most of their capabilities for the users had been shaped through applications since 1890. Hollerith devised only a few new features for which it is not possible to document the application that shaped them. The most conspicuous feature was the extension of the 36-column punched-card to forty-five columns, which offered the possibility for the processing of 25 percent larger records. However, the main accomplishment of the standardized punched-card system for statistics processing from 1907 was that it allowed simplification of the Tabulating Machine Company's business.

Although the introduction of the standardized punched-card system for statistics processing simplified the Tabulating Machine Company's production and maintenance, the expansion of business that this involved came to challenge Hollerith's personal management approach to technological development and sales. Even though the changing networks were a crucial dynamic factor in the history of punched-card systems up to 1907, Hollerith's technological system remained by and large unchanged. Hollerith ran the Tabulating Machine Company himself. He was the inventor and managed production and sales. From 1901 to 1905, Hollerith owned the Taft-Peirce Company, which produced his machines, but it was kept as a separate entity, and he eventually sold the company. The only change he made was to improve his innovation capability by engaging Eugene Ford in 1905. But Ford worked far away in his own workshop up in Massachusetts. Further development of the increasingly complex punched-card machines would require a development strategy involving a group of people in one location.

At the same time, basing the company on many business customers required a different approach from that needed for a few big customers. Hollerith preferred to have the contact with the customers himself. This was possible when he had few customers, but a large number of customers could not be reached by one person. Hollerith abhorred advertising and salesmen. He preferred advertising through word of mouth, as his machines were of an excellent quality. He supported this process by calling on customers and sending them letters, acting as his own salesman. This marketing strategy resulted in several customers. The foreign census orders from 1890 to 1896 were because of the satisfaction in the United States Census Offices, promoted by Hollerith's circulation of reprints of enthusiastic articles. The business market provides similar examples. The description in *Railway Gazette* in 1902 of the application at New York Central and Hudson River Railroad inspired the applications at Marshall Field and Union Pacific Railroad.[79] Enthusiastic punched-card installation managers wrote to colleagues and friends recommending the system. Soon the tried-and-trusted applications—railroad audit and statistics, shop operation statistics, and sales analysis—were taken up by new customers. Another example was the companies supplying electricity to customers; the use of punched-card statistics spread from the initial application at the New York Edison Company, in 1903, to several corporations in leading cities throughout the country, like Boston, Brooklyn, Chicago, Minneapolis, and Philadelphia.[80]

At the same time, old customers bought additional installations for other departments.[81] By mid-1907, several big companies were customers: Marshall Field, Eastman Kodak, National Tube, American Sheet and Tin Plate Company, Pennsylvania Steel, Western Electric, and Yale and Towne. In addition, Hollerith negotiated with Simmons Hardware, Heinz Pickle, Regal Shoe, Carnegie Steel, and several railways.[82]

The increase in the Tabulating Machine Company's business required several salesmen, but Hollerith did not want to delegate. This was a fundamental problem for the company, which grew more urgent as the business continued to expand.

THREE

U.S. Challengers to Hollerith

American offices in the turn of the twentieth century, as they do now, processed the information needed to produce, distribute, sell, and purchase products and services in the country's private enterprises and public organizations. They recorded sales and purchases in financial accounting, administered wages and salaries, and compiled information on the internal transactions and production in their company. The massive expansion of private enterprises between the Civil War and the 1930s was reflected in the growth in scale and scope of tasks that were carried out in more and bigger offices. The distinction between the growth of scale and of scope facilitates an analysis of the dynamics of offices and office machine production beyond the initial phase of mechanization in the late nineteenth century.

The research on office dynamics in this period has focused on the rising number of employees in offices, their changing educational background, gender, and the routinization and mechanization of jobs. The traditional American office before the Civil War was staffed by male clerks, who wrote everything by hand in pen and ink. Their work was done in bound volumes, like account ledgers, copy books of outgoing letters, and minutes of the board of directors' meetings, when the company became incorporated and a board was appointed. Bound volumes had the advantage of keeping things together, and they prevented fraud. Incoming letters were loose leaf, which was considered a problem. Some companies bound incoming letters, others kept brief records in bound letter journals, while a third answer was just to keep the loose-leaf letters stitched on strings or as a stack in a drawer or a box. Vertical filing systems were marketed for this purpose around 1900, and accurate filing was ensured by trusted clerks and through records in bound volumes.[1]

The growth in the number of clerical workers in the United States was striking; from 1870 to 1930 numbers rose from 81,619 to 4.2 million, that is, by a factor of 51. Women contributed significantly to this expansion. Only 1,910 women were recorded as clerical workers in 1870, but by 1930 the total had soared to 2,038,494, accounting for 49 percent of the total clerical workforce.[2] This growth came about through changes in the structure of the office and in the office workers' career paths. The small traditional offices had had few organizational levels; the new larger offices had multilevel hierarchies.

The male clerks in the traditional office had received an in-service, general training, and their positions formed part of a career leading to managerial positions. In contrast, a large proportion of the workers in the new-style offices were assigned narrow duties involving routine work, like typing, punching, or shorthand, based on training in these fields. Female office workers were often employed in this kind of routine job, and most of them remained in such positions. Routinization, formal hierarchies, and office machines became crucial components in the organization of big offices. An expert on railroad accounting in the United States found in 1919 that the value of office machines "lies chiefly in greater speed and volume of output, accuracy, and the assignment of women operators thereto, relieving the male clerks of general routine work and allowing them to concentrate their entire time on more difficult mental work. The assignment of women operators was particularly valuable during . . . [the First World War] and the resultant scarcity of clerical and other labor in general."[3] Office machines—and female office workers—were regarded as tools to enhance the scale of office work at a low price.

In addition to the increase in scale, the development of office machines between 1900 and 1940 underwent a growth in scope, which reflected a growth in the scope of significant office tasks. Industrial production of simple keyboard office machines emerged from skilled, almost craftsman-like, methods in separate industries for typewriters and adding machines. Typewriter print was more legible than most handwriting, and adding machines enhanced the operators' arithmetic skills. These were stand-alone machines with few organizational requirements beyond training the operator.

Remington of Ilion, New York, started to produce typewriters in 1874. Their success encouraged competitors to enter the market, so that by 1890 thirty typewriter producers existed in the United States, growing to eighty-five producers by 1910.[4] Similarly, the first reliable keyboard adding machines started to be produced in the United States by Dorr E. Felt and William S. Burroughs in the 1880s. In 1887, Felt began to market his nonprinting adding machine, in which the total was displayed on a visible result register. Subsequently, to produce his machine, Felt went into a partnership with manufacturer Robert Tarrant and founded Felt and Tarrant.[5] Burroughs' first adding machine design from 1885 also used a visible register. However, in 1887, he added what became his adding machines' outstanding accomplishment: printing the numbers as they were entered and the totals.[6] A printout enabled the operator to check an addition much more easily; previously operators had had to repeat their calculation until they got the same result twice. Furthermore, some businesses, like banks, needed a written record. Serial production of the printing Burroughs machine started in 1891.

By 1900, separate typewriter and adding machine industries had emerged, each based on a machine stabilized in one or a few versions, and each market segment was highly competitive. After the stabilization, the calculating machines remained more open to development than did typewriters. Calculating machines could add or multiply, be equipped with a printing capability, have a full keyboard or just ten keys—and additional features continued to emerge. In 1902, Felt and Tarrant launched a machine that could cross-tabulate, that is, add two or more columns of figures horizontally—as well as the more typical vertical addition.[7] Three years later, Burroughs launched a machine that could print identification numbers and dates in addition to the figures posted, enabling a bank, for example, to print the account number for each entry. Both machines had a wide carriage that allowed the use of wide sheets or forms. This eased the use of adding machine prints as pages in loose-leaf ledgers. Burroughs developed a machine that could handle both subtotals and grand totals in 1910.[8]

The emergence of a subtraction capability around 1910 greatly improved the adding machines' suitability for bookkeeping. Until then, credit-debit bookkeeping could only be carried out in one of two ways: Keeping debit and credit in separate registers, or entering the debits as complements (i.e., adding the "opposite" number rather than actual subtraction).[9] As complements require nontrivial computations, calculating machines were sold with a special keyboard for entering complements.[10] In 1911, Burroughs produced a machine that could subtract without entering complements.[11]

The integration of typewriters and adding machines emerged in several variants. One type consisted of invoicing machines based on combinations of typewriters and adding machines, for example the Ellis adding-typewriter and the Moon-Hopkins billing machine. They were complex designs and slow to use, as the operator had to type the full name and address for every invoice. The design of the Ellis adding-typewriter was finalized in 1906 and subsequently produced. Production of the Moon-Hopkins billing machine started in 1908.[12] An alternative was to use a set of separate machines to calculate, issue, and address invoices. For example, addressing could be performed by use of an addressograph, but then the alphabetic capability was only used for writing invoice specifications. Further, it was necessary to perform multiplications either by hand or by use of a separate multiplication machine.[13]

But it was not a simple matter for the various machine producers to decide how to improve the capabilities of their machines for bookkeeping operations to enhance their competitive position, except that this application required more than could be accomplished by use of a typewriter, a keyboard adding machine, or Hollerith's standardized punched-card system for statistics processing from 1907. Producers needed to answer several questions: For example, what was important for machine-based invoicing? How important were computation capabilities like subtraction, multiplication, or division, none of which were available on key adding machines or punched-card machines by 1900? How important was the ability of the machines to write specifications on an invoice or a wage statement? How about the capability to print the recipient's name and address on an invoice or a wage statement?

As early as 1901, Hollerith clearly intended to extend the tasks his machines could accomplish bookkeeping through the development of a system for wage administration based on punched cards, but he never attempted to implement this idea. He chose instead to develop a standardized punched-card system for statistics processing, which became a major tool in large-scale business statistics.

In the decade after 1905, three challengers to Hollerith emerged: the Census Bureau machine shop, John Royden Peirce's punched-card systems, and the Powers Accounting Machine Company. In 1907, the year Hollerith finalized his statistics-processing system, John Royden Peirce envisioned a punched-card system of mechanized bookkeeping that replaced the ordinary statement of accounts by printouts of the transactions on loose-leaf ledgers. In 1913, *Scientific American* broadcasted these ideas in an article, "Keeping Books by Machine: The Punched Card as a Saver of Brain Energy." The vision was a machine, that

> performed the remarkable feat of recording approximately eleven entries of a single transaction—eleven entries which would ordinarily have to be made with pen with eleven chances of making a mistake. In recording a sale, the amount is entered upon a sales check, to be entered upon a bill, again upon the segregated sales record, again into the sales ledger, through another operation placed to the credit of the sales person, then to some department, in addition to a record as to whether the package was delivered by mail, carried away or sent by regular delivery express. Instead of this constant juggling with the same figures through a maze of operations there will be an original entry upon a punched card, and this card will, through the medium of motor-driven machines, be automatically sorted into various divisions and subdivisions, and recorded item by item upon counters or wheel sets into adding mechanisms . . . After the [printed] ledger postings are complete, the sales checks will be passed through the machine again and listed according to departments.[14] (Listing was to print selected information from every card.)

This was a production-line form of bookkeeping system facilitated by punched cards, though the card remained a processing tool. No data were to be stored on punched cards beyond the next closing of accounts, and there was no vision of using the system to generate additional extracts of accounts or collection forms. Compared with the key-set bookkeeping machines built during the next couple of decades, the core advantage of the punched card was the need for only one data entry for several jobs. The cost of attaining this was higher standardization requirements and more rigorous organization.

The Census Bureau Machine Shop

The first challenger to the Tabulating Machine Company's prime mover position was the Census Bureau machine shop established in 1905. The machines that came out of this shop were intended to break the Tabulating Machine Company's monopoly and to find an alternative for the excessive prices charged by the company as a result of their monopoly. However, the scope of the new enterprise remained within the Tabulating Machine Company's emerging closure of punched cards for statistics processing. This, in turn, became a major reason for containing the machine shop within the Census Bureau—a situation that led to its slow suffocation.

The Powers Accounting Machine Company was the only full punched-card competitor to emerge during the 1910s. This brought about conflict over the Tabulating Machine Company's patents, a cornerstone of their prime mover position. Powers planned to go for bookkeeping jobs; from the outset he based his punched-card system on a numeric printing capability and called his machines "accounting machines." However, he imagined that a numeric printing capability would be sufficient to make his system applicable to bookkeeping jobs. The actual requirements for extensive bookkeeping applications proved more demanding, and the Powers company only reached a closure on a complete system of punched cards for bookkeeping in 1943.

In contrast to these conservative approaches to machine development, John Royden Peirce was the visionary contender who first proposed bookkeeping and alphanumeric systems using punched cards to gain access to business of much greater volume. He challenged the closure of punched cards for general statistics and did not even accept Hollerith's established punched-card standard. Peirce encountered severe problems in converting his visionary designs into well-functioning constructions and in establishing machine production. The stories of the three challengers illustrate a prime mover position's robustness and the demand in offices in the United States in the 1910s for advanced mechanization.

The Census Bureau machine shop was established by a Congressional appropriation in 1905, and subsequent appropriations ensured its existence for several decades. Establishing and maintaining this machine shop was a part of the federal government's growing influence in society since the 1870s, and it had two immediate contexts: First, the Census Bureau's endeavor to establish for itself a permanent role, which was used to analyze the break with Herman Hollerith. Second, the antimonopoly wave since the 1870s, aimed at the new big companies and all sorts of market manipulations. The general public believed that American industry was being monopolized, but a monopoly was no simple entity and the federal government granted patent monopolies. The legislature tried using the antitrust laws, the Interstate Commerce Act (1887), and the Sherman Act (1890) to protect the free market through bans on restraints of free competition.[15] But in 1905 Hollerith had no competitors and the punched-card market was small. In addition, the antimonopoly endeavor came up against the federal patent system that granted a seventeen-year protection for inventions, conditional on their subsequent transfer to the public domain. The purpose of the patent system was to encourage inventions, and similar systems existed in all European and North American countries.[16] The punched-card history illuminates how further technological development enabled the extension of a patent-based monopoly far beyond the initial seventeen years.

To obtain the best available system to process the returns from the census in 1900, the office of this census had invited tenders. The invitation was based on the processing of population returns, like ten years earlier, and three offers were received: Hollerith's well-tried system from the previous census competed with two systems submitted by Charles F. Pidgin, who also had submitted proposals for census processing ten years earlier. This time he offered a manual card system and a key-entry system. The office of the census in 1900 (which became the permanent Bureau of the Census in 1902) once again chose Hollerith's offer.[17]

The Census Bureau in 1904 adopted a strategy of establishing its own production of punched-card machines. The following year, the bureau obtained a congressional appropriation to establish a machine shop.[18] Hollerith immediately withdrew his tabulators to avoid their being used as models for the bureau's own machine production. The Census Bureau's gamble had put them in a position of no return, which placed heavy strains on their new machine shop. They had less than five years to build the reliable machines needed to process the next census in 1910, but they experienced three interrelated problems: their ambitions to build better machines than Hollerith's were confounded by technical problems, the constraints of the still-valid Hollerith patents, and the difficulty of hiring qualified people to build their machines.

The early Hollerith patents would expire in 1906, which enabled the Census Bureau to copy the machines that he built for the census in 1890. From the outset, the Census Bureau had no intention of infringing any patent or competing commercially with Hollerith.[19] Their machine building was a parallel to Gore's punched-card system built during the 1890s at the Prudential Insurance Company and, as such, was not a cause for litigation. It proved difficult, however, to observe their intention neither to compete nor infringe existing patents. As the Census Bureau's ambitions exceeded the simple sorting of the Gore system, Hollerith's still-valid patents proved difficult to circumvent, and the bureau could not prevent employees from resigning to compete with Hollerith, as James Powers did in 1911.

Further complicating the issue, Hollerith had improved his punched-card machines substantially after 1890. He had built a mechanized feed, brush reading, and an adding unit for his tabulator, a sorter and a keyboard punch, and his patents on these facilities and machines would remain valid well after the census in 1910. In particular, it proved difficult not to violate Hollerith's patents on sorting and punching.[20]

The Census Bureau Director, Simon N. D. North, planned to improve on Hollerith's system from 1890 and, probably, even on his contemporary systems. To attract mechanics capable of making the needed inventions and innovations, these were allowed to take out private patents on their inventions in government service—with the proviso that they should be freely available for government use.[21] This privilege was a reasonable concession to attract mechanics, but it initiated commercial competition with Hollerith, which he considered a threat.

The machine shop staff grew from two people in 1905 to twelve in 1907 and to sixteen in 1909. Four were former Hollerith employees, one of whom had been at Hollerith's company for twelve years, part of the time as a foreman.[22] Hollerith considered this a theft of expertise, but he had no arrangement to prevent his former employees from taking their new positions. However, their importance as agents of technology transfer was probably limited as Hollerith was a drawing board inventor, who had outsourced innovation for machine production and production. His company only maintained machines and printed punched cards.

One of the people who moved from Hollerith to the Census Bureau machine shop, Charles W. Speicer, incorporated the Speicer Tabulating Machine Company in Washington, D.C., in 1912. The company tried to sell punched-card machines in Britain, and Speicer was granted a patent on tabulating equipment. However, he was not successful as a punched-card challenger, and he filed a patent application in 1914 on an invention in a different field—indicating that he had moved away from punched cards.[23]

James Powers was the most promising inventor hired by the Census Bureau machine shop, and he had not been employed by Hollerith. He was born in Odessa, which was then part of Russia, and graduated from a technical school there. Then he was employed for a period in a precision shop making scientific instruments for the physical laboratory at the University of Odessa. James Powers immigrated to the United States in 1889. During the next eighteen years, he worked for several firms in New York and became a partner of a small experimental workshop in Los Angeles.[24] He carried experimental work on in several fields, including cash registers, typewriters, and adding machines.[25] This and his subsequent work showed that he was a fine machinist and craftsman, who preferred purely mechanical constructions. Powers had no need to rely on electric constructions, as Hollerith had originally done. Powers' experimental work and inventions were spread over many fields, and he had still not settled in one field when he was hired by the Census Bureau in 1907. Up until that point he had proven himself an inventor, not an entrepreneur. He had tried to get others to develop his patents into marketable items and then had moved on to new projects.[26]

Powers lived in New York and was working at the mechanics laboratories of Francis H. Richards when he was hired by the Census Bureau. The Census Bureau had an agreement with this laboratory for workshop facilities. The census shop in Washington, D.C., at that time focused on the development of tabulators, while Powers was employed to construct a sorter. This arrangement showed the census workshop's limited capacity. In 1909, Powers finished his basic machine design for a punch and a sorter and moved to work in the Census Bureau machine shop in Washington, D.C.[27]

The Census Bureau shop's punched-card plans were based on the 24-column punched card that had been used to process the population census returns in 1890 and 1900. This allowed for all the punches used for these censuses, but it prevented the use of the twenty sorters bought for the agricultural census in 1900 because they were built for a shorter card of only twenty columns. The original Census Bureau plans consisted of three machines: an improved counting tabulator that could print the results, a sorter and, less urgently, an adding tabulator.[28] The desire for a printed result was probably inspired by printing adding machines. If implemented, this set of three machines would have placed the Census Bureau's technology ahead of Hollerith's. The key to the bureau's success was the ability to design and build a printing numeric tabulator and a sorter.

The machine shop's starting point was the transfer to public domain of Hollerith's early tabulator patents in 1906. This allowed the bureau free copying of the old pantograph punch and Hollerith's tabulator with manual card feed from 1890,[29] as well as Hollerith's electrical reading method. The machine shop started to develop their own counting tabulator, which soon included two improvements: mechanical card feed and numeric printing. For the card feed, they started with plans for a fully mechanical feed but ended up substituting an electric button for the hand lever. With the Census Bureau's new system, the operator placed the card in the reader and pushed a button that released an electric motor turning down the press. This process required less energy from the operator, and the new card reader raised the speed of the tabulator by about half, but it still required manual handling of every card and it only operated at a third of the speed of Hollerith's newest tabulator. From 1909 to 1910, one hundred copies of this tabulator were built by the Sloan and Chance Manufacturing Company of Newark, New Jersey, who specialized in making precision machinery.

First in 1911, the Census Bureau machine shop succeeded in building a mechanical card feed using an electric pin box, which required the card to stop during the reading. Despite the intermittent card movement, this tabulator's speed was 50 percent higher than Hollerith's. Number printing was the most conspicuous tabulator improvement in the Census Bureau machine shop. For the first time, a tabulator could print the result on paper. This removed the time-consuming process of reading and writing manually from forty to sixty counters, an important source of error. The printing unit consisted of six rollers containing ¾-inch- (2.5-cm-) wide paper strips, which were carried under each of the six rows of counters in the tabulator. When the printing unit was released, the numbers were printed on the paper strips by hammers and ink ribbons. As the print of a set of results filled up 27 inches (90 cm) of paper strips, it was still necessary to transcribe the data to consolidation sheets before making any use of them in preparing tables. In 1912, the Census Bureau machine shop succeeded in building a tabulator unit that printed the figures from all the sixty counters on one sheet of paper, and this new tabulator was used to process returns from the census in 1910.[30]

James Powers had been employed to build a sorter, but during this work he conceived the idea of an "automatic" punch as a substitute to Hollerith's pantograph and key punches from the censuses in 1890 and 1900. The Census Bureau adopted this idea as an additional machine building project. The Powers keyboard punch from 1908 was "automatic" in the sense that an electromotor performed the punching operations, which significantly reduced the operator's work.[31] The Powers keyboard differed from its predecessors. Hollerith's original pantograph punch had one "key" that was moved to punch in all positions, and the card was placed on a carrier in his key punch and moved, column by column, under the same punches, as the eleven to thirteen keys were operated. The Powers keyboard punch had 240 keys, one for each of the twelve punching positions in twenty columns on the card.[32] The full keyboard had twelve keys in each digit position, as did the Burroughs and Felt and Tarrant adding machines. The celluloid keys held abbreviations for each punching position.

Powers' punch was unique because the operator keyed in all the information for one card, before he or she, by use of a special key, ordered the machine to perform all the perforations in one operation. An electromotor punched, moved the punched card out of the punch, and entered the next card. Previously any keying error had caused a card to be scrapped; now the errors could be corrected before the card was punched. This enabled the correction of any errors discovered while keying-in, although those detected later still required a new card to be punched. Further, this punch was easier to operate than Hollerith's key punch, as the operator did not have to perform the punch manually. However, all these capabilities meant that the new key punch had an extremely complex design.[33]

Three hundred Powers keyboard punches were produced for the

census in 1910, but the new punches were not reliable and were abandoned, making their three hundred copies a big test production.[34] To alleviate this problem, the Census Bureau reintroduced Hollerith's old pantograph punches from 1890 to punch a third of the information on the population census in 1910. The Powers punch was advanced compared with its predecessors and was designed to reduce the workload of the key punch operator. However, its construction was not essential for the Census Bureau, and its development and production diverted attention from building a much-needed sorter.

Building sorters was the second most important machine construction project originally planned for the census in 1910. James Powers was hired for this task in 1907, but the following year he refocused his work to design the keyboard punch. After completing the punch, he returned to his original project and developed a sorter. Whereas Hollerith's machine sorted on one column at a time, Powers' first effort enabled sorting in one operation using several punching positions spread out over the card.[35] However, Powers' sorter was a complex mechanical design that only could operate at a third of the speed of the Hollerith sorters. In addition, it was an infringement of Hollerith's still-valid sorter patent, which made the Powers sorter a dubious project.

The alternative for the Census Bureau was to rebuild the twenty sorters bought from Hollerith for the agricultural census in 1900. Rebuilding was necessary to use the 24-column card for the population census; the sorters had been built for the shorter 20-column card used for processing the agricultural census in 1900.[36] The rebuilding alternative was chosen, but Hollerith's Tabulating Machine Company brought a lawsuit claiming that the alterations to these machines infringed their patents. The case was dismissed by the court, as the alterations were not considered sufficient to constitute an infringement of the Hollerith's sorter patent, and also because Hollerith's basic sorter patent was filed after the sorter had been disclosed to the public by being supplied to the Census Bureau, which nullified the patent.[37]

An adding tabulator was the last of the Census Bureau's original machine building projects. Their machine shop worked on the tabulator from 1908 to 1909, but it never became operative.[38] This combined with the animosity toward Hollerith's company compelled the Census Bureau to abandon punched-card processing of the returned information in the agricultural census, which was accomplished by purchasing key adding machines and reducing the investigations.[39]

The Census Bureau's development strategy was based on a development group, in contrast to Hollerith's lonely inventor approach, which, in fact, relied on a set of subordinate technicians. The Census Bureau's strategy resembled the contemporary introduction of development departments at many big industrial firms. In the Census Bureau, this strategy proved successful for building the counting tabulator that could print, but building a keyboard punch and a sorter was assigned to one individual, James Powers, who worked in a way similar to Hollerith.

The Census Bureau machine shop accomplished designing and producing the first numeric printing tabulator for the census in 1910, but this accomplishment was not sufficient to keep the Census Bureau machine building project afloat. It failed to build two of the three planned machines, the sorter and, particularly, an adding tabulator. It simply was not able to attract a sufficient number of able people. Only two of the sixteen machine shop staff subsequently were granted patents. Charles E. Speicer was granted only one punched-card patent, and James Powers used his attained punched-card expertise to establish his own personal punched-card machine company. However, none of his sophisticated machines designed for the Census Bureau were reliable.

The census in 1920 witnessed a reduced role for the Census Bureau machine shop, as no new machine was built, and equipment was acquired from the Tabulating Machine Company. The problems with the counting tabulators from the census in 1910 were corrected. Further, for processing the census returns in 1910, the bureau only had the twenty sorters, which they originally purchased from Hollerith for the census twenty years earlier and later rebuilt. For the census in 1920, the machine shop built several additional horizontal sorters based on Hollerith's original design, without interference from the Tabulating Machine Company.[40] By then, Hollerith was no longer a problem. His basic sorter patent had expired and his patent on the horizontal sorters for the census in 1900 had been nullified in his lawsuit against the government in 1910.[41]

The Census Bureau never resumed their building of either the Powers sorter or the Powers punch from the census in 1910. Only the work on the adding tabulator continued during the 1910s, where the people in the Census Bureau machine shop encountered the problem that several adder designs were patented by Hollerith and by the adding machine producers. Then only the early simple designs had reached public domain. The Census Bureau machine shop preferred to base their design on expired patents, for example, they tried to use the design of the Burroughs adding machine. These difficulties combined with the problems of hiring capable engineers caused the project of building an adding tabulator to be abandoned, and tabulators were rented from the Tabulating Machine Company for tasks requiring addition.[42] This solution was then feasible, as all the main contenders in the battle from 1903 to 1911 had moved elsewhere, namely Hollerith and the census directors North and Durand.

During the censuses in 1930 and 1940, the Census Bureau gradually introduced the 45-column card for processing population census returns. The population census in 1930 was processed using the old 24-column card, but a column layout was applied and IBM punches were used. (The Tabulating Machine Company had been renamed International Business Machines, or IBM, in 1924.) The new column layout required changes in the various characteristics punched. One change was punching the age in tens and units that required two columns, instead of in five-year periods and their units that required two and a half columns, as in the census in 1920. Another change was that state of birth—both domestic and foreign—was punched by use of a two-digit code, which had to be looked up or memorized by the key punch operator. Finally, the old pantograph punches from 1890 were retired. In 1940, IBM's 45-column card was introduced, and for the new punched-card format the Census Bureau machine shop rebuilt its existing sorters and counting tabulators. IBM had introduced the 80-column card in 1928, but processing using 45-column cards was cheaper. Also, the Census Bureau got 45-column punched-card machines from IBM, such as punches and adding tabulators. Thus it took the bureau and the company twenty-five years to get together again.[43]

The outcome was a reduced Census Bureau machine shop. During the years around 1910, the focus had been inventions and innovations. Later, its main occupations had become producing, modifying, and repairing equipment, with only modifying and repairing remaining after 1930.[44] To the Tabulating Machine Company and IBM, the absence between 1905 and 1930 of big counting statistics orders from United States censuses caused all their tabulator development work to be focused on adding tabulators, which were operated at a lower speed than simple counting-based machines to process counting statistics. In contrast, IBM's British agency, the British Tabulating Machine Company, had built or rebuilt machines for processing population statistics since 1910. Only in the late 1920s, did the American parent company return to design a special punched-card machine for counting-based statistics.[45]

The Census Bureau did not accomplish breaking the Tabulating Machine Company's monopoly, but they provided James Powers with the technological expertise he subsequently used to establish his own company. And once Powers emerged as the first open competitor, the Tabulating Machine Company's patents forced him to cooperate, and it regulated its market, which exposed it to an antitrust prosecution.

Powers Accounting Machines

The story of the Powers Accounting Machine Company started in 1911, when James Powers resigned from the Census Bureau to establish a company to design and build punched-card machines for "statistical and commercial works."[46] His main jobs involved numeric bookkeeping, but he would also collect statistics jobs, which for many years had provided the foundation of Hollerith's Tabulating Machine Company. This strategy proved harder to implement than Powers had anticipated. He ran into several technical and managerial problems, and he only attained well-functioning machines through engineer William Lasker, who brought experience from office machine building.

James Powers had built a punch and a sorter while employed at the Census Bureau from 1907 to 1911, but both were failures, due their high mechanical complexity. Further, Powers had no part in building the successful printing tabulator at the Census Bureau during his employment. However, he gained expertise in punched-card machinery.

From the outset, James Powers based his commercial machines on the 45-column punched card, introduced by Hollerith in 1907, which made this card an industry standard that remained until 1929. Powers needed to build a punch, a sorter, and a tabulator to establish his business, and he completed prototypes of all three machines in 1912.

Powers first wanted a to make a punch, so he tried but failed to improve his complex design for the Census Bureau from 1908.[47] A major problem was his approach with a full keyboard. The Census Bureau only needed to punch twenty columns, which required 240 keys, but these punches never became reliable. Now, his choice of a 45-column card would require a full keyboard of 540 separate keys, more than double the number. After having realized the shortcoming of his full keyboard design, Powers simplified the design and replaced the keys with slides, one for every column on the card, which reduced the number of entry determinants to forty-five slides. Powers retained the ability to simultaneously punch all the columns from his Census Bureau machines, and his company's punches continued to be able to perform this function.[48]

Powers built a new horizontal sorter with two decks. Probably the motivation for the two-deck design was to save floor space to compete with the vertical sorter from the Tabulating Machine Company. The Powers sorter operated at similar speed as the contemporary sorter from the Tabulating Machine Company.[49]

A programmable and printing tabulator proved hard to build. The ability to print and program was essential to attract bookkeeping jobs, as several tabulator settings were needed in a bookkeeping installation. Powers' first tabulator was a prototype that was based on the technology of the printing Felt and Tarrant adding machines, which greatly eased his work, but he never was able to make these machines reliable. The Powers tabulators read the punched cards in a pin box while the card was at a standstill. The reading was performed as spring-loaded steel pins, one for each punching position, went up toward the underside of the card in the pin-box reading unit. When there was a hole, the pin went through and affected an adding unit through a "connection-box." This was an interchangeable unit consisting of a rigid frame, which held rows of thin steel pins that could move a fixed distance up and down. The pins transferred, by the use of bar pulling, the movements from the pin-box reading to the adding units, where the bars pushed the various keys on the units.

The connection box was a joint unit, which could easily be exchanged for another when the tabulator was switched to another job, or programmed for the other job using today's terms.[50] Generally speaking, a user had one connection box for every task. The advantage of the connection box was to enable a quick and efficient switch from one task to the next, an operation only requiring instruction for a few minutes. At the same time, connection-box programming was inflexible, as even a small

TABULATOR-PRINTER

Powers company printing tabulator from 1914. (*The American-Canadian Mortality Investigation, Based on the Experience of Life Insurance Companies of the United States and Canada during the Years 1900 to 1915, Inclusive on Policies Issued from 1843 to 1914, Inclusive*, New York: Actuarial Society of America, 1918, facing page 14)

change required the box top be rebuilt at the company's factory, which took time and cost money.[51]

The Powers tabulator's ability to print was a requisite for most bookkeeping tasks. Also, printing became a useful capability of actuarial statistics, which was the reason the Actuarial Society of the United States chose Powers equipment for their big mortality investigation, which was processed from 1916 to 1918.[52]

Like the machines that Powers built at the Census Bureau, all his subsequent machines were based on exclusively mechanical technology, the only electrical part in a machine being an electromotor to operate it. This minimized their dependency on the various electrical powers systems of the age, which had direct or alternating current and diverse voltage. A machine could be moved from one power system to another just by changing the electric motor. Also, the choice of mechanical technology enabled the use of design and components from the adding and accounting machine industry, such as the use of a Felt and Tarrant adding unit for Powers' first tabulator.

Mechanical technology was not backward compared with electromechanical technology but rather more appropriate. Poor punched cards and unstable power supply had less impact on mechanical than on electromechanical punched-card systems. Punched cards did not always have the best quality. They could hold electric conductive particles, metal, or carbon grains, which became false holes in electric reading. Small cracks or fallen out bits could cause false reading for both kinds of machines, but the conductive particles only caused problems on electromechanical machines.[53] In addition, the direct current applied for card reading in the electromechanical machines required a stable voltage, and any brief power break could cause holes not to be read.

The major advantage of punched-card-based printing was the automatic execution, but in advance of being printed, all cards had to be packed in a deck of cards. Further, the paper applied was advanced by rubber rollers, which had limited precision. These operational problems were the focus of subsequent machine development. Powers' first tabulator showed his realistic vision of extending the scope of punched-card applications to bookkeeping and his technical problems in implementing this vision.

His first punch, sorter, and tabulator of 1912 were improved in the years until 1920, when a verifier was built to expand the scope of punched cards and attract bookkeeping applications. During this process, the mode of machine development was extended from the lone inventor, James Powers, to a group of people, similar to his earlier experience in the Census Bureau. During his Census Bureau years, Powers had gained experience in developing machines in cooperation with engineers at Francis H. Richards's machine shop and with the other people in the Census Bureau machine shop. Soon, he engaged his own team of engineers, first and foremost William Walter Lasker, who brought expertise in typewriter design.[54] Also, his company profited from inventions by the Powers agencies in Europe.[55]

The most urgent development task was a reliable tabulator. Powers and Lasker built an improved tabulator with fewer parts to make production cheaper and to ease maintenance.[56] The Felt and Tarrant adding mechanism was replaced by an adding unit from the Dalton Adding Machine Company. This was the first successful printing tabulator that both listed information from punched cards on a sheet or paper roll and printed totals. The operation of the machine, either listing all cards or printing the totals, was selected by shifting a switch. The tabulators in the Census Bureau could only print totals, and the Tabulating Machine Company only supplied a printing tabulator in 1921. However, in spite of this technical advantage, the Powers company did not do that well, and the Tabulating Machine Company remained by far the biggest company.

From 1914 to 1919, Powers' company amended the bookkeeping capability of their line of machines through the introduction of a verifier and an improved tabulator. The first verifier appeared in 1914 and was a crucial improvement for bookkeeping applications. Formerly, two cards had to be punched independently, and they were then superimposed and compared by looking at the light peeking through the perforations. The second card was subsequently thrown away. Powers' first verifier resembled his slide punch. When the slides were adjusted, a mask was superimposed on the printed card and a light was lit to check that the two perforations were identical.[57]

At the same time, work started on producing a machine that could make successive totals. Often several bunches of vouchers were recorded in bookkeeping or statistics operations, which required a subtotal at the end of each bunch of vouchers and a grand total at the end of the last bunch. On a tabulator up until that time, this was accomplished by running each bunch of punched-card voucher copies as a separate job. When a job ended, the operator stopped the tabulator and copied the result manually on a tabulator from the Tabulating Machine Company or released the printing mechanism on a Powers tabulator.

For the operator of a tabulator from the Tabulating Machine Company, this appeared reasonable, as he or she anyway had to copy the result by hand. This was not the case on a Powers tabulator, which the operator only had to stop the machine and print. The key to bypassing these simple but tedious operations was devising a way to instruct the tabulator when to make a total. For this purpose, Powers introduced separate "total cards" and "stop cards," which William W. Lasker implemented. A total card was a blank card with a special perforation in such a location as to allow a plunger to go through, thereby causing the adding unit to print a total without stopping the machine. A stop card had a perforation in a different location and caused the tabulator to stop. A stop card was inserted in a stack of punched cards at every point at which a total was to be calculated.[58] However, this arrangement only allowed one level of totals, which was a shortcoming for many commercial users.

When processing several bunches of related punched cards, for example a bunch from each department, they needed both subtotals and a grand total at the end of the job. However, they had to choose between the subtotals and the grand total, as it was only possible to transmit information from a card to one calculating unit on the tabulator. Only in 1923 did Powers introduce a tabulator that could do subtotals and grand totals by superimposing a grand total unit on the usual adding unit. Printing numbers from the grand total required the operator's initiation.

The slide punch seems to have been reliable, but entering figures was cumbersome. Powers worked through several designs, but William Lasker developed the successful automatic key punch that was launched in 1916. It had a small keyboard with twelve keys, one for each row on the card, and control keys for skipping a column, and so on. Further it retained the Powers punches' simultaneous electromotor-powered punching when the operator had keyed in all information for a card, and the automatic transport of cards. These capabilities eased the key punch operator's labor compared with the punch from the Tabulating Machine Company.[59] For more than ten years, this new punch from the Powers company remained the best punch available. Lasker also built a one-deck horizontal sorter, marketed in 1919, that remained the basis for all sorters from the Powers companies in the interwar years.[60] It had a simpler design than its two-deck predecessor, but it occupied more floor space.

Thus by 1919 Powers' company completed its first well-functioning set of punch, sorter, and tabulator that was necessary to exploit its original competitive advantage of a punched-card system with number printing. But this process had taken eight years, and the company's business organization was not yet fit to make the best of its technical accomplishments.

The Powers Accounting Machine Company had started machine production in 1914, but it ran into several problems. To establish production, it needed a license for tabulating and sorting patents held by the Tabulating Machine Company. This was granted, but the conditions were harsh. Powers was authorized to produce mechanical punched-card machines, but the company had to pay about 20 percent of its revenues in royalties to the Tabulating Machine Company. This should be compared with the 5 percent royalty for a similar license in Germany, which was imposed through different legislation. The Powers company could reduce its royalty payments through the development and production of machines that were outside the scope of the Tabulating Machine Company's patents. A futile attempt to this end was Lasker's invention of a sorter using a circular movement in contrast to the linear concept in Hollerith's patent, but it was never produced.[61]

In addition, the new company ran into financial problems. Costs to establish production proved to be higher than James Powers originally had anticipated. Powers had raised money to incorporate his company in 1911, but that did not prove sufficient to finance his machine building during the next several years, until the new company started to earn sufficient profits through sale. As a consequence, the company was reconstituted by money from financier John Isaac Waterbury from Morristown, New Jersey, in 1913, and three years later, the company received additional capital through a big loan.[62]

Further, the company's sales were poor, which was revealed by the company only getting a separate sales manager in 1920.[63] However, this was just a visible aspect of the company's managerial problems. James Powers did not care to manage and resigned from the company in 1918. The announced reason was failing health, but Powers lived as an active inventor for another nine years, and he filed five patents that were granted.[64] Further, he assigned three additional punched-card patents to the company from 1924 to 1925, and he even tried, in vain, to get reemployed in his old company.[65] These acts also paint him as a restless character, which is substantiated by his patent record.

In contrast to Hollerith and Peirce, Powers never settled in a technological field. For example, while engaged in the Powers Accounting Machine Company between 1911 and 1918, he filed eight patents in six separate fields other than punched cards, and three of his five patents filed after 1918 were outside the punched-card field.[66] His patent record and his preserved correspondence portray him as a prolific and eccentric inventor, but not an able entrepreneur, who returned to his preferred occupation as an independent inventor in 1918. Only for his punched-card patents did he try the role as entrepreneur.

The Powers company's management improved as able hired managers came to replace James Powers. He was succeeded as chef engineer by William W. Lasker in 1918.[67] However, the company encountered economic problems for several more years. It went into receivership in 1920, became charged to the Chase National Bank of New York and was offered for sale. Two years later, the company merged with a producer of mechanical adding machines and was reorganized twice. The receivership was only terminated, as the company was sold to the new Remington Rand Corporation in 1927.[68]

Remington Rand

The Remington Rand Corporation emerged as a general supplier of office machines and other office equipment through a broad merger in 1927. This merger was guided by James Henry Rand of Rand Cardex Services and had substantial financial backing. Rand Cardex Services was one of the country's most important producers of record control systems, and it merged with the Remington Typewriter Company, the country's leading typewriter producer, and the Dalton Adding Machine Company. Also in 1927, Remington Rand acquired the Powers Accounting Machine Company, which became the Powers Accounting Machine Division.[69] Over the next few years, the scope of the new company expanded through the acquisition of several companies producing loose-leaf ledgers, vertical filing systems, and safe deposit boxes.[70] The Remington electric shaver followed in 1938.[71]

The main advantage of Remington Rand was its ability to improve the use of the production capacity of its many constituent companies. Simultaneously, the wide scope of activities was the company's weakness, which resembled the Computing Tabulating Recording Company's situation in 1911 (see chapter 4). The acquisition of competitors, including three additional typewriter producers, strengthened the company's position by accumulating expertise and patents, but there was little interaction between the producers of typewriters, key adding machines, punched-card machines, loose-leaf ledgers, and vertical filing systems, not to mention electric shavers for the consumer market. However, in spite of the low level of integration, it proved difficult to get all the old companies to feel comfortable, which had an adverse effect on the creativity of inventors and innovators.[72]

Remington Rand proved more vulnerable to the economic crisis in 1929 onward than did IBM, due to its low integration and the high level of debt incurred during the merger. Remington Rand experienced a sharp dive in sales from 1930, causing net losses in fiscal years 1930–1931 and in 1931–1932. An injection of new capital reconstituted the company in 1936, but it proved difficult to reach the revenues and profits that had been attained in the last year before the crisis in 1929.[73]

The new company had inherited the patent license agreement with IBM from the Powers Accounting Machine Company.[74] This agreement expired in 1929, and $350,000 in royalty payments to IBM was due. The patents constituting the basis for the agreement in 1914 had expired, but Remington Rand still needed Hollerith's automatic group control patent. Hollerith had filed a petition for this in 1914, but it was only granted in November 1931.[75] Eight months earlier, a new five-year patent license agreement had been reached with IBM. Remington Rand was to pay the amount due, and in addition $25,000 in each of the five years. Further, the two companies agreed on pricing, which the following year caused an antitrust suit to be filed by the Department of Justice against the two companies.[76] The burden of the royalty payments was lower in the new agreement, but the very fact that Remington Rand entered the new agreement displays the difficult position of a challenger. Remington Rand still had to accept the conditions for the punched-card trade that were laid down by IBM as a prime mover.

Numeric printing had been the Powers company's original competitive advantage, but only in 1919 did it attain a full set of reliable punch, sorter, and tabulator machines to make the best of this advantage—which they then lost two years later when the Tabulating Machine Company introduced their numeric printing tabulator. These problems with their original competitive advantage combined with the restraints of the receivership between 1920 and 1927, restricted the Powers company to a strategy of developing its mechanical machines to compete with the Tabulating Machine Company on the markets for statistics processing and numeric bookkeeping. However, the Powers company twice attempted to develop new and exclusive punched-card application fields. First, the Powers Accounting Machine Company worked to develop bookkeeping applications requiring letter printing in 1924. Second, Remington Rand attempted to build special machines and cards to develop bookkeeping automation for the expanding department stores in 1932 and 1933.

Punched-card-based letter printing was first marketed by the British Powers company in 1921. However, this feature could only manage a reduced alphabet, and letter printing was restricted to separate printing positions, which could not print numbers. The British Powers company developed this feature at the instigation of the Prudential Insurance Company in London that wanted names as well as amounts to be represented in the perforated cards.

As early as 1916, the British Powers company filed a patent application on letter printing, but at the same time the American company refused to adopt this facility. The American company first embraced alphabetic printing in 1924. Lasker implemented the British design by developing a new tabulator that could print letters and numbers in separate printing positions like the British model. Lasker's tabulator was based on a reduced alphabet of twenty-three letters, like the British alphabet printing system. Further, to facilitate letter printing, a new model of the American company's key punch was built, furnished with alphabetic in addition to numeric keys. [77]

The first American alphabetic printing tabulator was installed in the Metropolitan Life Insurance Company in New York in 1925.[78] This installation replaced Metropolitan's cooperation with John Royden Peirce, which had existed since 1918, to develop punched-card machines with alphanumerical capability. However, Peirce had not proven able to build the promised machines and had been hired by the Tabulating Machine Company in 1922, which by then was not interested in this capability. In 1925, the Powers company's system with a reduced alphabet was the only other punched-card system with letters available in the United States. However its success was very limited, and Remington Rand chose not to implement it on the 90-column punched card, when it introduced this card as its standard between 1929 and 1935. This history indicated that the Powers company and Remington Rand perceived a very weak demand for letter printing until the mid-1930s.

The second attempt to develop a new punched-card application field was when nonstandard machines and cards were developed for bookkeeping automation for the expanding department stores in 1932 and 1933. This system was exclusively numeric and showed how far Remington Rand went to build nonstandard punched cards and equipment to gain a new market segment, in which it would be a prime mover.

The American department stores' sales had grown by 39 percent between 1919 and 1929, to $4.4 billion from 4,221 department stores, and constituted 9 percent of total retail sales in the country.[79] Further, credit sale made up a growing share of the department stores' sales. Punched cards were well suited to administer credit sales, as they facilitated distributing information on every sale to the customers' accounts and processing frequent statements of all customer accounts for the purpose of monitoring them. Punched cards offered mechanical sorting and printing of lists, and since this bookkeeping operation only required processing of numbers, punched cards were eminently suitable.

In 1932 and 1933, Remington Rand's punched-card division and a company specializing in electric automation designed a central record system for the Kaufmann's Department Store in Pittsburgh, Pennsylvania. It introduced up to two hundred fifty counter type sales transmitters, which were wired to transmit punched information to a common credit and authorization office with fifteen teleprinters, and to a central accounts receivable office equipped with twenty on-line recording and card punching machines.

Various types of nonstandard equipment were applied to handle the nonstandard punch tokens, which were small cardboard cards holding price and other sales information as perforations. Also, a recording punch was developed for the accounts receivable section to print and accumulate the same information on paper as was punched.[80]

It was planned to use punched cards in issuing customer statements, as opposed to address plates that were usually used for this purpose in the Remington Rand applications at that time. The addresses were to be produced by use of the reduced Remington Rand alphabet with twenty-three letters and three cards for every address, each corresponding to one of the lines: name, street number and name, and city and state. This application was to be based on the old 45-column card, as the new 90-column card could not yet hold letters. This project was never implemented, and Remington Rand did not pursue a comparable integrated system until the end of the Second World War. Mistakes due to technical imperfections were highly probable, but Remington Rand's subsequent nonpursuance of this promising application indicates a lack of demand rather than technical difficulties.

The main efforts to improve the Powers company's and Remington Rand's punched-card systems between 1920 and 1935 were responses to two significant improvements by the Tabulating Machine Company and IBM. One was that the Tabulating Machine Company's initial numeric printing tabulator from 1921 was also the first to offer automatic group control, which was a significant competitive advantage compared with the tabulator from the Powers company. Automatic group control rendered superfluous the cumbersome and time-consuming insertion and removal of total cards. However, the Powers company only managed to design and implement the ability to perform this task on its machines in 1927. The new mechanism allowed the tabulator to sense the change of designation on the cards and, without requiring action by the operator, to record the total at the end of each card group (for example, a customer number) and to begin processing the next group of cards without a pause.

This capability was attained through substantial and protracted labor. Eight years earlier, James Powers had filed a patent describing this mechanism. However, neither Powers nor Lasker was able to implement this mechanism, which was only achieved after they had hired Robert Edward Paris.[81]

The other significant improvement was that IBM changed the focus of punched-card machine development in 1928 from improvements of the numeric capability to increased card capacity through the introduction of a new card with eighty columns and matching machines. The new card was the same size as the old 45-column card, and IBM accomplished this increase of 78 percent by introducing rectangular perforations to avoid weakening the cards when inserting more columns. IBM patented the new card to prevent others from doing the same, which made it proprietary. This card became known as the IBM card.

Remington Rand accepted the challenge of creating a card with increased capacity. The rectangular perforations were no advantage to the mechanical Remington Rand machines. Therefore, their researchers decided to squeeze more information onto the same card either by applying smaller holes and more columns—or by introducing a more compact proprietary punch code. In contrast to the British Powers company, Remington Rand chose a more compact punch code and introduced their proprietary 90-column card in 1929, a double-deck card also based on the existing standard 45-column card. It kept the old card's layout of perforations, but redefined their meaning, by splitting the card into two decks of every six rows. On each deck, row number 0 was used for control information, like the negative sign in a debit figure, which left five rows to represent the ten digits.[82] This design could be extended into an alphanumeric representation, as six rows can hold up to sixty-three different characters, which could accommodate both a full alphabet and the ten digits with control information. However, this possibility was not pursued for the next nine years.

The numeric double-deck card became the basis for Model 2, Remington Rand's new line of punched-card machines, introduced in 1929.[83] However, this equipment only became available in 1931, as all the machines had to be redesigned to accommodate the new card and production established. Though the new Remington Rand card had capacity to become alphanumeric, the first new machines only served the double-deck card with numeric representation. The American Powers system of reduced alphabet representation from 1924 had not been a sufficient success to be included into Remington Rand's new line of machines. However, it was possible to use a part of a punched card for letters to be printed, by using the old standard 45-column positions, and the rest of the card for the numeric double-deck standard. But such an arrangement reduced the card's capacity and required both machines for 90-column cards and machines for 45-column cards.

The difficulties of visual reading of the numeric double-deck standard, distinguished it from the old numeric 45-column standard. To alleviate this problem, Remington Rand produced an "interpreter" that printed the meaning of the punched holes on the card.

During the period from 1932 to 1935, the Model 2 line of machines was extended by the introduction of a new verification scheme, a summary punch, and a multiplier. The new verification system introduced in 1932 was composed of an attachment for the key punch and the "automatic verifier." The verifier attachment to the key punch elongated or offset the original perforation when the card was punched a second time by a separate operator. The automatic verifier searched these twice-punched cards.

```
  A B C D E F G H I J K L M N O P Q R S T U V W X Y Z    0 1 2 3 4 5 6 7 8 9
0   O O O       O    O      O O O       O O O O        O
1 2 O O      O        O       O O   O O           O       O O
3 4      O O   O O O O        O O O      O  O O   O        O O
5 6 O O   O    O  O O O   O O     O    O  O          O O
7 8   O     O O O            O O O O O O    O O  O           O O
9 + O      O        O O   O         O  O  O O O        O  O  O  O O         45
         O O O          O O O           O O O O         O
1 2 O O      O        O       O O   O O           O       O O
3 4      O O   O O O O        O O O      O  O O   O        O O
5 6 O O   O    O  O O O   O O     O    O  O          O O
7 8   O     O O O            O O O O O O    O O  O           O O
9 + O      O        O O   O         O  O  O O O        O  O  O  O O
46                                                                         90
```

The Remington Rand 90-column numeric punched card introduced in 1929, a double-deck card derived from the existing standard 45-column card. The perforation was in Remington Rand's alphanumeric system from 1939. (H. L. Tholstrup, "Perforated Storage Media," *Electrical Manufacturing* December 1958, 58. Measurement details deleted.)

When a circular perforation was detected on a card, it indicated an error, and a signal card was inserted in the deck of cards on this card.[84]

Remington Rand marketed a punched-card multiplier in 1935 that multiplied two figures punched on a card and punched the outcome. This design was based on a patent of the United Accounting Machine Company, which had been acquired by Remington Rand in 1934.[85] This multiplier completed Remington Rand's development of punched-card machines for numeric bookkeeping before 1945.

Only in 1938 did Remington Rand introduce a new line of alphanumeric punched card machines, which broke its focus after 1924 on the numeric capabilities. The new line of machines was called Model 3 and was based on extending the numeric representation on their double-deck card to become alphanumeric.

Demand from public utilities and public administrations provided the major impetus for Remington Rand to develop this system. Remington Rand had suffered the painful experience in 1936 of having their punched-card system turned down by the Social Security administration because of its lack of alphanumeric capability. The Social Security administration was looking for a system to handle the compulsory old age savings by many millions of Americans, which would have provided extensive business for the company awarded the contract, which IBM won.

Addressing items was the core functionality of alphanumeric systems. Addressing was made easier in 1943 by a new tabulator that could print the three lines of an address from the same card. Before, only one line could be printed from a single card, meaning three cards for each address, which was extremely cumbersome to handle.[86]

John Royden Pierce's Punched-Card Systems

John Royden Peirce was born in Maine in 1878. His father headed a construction company and directed several large construction projects in New York City. With this background it is not surprising that John Royden Peirce graduated an engineer—though a mechanical engineer—from the Stevens Institute of Technology in Hoboken, New Jersey, in 1900 at the age of twenty-two.[87] The Stevens Institute of Technology had opened in 1870, and within a few years it was widely recognized as a premier academic institution for training mechanical engineers. Like Columbia College (later Columbia University), the teaching at Stevens emphasized mathematics and sciences rather than machine shop experience.[88] However, Peirce chose to base his constructions on mechanical technology, in contrast to Hollerith who was trained at Columbia College.

After graduation, John R. Peirce worked for four years as an estimating and costing clerk in New York with two construction companies. During this period, his inventive skills found outlet within construction work and office jobs. First, he invented several machines to improve basic construction techniques, the first in 1904, which resulted in his being awarded several patents.[89] Second, Peirce became involved in office mechanization. He conceptualized a mechanized punched-card system for bookkeeping in 1906, based on mechanical technology. Each of his punched cards held the items of information required in a bookkeeping entry, and he used the cards to print the entries on loose-leaf ledgers. To enhance the quality of these printed lists, he introduced letters into his punched-card system. His punched card had forty-three columns and was tailored to the application.[90] Peirce's tailoring of the card to an application indicated that he planned to design different punched cards for various applications, which would have required different machine models. He planned to build a new punch and an adding tabulator, since his cards were different from

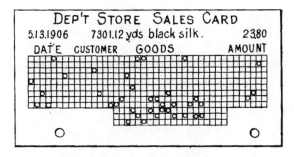

A John Royden Peirce punched card for department store bookkeeping, 1907. (J. R. Peirce, Perforating machine, [U.S.] Patent 998,631, filed and issued 1911. Reference numbers deleted.)

Hollerith's card, which prevented his using machines from the Tabulating Machine Company.

During the following year, Peirce designed a tabulator and a key punch, which simultaneously typed a keyed-in character on the card and punched it.[91] The new punch applied a modified standard typewriter key board with thirty-seven keys. He now had a punch code using twelve punching positions that represented each character by punching one or more holes in one column. This system was most ingenious as it introduced letters on punched cards, but Peirce applied separate representations for numbers and letters that required printing digits in some positions and letters in others. He had nine rows or punching positions for digits (zero was not represented) and twelve rows for letters.

Peirce developed his early designs for several applications over the next six years. He designed applications to produce restaurant invoices, to record consumption in electricity and gas utilities, for bank accounting, for stock bookkeeping, and for processing sales in a department store.[92] Wherever possible, punching the card was carried out within an existing procedure to save labor and reduce errors, in contrast to a subsequent isolated punch operation. For example, the card used for reading an electric or gas meter was entered into the meter that mechanically punched the reading.

For all the applications considered by Peirce, the card was the original form that recorded the transaction. This shows that Peirce saw punched cards in a different role than his competitors did. He was the first to realize that a punched card could contain the original entry. In contrast, Hollerith, the Census Bureau, and Powers had the card as a copy of the original form

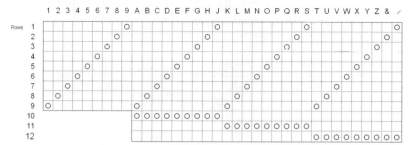

Peirce's punch code, 1907. Zero was not represented, and the letter *I* was represented as a numeral 1. Except for zero, the numeric code followed Hollerith's contemporary code. (Drawing based on J. R. Peirce, [U.S.] Patent 1,236,475, filed and issued 1917)

or invoice. Anyway, there is no evidence that any application was developed in cooperation with a real company or institution. The applications just served as a basis to implement Peirce's original concept.[93]

Peirce shaped his punched-card systems as he worked to implement his ideas. The letter representation disappeared and the punched card became smaller, as reading letters using two holes in each column was difficult to implement. Furthermore letters did not prove crucial for the chosen applications and offered no scope beyond the contemporary numeric key bookkeeping machines. The outcome was several small punched cards and a compact character representation, only comprising the digits.[94] Further, Peirce's small cards enabled him to design compact machines. The choice of basing his design on several applications resulted in various punched-card sizes and plans for specially designed machines, like a cash register perforating the amounts entered and an electric meter equipped with a small perforating attachment to facilitate punched-card-based reading of the meter.

The next problem for Peirce was implementing his designs, which he only seems to have attempted after several years of designing. A prototype tabulator was completed in 1912 that displayed his technical approach. Like Powers, Peirce used exclusively mechanical technology for card reading, data transmission, and adding and printing, and he also applied an electric motor to operate the machine. The adding units printed only totals and were based on Burroughs adding machine wheels. Their patents had expired and a former Burroughs engineer supervised the machine building, which can explain why Peirce succeeded in using this approach in contrast to the people in the Census Bureau machine shop.

The electromotor operated the machine feed of punched cards, and the prototype tabulator had a processing speed of about half that of the contemporary tabulators from the Tabulating Machine Company, but Peirce envisioned a higher speed for the planned serial produced machines. Peirce's tabulator used a manual feed of paper for printing, as did several key accounting machines, which circumvented machine feed problems and simplified exact printing on forms. Mechanized location of the print on the paper required precise horizontal and vertical control of paper movement, which was crucial when using preprinted forms. Peirce planned to sell the tabulator at a price from $1,000 to $1,500, depending on the number of adding positions.[95] This was cheaper than renting a tabulator from the Tabulating Machine Company for two years. The lower price and modest ambitions of his individual application indicates a business strategy of selling to medium-size firms.

John Royden Peirce moved to incorporate a production company in 1912 based on the prototype tabulator and his ideas for applications. Until then, he had worked with his inventions and development as an individual; now, he needed to raise $400,000 to establish a company to produce and sell his machines. He only found $100,000, and, consequently, had to find additional funding for development and to establish production. In the prospectus used for raising money for the incorporation in 1912, his machines were recommended by nineteen statisticians and accountants, of which fifteen were from big companies or important federal organizations.[96] This could indicate a change in Peirce's business strategy toward big business and major federal organizations, who could be partners in his development and production of punched-card machines, in contrast to the small and medium-size firms apparently targeted in his development strategy thus far.

One of the recommendations in his prospectus in 1912 came from the comptroller of the Mutual Benefit Life Insurance Company in New York City, a moderately large company.[97] Peirce supplied this company with punches and sorters to process their insurance statistics in 1914. This, together with the several tabulator versions on the drawing board at that time, indicated that Peirce's tabulator design was not yet finalized.[98]

The contract with the Mutual Benefit Insurance Company supports the impression of a change in business strategy to pursue big customers. This is further substantiated by a contract he had been awarded, in 1913, to develop punched-card machines for the Metropolitan Life Insurance Company, in New York City, which was the largest insurance firm in both number of policies and value of insurances in the United States. This contract tied Peirce to a large company that could finance his machine development. However, he had little choice, as his original business strategy of selling to small and medium-size companies required that he was able to finance his development work and the establishment of production—which he had proven unable to do. By then, his development work had lasted for six years, and he had not sold a single machine.

The Metropolitan Life Insurance Company contract in 1913 entailed a system for compiling and printing five standard numeric reports used to control the transactions in the big company, like lists of policy numbers, possible disabilities, premiums to be paid and whether the policies were paid annually, semiannually, or quarterly. The tabulator was to be a substantial improvement on his prototype tabulator from 1912, as it would both list information from cards and print subtotals and grand totals, without intervention from the operator.[99] Formerly such lists had been compiled by fifty-four clerks.[100] Now Peirce had an order, but he experienced problems in providing the machines promised within the contractual time limit in 1914. The machines were only completed in 1916, with the tabulator costing double the originally agreed price; but in addition to the original stipulations, the Metropolitan Life Insurance Company received automatic group control, which eased processing of punched cards. The delay and additional machine development created financial problems for Peirce, and Metropolitan had to pay every invoice he received from the builder of his machines, the DeCamp and Sloan Manufacturing Company in Newark, New Jersey. When the machines were completed in 1916, Peirce claimed they had cost his company $21,000 in excess of Metropolitan's payments.[101]

However, the Metropolitan Life Insurance Company did not consider Peirce's problems more serious as it, in 1916, contracted for additional Peirce machines that were not yet developed. Already during the negotiations in 1913, Peirce had proposed a system based on a "master card" on every policy that held punched information of the policyholder's name and address along with the relevant numerical information indicating amount insured, amount to be paid, date of payment, and other relevant information. The punched information was also to be printed on the card, which would be used to generate invoices, receipts, and other transaction documents.[102]

This was an ingenious concept, but its implementation would be protracted. The master card revived Peirce's original vision of punched-card systems with letters, but no one had any experience using letter codes in punched cards in 1913. A suitable code was needed, together with a punched card holding sufficient information. After much consideration, Peirce settled for a code using four punching positions, or rows, for numbers and six rows for letters. His new master card held an ambitious total of one hundred fifty characters, which should be compared with the forty-five digits on the cards from the Tabulating Machine Company.[103] In addition, Peirce's new machines remained to be built. Considering the technical ambitions among the other punched-card producers in the 1930s this was extremely ambitious, and, once more, Peirce encountered problems implementing his concepts in the machine shop. Regardless, by 1916 he submitted a detailed proposal to Metropolitan to build customized punched-card machinery to prepare and address premium notices, receipts, and notices that would be mailed to the policyholders. Further, Peirce proposed preparing various internal records, including a register of policies issued and an agent's list of notices. The planned alphanumeric punch was a modernized version of his punch design from 1907 and resembled a typewriter.

The proposed system also included a listing machine and a machine duplicating information from one card to a new card, which relieved the wear and tear on the master cards. Peirce made additional elaborations of this plan, and he concluded a contract with Metropolitan Life in 1918 to build customized punched-card machines for administering their policies. By 1922, this system included a prototype tabulator that implemented the first alphanumeric representation in punched cards, needed for addressing letters to policyholders.[104]

To improve his revenues, Peirce got additional orders in 1918 to supply punched-card machines to the Prudential Insurance Company in Newark, New Jersey, and the Bureau of War Risk Insurance. The Prudential Insurance Company's contract had a limited scope of application, compared with the Metropolitan Life's contract, and it only comprised two tabulators and two sorters.[105] In contrast, the Bureau of War Risk

Insurance entered a contract for the development of systems for both actuary and policy administration. This contract resembled the contract with Metropolitan Life.[106] The Bureau of War Risk Insurance in the Department of the Treasury insured members of the armed forces after 1917.[107] In addition to these contracts, Peirce tried to capitalize his foreign patents to raise money.[108]

However, in 1922 he had not yet fulfilled any of his three insurance contracts from 1918, and he sold his patent rights to the Tabulating Machine Company and became a development engineer in that company.[109] The three insurance companies had received some of the contracted machines, but the three contracts ceased to exist one by one over the next few years, without any known regress from the insurance companies. The Metropolitan Life contract was the last to be terminated—in 1926.[110]

Since 1918 Metropolitan Life had paid $1.1 million to Peirce for developing and building punched-card machines. However, by the termination of the contract from 1918, the total cost of the machines in the contract was estimated to be about $3 million, thus there remained about 63 percent of the machine development and production.[111] Still Peirce had problems completing his machines. The $1.1 million paid also disclosed the importance ascribed by Metropolitan Life to a system to mechanize their policy administration. Eight years had elapsed since their first contract with Peirce and a usable Peirce system was still estimated to be several years away, substantiated by the fact that Peirce had not yet settled on a standard for a letter representation in punched cards by 1922.[112] But Metropolitan Life still had no other option for punched-card based policy invoicing and payment control. The reduced alphabet system developed by the British Powers company and marketed by the American Powers company in 1924, was not a viable alternative. Further, IBM did not yet offer letter representation in punched cards.

A Challenger's Possibility

Alfred D. Chandler analyzed the position of a company challenging another company's first mover position in terms of certain internal features of the challenger including its organizational capabilities, which he defined as the company's total physical facilities and human skills.[113] The three challengers to the Tabulating Machine Company's first mover position in the United States were the machine building in the Bureau of the Census, the Powers Accounting Machine Company, and John Royden Pierce's work before he was hired by the first mover. However, to understand the possibilities of these challengers, the analysis needs to be extended to encompass the patent system used by the first mover to enforce his position and the technology. The technology provided possibilities to challenge the prime mover, but patents were a powerful tool to restrict their options.

The prime mover, the Tabulating Machine Company (later IBM), originated in the inventions and innovations of Herman Hollerith in the 1880s. In 1889, Hollerith was granted patent protection on his original equipment, lasting for seventeen years.[114] Hollerith's improvements in the 1890s were the basis for additional patents granted in 1901.[115] This renewed his patent protection, which enhanced his prime mover position. The second renewal was his patent on "automatic group control," which he filed in 1914, while James Powers' less comprehensive patent application was filed the following year and granted in 1917. These two applications ran into a lengthy patent conflict in the United States both with each other and several other patent applications. Therefore Hollerith's important patent was only granted in November 1931, making it valid from 1914 to 1948.[116] Consequently, the prime mover succeeded in achieving substantial patent protection for about sixty years based on five patents. The stories of the challengers in the United States document the importance of these patents in shaping the punched-card industry.

The people in the Census Bureau machine shop envisioned their development of punched-card equipment within Herman Hollerith's closure of punched cards for general statistics, from 1907. They worked to establish what today would be called a clone production of equipment simply to reduce the bureau's costs. In addition, the existence of the machine shop enhanced the bureau's position in the federal government. Their objective was a punched-card system with a number-printing tabulator, which represented a significant improvement on the machines then available for general statistics. This objective was accomplished, but they failed to build other needed machines. The project at the Census Bureau was foiled by a combination of internal and external factors: the inability of the engineers in the Census Bureau machine shop to design the needed machines, the containment of the project within the Census Bureau with its yearly struggle for appropriations, and the restrictions on the machine-building possibilities imposed by the patents granted to the prime mover.

The longest lasting importance of the punched-card endeavor in the Census Bureau was through its offspring, the rival company founded by James Powers. His company was the only full punched-card competitor to emerge in the United States, which brought about conflict with the Tabulating Machine Company's patents. The outcome was a costly license that had a lasting negative impact on the business potential of the Powers company. From a technological perspective, the provisions of the license excluded it from applying electromechanical technology for two decades.

James Powers' punched-card equipment originated in Hollerith's closure of punched cards for general statistics. However, he believed that it was possible to extend the technology within this closure to encompass bookkeeping tasks through building numeric printing tabulators. The requirements for extensive bookkeeping applications proved more demanding and the Powers company only achieved a closure on punched cards for bookkeeping in 1943. This raises the question of the delineation of technological capabilities within a closure, which will be addressed at the end of the next chapter. A most amazing part of the Powers company's punched-card history was the robustness of the Tabulating Machine Company's first mover position, demonstrated by its ability to its position in spite of its abstention from the bookkeeping market until 1921.

In contrast to the two other challengers, John Royden Peirce was a visionary and from the outset worked for punched-card-based bookkeeping and alphanumeric systems to gain access to business of much greater volume. By doing this, he opened a renegotiation of the very nature and purpose of punched cards and which abilities punched-card technology should have. However, Peirce ran into problems caused by his inability to implement his ideas and designs and to produce reliable machines, the well-known problem of an inventor who becomes an entrepreneur, a transition accomplished only by Herman Hollerith in the United States punched-card industry. Peirce's brilliant conceptualization led him to devise a clever version of the "automatic group control" patent, which IBM acquired in 1922 and thus continued to control the industry through patents. The acquisition of the patent also involved hiring John Royden Peirce for the Tabulating Machine Company's machine development.

FOUR

The Rise of International Business Machines

The Tabulating Machine Company had its origins in Herman Hollerith's punched-card business in the 1890s and became the prime mover for the use of punched cards in statistics processing. As punched-card systems were developed to handle bookkeeping between 1907 and 1933, the Tabulating Machine Company and its successor company, the International Business Machines Corporation (IBM), retained the prime mover position and became one of the most influential companies in the United States. This was achieved through the company's shaping of punched-card-based bookkeeping and in spite of the emergence of several challengers. The successful transformation of the company's primary application field from statistics processing to bookkeeping operations was facilitated by fundamental changes in the company's sales and machine developments.

The Tabulating Machine Company merged into the new Computing Tabulating Recording Company in 1911, which was renamed IBM in 1924. Thomas J. Watson Sr. joined the company in 1914, organized its sales to be the most efficient among office machine producers, and established a machine development department. The machine development staff first managed to bring the company abreast of the Powers company by 1921, and then they went on to shape a punched-card system for bookkeeping that became the industry's de facto standard.

During this process, several conflicting paths of technological development were pursued in the search for not-yet-defined goals. The core of this process was identifying and interpreting a new prime application in bookkeeping.

The Computing Tabulating Recording Company

In 1911, Herman Hollerith and his directors sold the Tabulating Machine Company to a conglomerate established by the entrepreneur Charles Ranlett Flint. Company growth through mergers and acquisitions had become increasingly widespread in the United States since the 1880s. Charles R. Flint was a prominent promoter of mergers, and among his more important promotions were the United States Rubber Company (1892) and the American Woolen Company (1899).[1]

There were two main reasons for the sale of the Tabulating Machine Company. First, the latent reason, that Hollerith's one-person management style could not control his fast-growing company much longer. The company's revenues had grown on average 50 percent per year from 1908 to 1911. Second, Charles R. Flint's great offer was hard to turn down and served as a trigger. Flint's offer of $2.3 million was about twenty times higher than an offer made three years previously that had been turned down. Hollerith claimed failing health as a reason to accept the offer.[2] He might have had health problems, but he was only fifty-one years old and would live for another seventeen years.

The newly merged company was named the Computing Tabulating Recording Company (CTR). In addition to the Tabulating Machine Company, the components were the Computing Scale Company of Dayton, Ohio, the International Time Recording Company of Binghamton, New York, and the Bundy Manufacturing Company of Endicott, New York. The Computing Scale Company produced scales equipped with a chart enabling clerks to determine an item's price as it was being weighed. The Bundy Manufacturing Company and the International Time Recording Company produced machines that recorded when workers entered or left their workplace.[3]

Charles R. Flint later described CTR as his most successful trust construction,[4] but the shape appears curious. It was neither a horizontal consolidation of concerns of similar or competing products, nor was it a vertical combination. It was a conglomerate of four producers and manufactures of fine mechanics-based business equipment, which largely supplemented each other. The main advantage sought seems to have been the consolidation of their facilities for manufacturing and distribution. Of the four companies, only the Tabulating Machine Company did not produce its own machines. It is true that in 1901 Hollerith had bought a machine shop, the Taft-Peirce Company of Woonsockett, Rhode Island, but it had been sold in 1905. In 1911, the Tabulating Machine Company moved production to the amalgamated company's various production facilities.[5]

Another advantage of the merger was the use of skilled engineers for inventions and for developing projects across the borders of the old companies.[6] However, the new company formed a weak umbrella organization for the four merging companies, which remained largely independent. Therefore, the Tabulating Machine Company kept its former style regarding customer contacts and patents.

In 1911 the four parties to the merger seem to have been of comparable sizes, but as time went by, punched cards became the most prominent activity in the company. In 1918, the Tabulating Machine Company earned half the total revenue of the merged company, and adaptation of punched cards for bookkeeping purposes offered a far larger sales potential than did smart scales or time recording systems.[7] In 1935, the company sold its scale interests, and the time recorders followed in 1958.[8]

From the outset, the Computing Tabulating Recording Company's headquarters was in New York City, where the Tabulating Machine Company's management moved from Washington, D.C.[9] After the merger, Hollerith resigned as general manager and later left the board.[10] This freed him from executive responsibilities and enabled him to concentrate on machine development, the work he always enjoyed most. He stayed with the company as a consulting engineer for a decade, and he was very influential until 1914 because all proposals for change in machine design were submitted to him for approval.[11]

Hollerith was far from burned out and made important improvements to his system and filed eight patent applications between 1911 and 1914. His main collaborator was still Eugene A. Ford, who continued to work up in Massachusetts.[12] Machine development in this period was influenced by the new strength of the Powers Accounting Machine Company, and the management of the Tabulating Machine Company noticed the emergence of John Royden Peirce as a punched-card producer.[13]

At that time, Hollerith was not interested in extending the scope of applications to encompass bookkeeping. He worked instead on four machine improvements, of which two were influenced by competition from the Powers company. Electric brush reading of cards was susceptible to error due to tiny holes or specks of conducting material in the cards that could cause the machines to register incorrect figures. Hollerith's early pin reading, which James Powers had adapted for his mechanical punched-card system, was less sensitive because the reading only took place as a pin passed through a perforation on a waiting card. To solve the problem of increased errors from brush reading, Hollerith invented a speck detector in 1914 that located flaws in card stock using an electric test current.[14] The date indicates an influence from Powers' system, as brush reading had been on the market for more than ten years.

The start of Powers' production of a printing tabulator in 1914 also influenced Hollerith's considerations. In 1908, the first printing tabulator in the Census Bureau had demonstrated the feasibility of such a design. This printing tabulator required manual release by the operator for each print and only printed totals. Another possibility was to "list" cards, that is, to print selected information from every card. In statistics processing total printing was an advantage as few results were produced compared to the number of entries. In contrast, in bookkeeping the number of results was high compared to the input, and listing data on the cards was also important. By 1911, the listing and total printing Burroughs adding machine had been a great success for two decades. Furthermore, a listing and total printing tabulator was a major feature of John Royden Peirce's punched-card system, marketed in 1912.

Hollerith had worked on a total printing unit as early as 1899, and he filed patent applications for this capability in 1905 and 1913.[15] In 1913, he also filed a patent application for a listing tabulator.[16] However, Hollerith stuck to the view that only a total printing tabulator should be built; his vision of punched-card applications only encompassed statistics processing. He even resisted an attempt to build a listing tabulator as late as 1917.[17] Since the 1880s, Hollerith had been the imaginative designer of punched cards for statistics processing but never appreciated the vast business and technical potentialities of punched card for bookkeeping. It was true that in the spring of 1912 the Tabulating Machine Company collected information about the emerging John Royden Peirce punched-card system for bookkeeping, but the initiative rested with the management succeeding Hollerith, and he was only informed subsequently.[18]

Hollerith's first machine improvement of his closure for general statistics processing aimed at improving processing cards with information on many small groups, for example, for processing statistics on the age distribution of women in a county. For that task, the cards were sorted according to age, and then the cards for each age group were taken to be run separately on the tabulator. The operator had to stop the tabulator each time, copy the results from a Hollerith tabulator by hand—or release the printing mechanism on a Powers tabulator. Subsequently, the operator removed the small punched-card stack, put in a new stack, and reset the tabulator. Hollerith filed patents on batch processing of several individual stacks of punched cards separated by introducing "stop cards" in 1912, as did James Powers shortly afterward.

These special cards were introduced into the punched-card stacks manually or through sorting, in the above example, after the cards on each age group. The tabulator came to a halt when it reached a stop card, and the operator either copied from the counters or released the printing mechanism before the next group was processed. In the Hollerith system, the stop cards were blank cards provided with a particular cut out. When a stop card was hit by the seizing arm, the tabulator stopped and the sum of the group could be read. The adding unit was reset manually and the tabulator restarted.[19] The Hollerith company introduced stop card control in about 1914.

The use of stop card control was unsatisfactory, as inserting and removing cards was time-consuming. Hollerith devised a way to control processing separate groups without stop cards using information punched on the cards, read twice in two separate reading stations. The simultaneous reading of two consecutive cards enabled the tabulator to compare without storing any information. When a change occurred in a selected field, the "group indicator," the machine stopped. In the age statistics example, the age could be used as the group indicator. The technique was called "automatic group control" and was the earliest example of conditional programming of a punched-card machine, that is, control of processing of punched cards based on information on the cards. This technique significantly eased processing, and it became essential for developing efficient programming of electronic computers after the Second World War. Hollerith's patent application ran into a lengthy litigation in the United States against James Powers and several other inventors. Hollerith's important patent was only granted in 1931.[20] Automatic group control was first implemented by the Tabulating Machine Company in 1921.

Hollerith's second machine improvement of the closure in general statistics processing was the design of an easy touch key punch. The pantograph punch had been lever-based, which eased punching, and the Powers key punches used an electric motor for punching holes in the card. Hollerith's idea was that key action would energize an electromagnet that would then perforate the card. A patent was filed in 1914, but the "electric key punch" was first marketed in 1917.[21] This launch date might have been influenced by the improved Powers punch, the "automatic key punch," which emerged in 1916.

Back in the years around 1911, the Tabulating Machine Company enjoyed fast-growing revenues, and the sales volume is apparent from the company's having distributed 113 tabulators and 97 sorters to customers in 1914.[22] However, the Tabulating Machine Company suffered from operational problems that needed to be solved. They had unsatisfactory sales activity, problems with technical development, and the newly amalgamated company experienced hitches in coordinating production.[23]

Of these, unsatisfactory sales activity was considered the major problem.[24] To enhance sales, the company established branch offices. Prior to 1911, Chicago had been the only office outside Washington, D.C. Over the next few years sales activity was reorganized, based on a general sales office in New York with sales offices located in key cities throughout the United States.[25] At the same time, the use of the various production locations within the amalgamated company required coordination, which did not always materialize.[26] Furthermore, the new company was heavily in debt, having financed its own merger.[27]

The board addressed these problems in 1914 by engaging Thomas John Watson Sr. as general manager for the amalgamated company. The following year he became president of the company, and he retained this position for forty-one years until 1956.[28] During his appointment, Watson proved a paternalistic entrepreneur who managed to turn the company into one of the most profitable in the United States. Thomas J. Watson was born in 1874. He attended one year of business school and became a salesman for the National Cash Register Company (NCR) in 1895. He learned the profession well and became sales manager in 1910.[29] The extensive development of cash registers at NCR demonstrated to him the importance of a development department.

In 1912, John Henry Patterson, Watson, and twenty-eight other NCR officials were indicted and brought to trial for antitrust violations. Reportedly NCR had over 90 percent of the cash register business. A year later, the jury reached the verdict that all defendants were guilty except one. Watson, Patterson, and one more of the defendants were given the maximum sentence of one year in prison. An appeal was granted in 1914, and the case ended with NCR signing a consent decree to clean up its business practices. During the appeal, tensions rose between Watson and Patterson, and Watson resigned. Indications exist that these tensions were a result of the antitrust case, but Patterson's biographer found it but a sign of Patterson's eccentric way of running his company. Seen from this perspective, Patterson fired Watson, along with several others, because he had become too powerful.[30]

Watson brought extensive knowledge from NCR of how to run a business machine producer, especially sales. Over the next several years, he gradually reshaped his new company, building an efficient sales organization, establishing a development department, and creating a corporate spirit. When Watson entered the company, the foundations of a sales organization had been established, but he made fundamental improvements. Each salesman was assigned his own exclusive sales district where he had the responsibility to develop sales, knowing that no other salesman from his company would try to poach his customers. The salesmen were considered the cream of the company, well-paid through generous commissions and smartly dressed as the company expected. Systematic selling was taught within the company.

The company's sales strategy was first to analyze the customer's problem, to demonstrate how IBM punched-card equipment could solve the problem, and finally to sign a contract that had the same rental rates for all customers. To monitor and improve sales, a quota system was introduced. The initial sales quota was based on the population and the business structure of the district, and the quotas for the following were adjusted according the sale in the district. When a salesman reached his quota, he became member of the "One Hundred Percent Club" of that year. Every year all members of the club were invited to and applauded at sales conventions. Watson organized the first One Hundred Percent Club in 1916 in each of the constituent companies, and they became a united event after 1922, which indicated the ongoing integration of the company. Over the next few years, these events evolved into huge conventions, which were held at impressive hotels like the Waldorf-Astoria in New York.[31]

Sales conventions became a principal component of the corporate spirit that included all employees. They met at various big gatherings, listened to talks, and sang company community songs, for example, "Ever Onward." The lyrics were written to a stirring tune from a Sigmund Romberg musical. Community songs were typical of the age, as reflected in the distinctive songs of various political movements of the interwar years. The corporate spirit was also promoted through slogans, foremost "Think." Other examples were "There Is No Such Thing as Standing Still," "Time Lost is Gone Forever," and "Beat Your Best."[32]

When Thomas J. Watson became general manager of the Computing Tabulating Recording Company in 1914, the company produced four punched-card machines: a key punch, a vertical sorter, a nonprinting tabulator that could operated by using stop cards, and a gang punch, which in one operation could punch one perforation in several cards. The key punch exhausted the operator, as it required substantial force to operate.[33] The tabulator was reliable, but the absence of a printing capability was an increasing disadvantage as the number of printing Powers tabulators grew.[34] Hollerith had worked to improve the punch and to build a printing tabulator for three years in Washington, D.C., in cooperation with Eugene A. Ford, who worked in Massachusetts. But no solution had been found.

Thomas J. Watson came from eighteen years at NCR, which had had a highly active development department since 1888. Within a few months of becoming general manager, Watson made Eugene A. Ford move from Massachusetts to New York to establish the first development department for punched-card machines. Soon a model shop was equipped and ten model makers were hired.[35] Hollerith never traveled to New York to contribute directly to the development efforts, but he contributed through an extensive correspondence for several years.[36]

The first new senior technical person to be hired for the development department was Clair Dennison Lake in 1915. He was an automobile designer with no previous punched-card experience. Fred Merchant Carroll was hired the following year. He was an experienced inventor who previously had filed patent applications for printing adding machines and cash registers.[37]

In 1917, Watson hired James Wares Bryce as supervising engineer of the time recording division of the company. Bryce studied mechanical engineering at the College of New York, but he quit after three years to take up a position as draftsman and designer. Since 1905, the patent rights for his time recording developments had been assigned to the International Time Recording division, which became a part of the Computing Tabulating Recording Company in 1911. In 1922, Bryce was appointed chief engineer of the whole Computing Tabulating Recording Company, which put him in charge of the development of punched-card technology.[38]

The approach to punched-card development at the Computing Tabulating Recording Company—and later IBM—was decided upon by Watson and diverged from the development at the challenger companies. In several cases the company pursued competing project paths, and only at the end of the projects did Watson choose the design from one of the projects for production. In addition, the Tabulating Machine Company and IBM bought inventions from individual inventors and they, in fact, hired several of these inventors. This supplied the management with alternative solutions. But it was a costly approach that only this punched-card company could afford.

The development of a printing tabulator was the first instance in which the Tabulating Machine Company pursued several competing development paths, but the scope of what inventors were developing remained limited to the capability to print numbers and the implementation of automatic group control. Watson assigned both Clair Lake and Fred Carroll the task of independently developing printing tabulators.[39]

Clair Lake's development group developed printing units for the existing tabulators, and their tabulator could both list and print totals. First, they worked on a tabulator printer with printing wheels, which did not succeed.[40] Then, they developed a printing unit that used type bars and completed a prototype in 1917.[41]

Simultaneously, Fred Carroll's development group pursued the design of a completely new tabulating machine, which incorporated a novel rotating drum printer, a very complex design that only printed totals.[42] Watson selected Lake's design to keep development and manufacturing costs low. However, the printing tabulator was not introduced until 1920 and was supplied to customers the following year. The delay was probably caused by reliability problems.[43] Though the development of a numeric printing capability became protracted, the company desisted from buying alternative designs, which were offered by independent inventors.[44]

The Lake tabulator supplied from 1921 was the first commercial tabulator with automatic group control, which was based on a patent application by Herman Hollerith in 1914. The new tabulator brought the company into conflict with patent applications from James Powers, John Royden Peirce, and Charles A. Tripp. The Patent Office decided to give Hollerith's patent application priority over Powers' application in 1919. The conflicts with Peirce's and Tripp's patent applications developed less satisfactorily for the Tabulating Machine Company, as these applications were assessed as being superior by the Patent Office. This caused the Tabulating Machine Company to buy Peirce's patent in 1922 and Tripp's patent two years later.[45]

The key punch was the second reverse salient of the Hollerith system, to use Thomas P. Hughes's concept. The problem was that the operator had to punch the card manually, which was extremely tiring. In addition, the key punch became an issue of competition when the Powers Automatic Key Punch appeared in 1916. Even the Tabulating Machine Company's own engineers considered the Powers punch to be superior.[46]

Again several paths of development were pursued to improve the punch. First, an electrically driven key punch was built by Hollerith in 1913 but only supplied from 1917.[47] Second, a key punch developed by independent inventors was purchased in 1914, but it never reached the market.[48] As with the tabulator, the electrically activated punch brought the company in contact with another independent inventor, John Thomas Schaaff, who claimed patent infringement. The Tabulating Machine Company settled his claim by acquiring his patents and employing him as a development engineer.[49] The Tabulating Machine Company marketed a new punch with electrically driven keys in 1923, but the operator still needed to enter and remove each card manually. Only in 1929, did IBM get a key punch with motor-driven feed and ejection facility, which completely equaled the Powers company's punch from 1916.[50]

The Tabulating Machine Company introduced a "verifier" in about 1917 as a means of checking the work of key punch operators. An operator used it to repeat the original key punch operator's work, and if a keying discrepancy occurred the verifier signaled this by locking its keyboard. If an inspection revealed the questionable card to be incorrect, the verifier operator prepared a correct card with an ordinary key punch.[51]

By introducing the number printing tabulator in 1921 and the electrically enforced key punch in 1923, the Tabulating Machine Company now produced a total line of machines that were ahead of the Powers company's products. This number printing tabulator became IBM's basic tabulator during the 1920s and remained in service into the 1940s.[52]

The Patent License to Powers

Shortly after joining the Computing Tabulating Recording Company, Thomas J. Watson was approached by representatives from the Powers Accounting Machine Company requesting a licensing agreement to use some of the still-valid Hollerith patents. "You are in a position to put us out of business," Watson several decades later recalled one of them saying.[53] The Computing Tabulating Recording Company decided to grant the Powers Company a license to use these Hollerith patents to produce mechanical punched-card equipment for 25 percent royalties on the gross receipts from the rentals of tabulators and sorters and 18 percent on gross receipts from card sales. This accounted for about 20 percent of their total revenues, as the Powers punches and verifiers did not depend on Hollerith patents.[54] The license arrangement enabled Powers' production but placed the company in an inferior competitive position.

Thomas J. Watson's biographers saw this agreement as an example of Watson's "clean, high-grade and square dealing[s]."[55] However, the approach from Powers in 1914 gave the Computing Tabulating Recording Company two additional alternatives: They could reject the approach, forcing the Powers company out of business, or they could buy the Powers company (in the following decade they acquired patents from many individual inventors and bought John Royden Peirce's punched-card company). Further, the royalty of about 20 percent in the United States should be contrasted with the royalty of 5 percent in the similar, but court-imposed, license in Germany.[56]

No record exists of the deliberations at the Computing Tabulating Recording Company, but Thomas J. Watson was still facing criminal antitrust charges for his activities at NCR.[57] If James Powers' request for a license had been rejected, or if the Tabulating Machine Company had purchased the Powers company, Watson would have reestablished Hollerith's punched-card monopoly. This would certainly have harmed the reputation of Watson's new company at a time when NCR's antitrust case was not yet settled, and it could have attracted the attention of the Department of Justice to investigate possible antitrust violations.

The Powers company's problems between 1918 and 1927 prevented them from paying their royalties to the Tabulating Machine Company. Their debt accumulated, and they got the royalty rates reduced by half in 1922.[58] During the 1920s, the legal position of the conflicting patent applications on automatic group control remained unsettled until Hollerith's patent was published in November 1931. Well before the license agreement from 1914 would expire in 1929, IBM filed a patent infringement suit against the Powers company's successor, Remington Rand. To ward off this suit, Remington Rand entered a new five-year license agreement in March 1931, which also repeated the previous agreement's market arrangement.[59] The agreement between IBM and Remington Rand became the basis for the antitrust suit the following year against the two companies.

This history of the patent licenses to the American Powers company illuminates the different views on monopolies in Germany and in the United States. Alfred D. Chandler compared the cooperation among German companies with the competitive tradition in the United States, which was enforced through the antitrust legislation.[60] However, the patent system in Germany comprised court-enforced licenses at a moderate rate, which facilitated competition between a first mover and challengers. In contrast, the United States had ultimate patent protection and antitrust enforcement. The patent protection forced a challenger to obtain a license from the first mover on the first mover's conditions, which were hardly positively inclined toward the challenger's business. This strengthened the patentee's monopoly and weakened antitrust legislation.

Shaping Equipment for Bookkeeping

Before the introduction of the printing tabulator in 1920, the Tabulating Machine Company established a growing business with many manufacturers, railway companies, and insurance companies. Using their nonprinting tabulators, they were able to process business statistics with a limited number of outcomes, as the operators had to copy the results by hand. The printing tabulator eased this problem, but bookkeeping assignments were more complex to mechanize than processing statistics.

Access to a substantial extension of the punched-card business could be found either through the development of punched-card systems for core bookkeeping or by introducing less costly machines that could extend punched-card applications beyond the existing big company customers to encompass small and medium-size concerns. Both options were pursued by the Tabulating Machine Company. By 1920, core bookkeeping was carried out manually or by use of key office and address plate machines. Key office machines could subtract, multiply, and print letters. Address plate technology was a well-established, though cumbersome means to address letters and other correspondence.

The transition to bookkeeping was more complex than the development of the printing tabulator, produced from 1921. The development of the printing tabulator focused upon three tasks: list printing, total printing, and automatic group control. After the introduction of the printing tabulator, four more general elements were involved: subtraction, multiplication, letter printing, and an enlarged capacity card. How these were put in order of priority for implementation depended on the kind of bookkeeping application selected in the development process. Further, it was a very open-ended process, where even the basic punched card of twelve rows or punching positions was contested.

In 1922 the Tabulating Machine Company acquired all the patents held by the engineer John Royden Peirce, who was a small, independent punched-card producer, to get control of his important group control patent. It also hired Peirce. He brought with him his machines, which were different from the Tabulating Machine Company's electromechanical punched-card machines. Peirce's machines applied exclusively mechanical technology with pin reading, and he applied a double-deck 86-column card, quite different from the standard 45-column card.

Peirce had strong and attractive conceptual ideas about punched-card-based bookkeeping. He also had experience with printed lists of numbers and letters for policy administration in insurance companies. But for ten years he had been struggling with the problem of building reliable machines. For several years after he had joined the Tabulating Machine Company, his double-deck 86-column card remained the basis for policy administration at the Metropolitan Insurance Company in New York City and in the Prudential Insurance Company of Newark, New Jersey, and his new company built additional machines for this distinctive punched card.[61] In spite of these shortcomings, Peirce was considered so essential for his new company that it paid an amount for his patents equaling 25 percent of its total net earnings in the year he was hired, plus his salary.[62] The high price paid to hire Peirce and the conflicting paths of development pursued until 1930 indicated a company in jeopardy with respect to finding a path of development for their punched cards.

At the Tabulating Machine Company, the development process consisted of building several machines that contained various combinations of the four general elements of punched-card-based bookkeeping. There were up to four parallel machine-building projects at the same time. The two major lines of machine development were carried out by engineering teams headed by John Royden Peirce and Clair D. Lake.

Peirce's group worked to develop machines capable of handling alphabetic as well as numeric information. Their work was based on Peirce's experience from before 1922 and on his nonstandard double-deck punched card with 86 columns. Before joining the Tabulating Machine Company, Peirce had built a "notice writing" machine for the insurance division of the United States Veterans Bureau (Bureau of War Risk Insurance until 1924). This machine issued and addressed notices to be sent to policyholders, with a punched stub to be detached when paid to facilitate subsequent bookkeeping. After Peirce had joined the Tabulating Machine Company, a similar machine was built and supplied to the Metropolitan Life Insurance Company in New York.[63] During the 1920s Peirce continued this line of development at IBM, and it was only terminated in 1930 after an additional tabulator had been supplied to Metropolitan. This tabulator could print numbers and letters in separate printing positions and it used Peirce's double-deck 86-column card.[64]

Clair D. Lake had been at the Tabulating Machine Company's development department since 1915, and his group based their machine development work on improvements to the existing machines. They focused on bringing the existing tabulator to perfection until 1925 when they turned to a phased extension of its capabilities to encompass bookkeep-

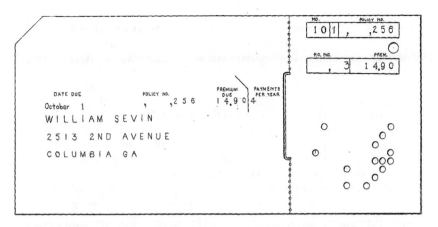

Peirce notice to insurance policyholders in 1922 with punched stub for subsequent bookkeeping. (J. R. Peirce, [U.S.] Patent 1,506,382, figure 9. Explanatory numbers removed.)

ing: first an increased-capacity card, then a subtracting tabulator, and—finally—letter printing. For the increased capacity card, Lake and his group chose to keep the existing numeric punch code on the 45-column card, and they squeezed the columns together to achieve an enlarged capacity. This required building new punches and card readers, but unlike Peirce's development approach, the basic numerical operations were not affected.[65] The available material does not disclose if this group had a prime application field in the 1920s.

Watson selected Lake's simpler design that used the 80-column punched card, and the machine was introduced in 1928. This design was chosen even though Lake's prototype alphanumeric tabulator was not yet completed. Lake's design was based on the representation of each character by up to two perforations, which was much simpler than Peirce's representation. The ability to store alphabetic information in addition to numbers was judged, as yet, to be not sufficiently important to the customers. The 80-column card offered 78 percent more storage capacity than did the old standard 45-column card, and the rectangular perforations were its most explicit characteristics.

The new cards were rectangular instead of round, because this facilitated electric reading. Further, cards with rectangular perforations proved sturdier and made the new design patentable. The rectangular perforations

The "IBM Card" of 80 columns introduced in 1928 with the standard English alphabet (four zone) representation from 1933. (IBM Corporate Archives, Somers, New York)

provided the card with a distinct appearance that could not be copied by others.[66] The age of the standard 45-column card was over and was about to be succeeded by proprietary cards. IBM protected their "IBM card" through litigation in France in the 1930s. In addition to handling the new card, Lake's new tabulator had the capability to be wired to process the old 45-column card, which eased IBM's service to its remaining 45-column-card customers.

A "reproducer" was introduced to punch 80-column cards from existing 45-column cards to help customers to convert their records to the new 80-column card. The importance of being able to reproduce cards became evident through this experience, and copying from 45- or 80-column cards became a standard feature.

The second improvement from the Lake group was a tabulator with subtraction, which also was introduced in 1928. For this tabulator, negative numbers were identified by a perforation in row 11. Previously, positive numbers (credit) and negative ones (debits) were punched in separate fields on the card and separately totaled by the tabulator. Then balances were determined manually. Alternatively, the complements of negative numbers were manually determined and then keypunched in the same field as the positive numbers so as to achieve subtraction through addition of the complements.[67]

The Lake group also completed a prototype alphanumeric tabulator in 1928. This alphanumeric tabulator had a slightly reduced alphabet and enabled writing a full address, but it was not reliable and the company discontinued its further development.[68] The dismissal of the Lake group's work on an alphanumeric tabulator in 1928 and the termination of Peirce's alphanumeric system in 1930 disclosed the IBM management's low assessment of the importance of this feature for its customers. It reckoned that the customers primarily wanted increased capacity punched cards and subtraction.

The business opportunities based on these capabilities were increasing, as was indicated by the emergence of the new large application of public utility billing, that is, calculating the price for electricity, gas, and water based on meter readings and issuing invoices to be mailed to the customers. For this application, IBM modified a standard tabulator that came to perform automatic extra checks on its computation. Public utility companies were extremely sensitive to invoicing errors, as the invoices were produced on large numbers and went directly to the public. In this application, the result was either printed on forms or on postcards to be mailed to the customers, and a paper feed was built to enable printing on forms. However, public utilities only required their invoicing process to encompass numbers, while addressing was accomplished by the use of address plates.[69]

A new drive for alphabetic tabulator capability started in 1930. While Peirce had built punched-card machines with letters for insurance companies between 1916 and 1930, IBM now focused on chain stores that were undergoing a roaring expansion. The number of parent companies of chain stores had grown from 2,030 in 1914 to 7,046 in 1929, simultaneous with a rise in the number of store outlets from 23,893 to 159,826.[70]

In the mid-1920s, IBM had cultivated the chain stores by designing punched-card-based systems to control goods stored in their central stores. Every box in the store had a punched card stating its particulars, including product number and price. These cards provided a representation of the whole stock, which eased control. When an item was taken from the store to be dispatched to an outlet, the matching punched card was removed and used to produce the invoice.[71]

To extend this chain store application, in 1930 IBM started to develop a reduced alphabet tabulator to print invoices on shipments of goods from a central store of a chain store to the outlets. Reduced alphabet specifications were easier to verify by senders and recipients than product numbers. This development was based on the invention in 1927 at the Danat Bank in Germany of a simple reduced alphabet printing unit, which substituted the ten numbers on a numeric type bar with ten selected letters. This German invention had been acquired by IBM in New York by 1930.[72] First, IBM implemented a reduced alphabet by substituting the ten digits on several type bars in a tabulator with ten selected alphabet characters. A few copies of this design were supplied to chain stores in 1931 and 1932. Then this design was developed to an extended type bar of twelve characters corresponding to all twelve rows on the punched card. The most used letters were spread over two adjacent type bars, and the machine was used to produce in-company invoices. The first tabulators of this design were shipped in 1931, and a total of 255 were produced. This tabulator remained on the market until the 1950s (though the company stopped soliciting orders for it in about 1937) as its features sufficed for in-company invoicing for chain stores and it was cheaper than the later full alphabet tabulators.[73]

Although a reduced alphabet worked for in-company invoice specifications, insurance companies needed a better representation to write lists of policyholders or to address letters. Around 1930, insurance companies typed addresses or printed them using address plates; they do not appear to have cared much for addressing with punched cards. When IBM terminated the production of equipment for Peirce's double-deck, 86-column card in 1930, his engineering group switched their efforts to developing an alphabetical capability on the 80-column card that had been introduced in 1928.

From 1931, IBM produced a machine from this engineering group that provided for a reduced alphabet of twenty-one letters, with the remaining five letters were substituted by digits. The product was acceptable for lists for internal use in an insurance company or a bank, but not for letters to customers. The Twenshe Bank in Amsterdam, Holland, received the first copy of this tabulator in 1931, and the next year the Prudential Insurance Company in Newark, New Jersey acquired the second copy. A historian explained the launching of this reduced alphabet tabulator as a countermeasure against Remington Rand's announcement in 1929 of their 90-column punched card.[74] However, Remington Rand had chosen not to implement the American Powers company's reduced alphabet from 1924 on its new punched card due to its lack of success.

Consequently, the reason for developing and producing this reduced alphabet tabulator was more likely a demand conceived by the IBM sales organization, and it was perhaps also a measure to curb discontent in Peirce's engineering group, who had had its proposal for a new card standard of a double-deck card turned down in 1928 and the production of its machines for the double-deck card terminated in 1930. The IBM tabulator with a reduced alphabet was a limited success, as only about ninety copies were produced, but it remained in use for several years just as the chain store tabulator was.[75]

The first of a series of full alphabet tabulators followed two years later, in 1933. A perforation in one of the top three rows combined with a perforation in one of the bottom nine rows were used to designate letters, which supplemented the old standard for representing digits on punched cards. This alphanumeric code provided for ten numeric and twenty-six nonnumeric characters, and it survived to the end of the punched-card era.[76] This was sufficient for the English alphabet, but the German alphabet, for example, has several additional characters, which caused problems. Later types of tabulators enabled twenty-eight alphabetic characters in addition to two special characters, for example, & and * and were sufficient for most national alphabets.[77]

Addressing letters later became an important punched-card task, but IBM did not perceive sufficient demand for this application in the United States in the 1930s to prioritize production of this feature. The first alphanumeric tabulators could only print one line per card. This was not adequate, as three lines were needed for an address and one card of eighty columns could hold all the information needed for an address. In 1937, IBM began production of a tabulator for the French market, which could print two lines from a single card, possibly to attain a contract on a conscription and mobilization register. However, a tabulator that printed three lines from a single card only appeared on the American market in 1941.[78]

Extending the application of punched cards to encompass bookkeeping called for an improved multiplication capability. For invoicing, it was crucial to be able to compute the price of a commodity given the item price and the number of items. Further, attracting bank accounting applications would require the ability to compute interest rates. When using the existing IBM tabulators, multiplication was achieved through successive additions, through mental calculation, or by use of a separate key calculating machine.

As early as 1926, IBM had decided not to base punched-card multiplications on an improved tabulator but to build a separate nonprinting machine that could read figures from a punched card, perform the required arithmetic operations, and punch the outcome on the same or a successive card. In contrast, Deutsche Hollerith-Maschinen Gesellschaft (Dehomag), IBM's German subsidiary, built a multiplying tabulator between 1931 and 1935. In the United States, James Bryce filed patents on a separate punched-card multiplier in 1928, and a prototype was built. At that time, a study identified applications for such a machine but did not find enough demand to start production. Two years later a big telephone company approached IBM. The telephone company wanted a punched-card multiplier to process invoices to their customers. This made IBM start production of the Bryce machine from 1928, which was marketed in 1931 (IBM Type 600). It could read two factors from a single punched card, multiply them, and punch the result onto a blank field on the same card. Two years later it was superseded by an improved model (IBM Type 601), which could calculate combinations of multiplications and additions.[79]

In 1928, during Bryce's work on his inventions for the first punched-card multiplier, Dehomag brought attention to the Austrian inventor Gustav Tauschek's patent application for a punched-card multiplier, filed in Germany in 1926 and granted in 1928.[80] Tauschek's German patent contained very broad claims, and the IBM patent department was troubled by its consequences; an application in the United States would have priority from the date on which the German application had been filed. In order not to reveal to Tauschek the assumed value of one of his patents, IBM bought all of his patents in 1930 and employed him in the United States for five years starting in 1931. This solution was similar to the purchase in 1922 of patents held by John Thomas Schaaff and John Royden Peirce. However, the United States Patent Office turned down Tauschek's patent application in 1935, as it found his multiplying invention to be inoperative, and Tauschek's contract with IBM was not extended beyond the original five years.

A historian regarded IBM's purchase of Tauschek's patents and his employment as merely an attempt to terminate a competing punched-card producer.⁸¹ This might have played a role, but IBM's action was a repetition of the actions that ensured the granting of Hollerith's crucial group control patent. It worked to prevent Tauschek from gaining control of punched-card multiplication in the United States in the event of his application's being granted. In fact, IBM applied six out of Tauschek's twenty-three patents in the United States, including a patent for the improved punched-card multiplier launched in 1946 that also could divide.⁸² As an IBM employee, Tauschek worked on design in several fields, as recorded by his patents, including multiplication and division.⁸³ He also developed ideas beyond applications that IBM found relevant and filed applications for his own nonassigned patents.⁸⁴

Reshaping Punched Cards

Reshaping punched cards was a prolonged process. The basic idea of using punched-card systems for bookkeeping tasks was conceived in 1906 by John Royden Peirce and published in an article in *Scientific American* in 1913.⁸⁵ However, actually realizing this concept only took place in 1933. This process provides insight into the opening up of an established closure and demonstrates how a new closure is established.

In 1906, Herman Hollerith was still working to complete his closure of punched cards for processing general statistics. As his Tabulating Machine Company had a complete monopoly, he controlled this process, which included his customers and outsourced machine production. The reopening of this closure and the subsequent new closure were established through a more complex process as the users' requirements grew extensively and challengers emerged.

Based on observations from several technologies, Thomas P. Hughes introduced a distinction between radical and conservative innovations. While radical innovations are system originating, conservative innovations are system improving. Hughes also found that independent inventors invented a disproportionate share of radical inventions, while organizations established around established closures, preferred conservative or incremental inventions.⁸⁶ These findings assist the analysis of the reshaping of punched cards.

John Royden Peirce was an outsider to the producers of punched card for statistics processing and the related network of users, which confirms Hughes' observation. In contrast, the two other challengers originated in the statistics-processing closure, which curtailed their imagination. The Census Bureau machine shop remained within the established closure and worked only on incremental changes. In spite of its name, the Powers Accounting Machine Company had the same origin, but by 1915, it was trying to attract bookkeeping tasks that would use statistics-processing equipment in which the only improvement was a number printing tabulator. It also had pin-box programming, but that was only a mechanical implementation of flexible programming, first implemented by Otto Schäffler in Vienna in 1890. Compared with the machine-building activities in the Census Bureau, the Powers company's decision to market beyond statistics processing in the federal government provided the dynamics, which gave the company opportunity to contribute to the shaping of punched cards for bookkeeping and explore this market.

Hollerith's negative reaction to printing capabilities confirms Hughes' observation. Hollerith had shaped the previous closure. Further, his successor, Thomas J. Watson's, performance between 1914 and 1921 is notable. Only first in 1921 did his company manage to produce a number printing tabulator. This observation supports Thomas P. Hughes' observation of the preference for incremental innovations in an organization, in this case specifically an organization embedded in an established technological closure. Simultaneously, Watson improved the sales organization substantially, which increased the company's dependence on the statistics processing closure. For the Tabulating Machine Company this dependence was based on the expertise and machinery in its production facilities and, especially, on its business strategy of leasing equipment that remained the company's property.

The Tabulating Machine Company, and its successor IBM, was most influential in the shaping of punched cards for bookkeeping, as it was the biggest company involved in this field. It enhanced its momentum in 1922 by hiring visionary John Royden Peirce, but he brought prototype technology with different characteristics than IBM and for several years this threatened to displace the company's basic standards, the electric reading of cards and the 45-column card layout. This shows the extent of the negotiation of the characteristics of punched cards for bookkeeping. The winning characteristics were developed for actual or imagined users, and

IBM chose them because of their demand among the users. The number printing tabulator had given the Powers company a competitive advantage until 1921. It was, however, not able to exploit this because it lacked the ability to design and produce reliable machines—and because demand for their machines was poor. Further, the Powers company and Remington Rand were only able to adapt to the subsequent shaping of punched cards for bookkeeping to adapt to the new closure, as it was established by IBM.

FIVE

Decline of Punched Cards for European Census Processing

In the spring of 1889, Hollerith was struggling to attain the order to process the United States census in 1890.[1] He eventually succeeded in this endeavor in late 1889, but in the spring of that year his limited income derived from his only customer, the Office of the Surgeon General of the Army. Late the previous year, Hollerith had postponed his wedding as he was not able to afford to establish a family.[2] However, this did not prevent him from traveling to Europe from April to May 1889 where he visited Paris and Berlin. By doing this, he entered on a well-established route of expanding his technology-based business in the United States to other countries. He traveled to find customers for his census machine, not to find inspiration for additional capabilities or applications. He was looking for more scale in his business, not greater scope. Hollerith considered punched cards as a tool to process population census returns, and he saw the international statistical community as an extension of the United States' statistical community, where punched cards originated. He sought out the various European national statistical offices, in the same way that he tried public statistical offices in the United States. This strategy was based on the assumption that the problems and opportunities in the United States also applied in other advanced nations.

Hollerith had prepared for his visit to Europe by filing patent applications in several countries in Europe, as he wanted—like numerous other inventors—to protect his invention on foreign markets.[3] Even before this, he had filed two patents on railway brakes in Germany, which indicates that it was an established practice by 1890 to file patents in Europe, even for independent American inventors.[4] These patents were filed by local patent attorneys, which show that international patent attorney networks already existed.

Foreign patent protection made it possible to adopt the low-cost strategy of producing the machines at home and relying on patent protection on foreign markets. But a foreign patent was neither an easy nor an adequate tool. Obtaining patent protection in Europe required patents in all the countries, and patent laws varied. For example, Austria, Belgium, and France required local production to make a patent valid.[5] The alternatives for Hollerith had been either to organize foreign marketing through independent agents or via subsidiary companies.

Hollerith received a guarded reception in Europe in the spring of 1889, and he returned empty-handed to the United States. Over the following years, he gained orders to process censuses in Norway, Russia, and France, but most European countries did not embrace his census-processing equipment. Hollerith's foreign business became a substantial part of his activities only about twenty years later. This raises the question of the reasons for his limited initial success in contrast with the success of his second campaign that started in 1902.

Early Punched-Card Users in Europe

While in Europe in the spring of 1889, Hollerith exhibited his tabulator in Paris at the Great Exhibition commemorating the centenary of the French Revolution. At this exhibition, he was awarded one of the five gold medals that were given to American exhibitors.[6] The first Great Exhibition had taken place at the Crystal Palace in London in 1851 and was a huge success. Scores of Britons and visiting foreigners enjoyed the great entertainment that was intended to display Britain's world supremacy. Several European countries and the United States contested this position, and a succession of "great exhibitions" followed at only few years' interval. Domestic and foreign inventors displayed the products of their endeavors, and the great exhibitions became important fora for technological exchange. The Paris gold medal demonstrated that Hollerith was noticed. He exhibited the same machine in Berlin, probably because Hollerith's parents were German immigrants.

Though Hollerith had prepared for his visit by filing patent applications in several European countries, he still needed local support. Punched-card machines were precision devices requiring qualified technicians. From the outset, Hollerith exclusively rented out tabulators, which shows his early appreciation of the importance of support and of the income from card supply. When installations became scattered over a wider area in the United States, even the ability to provide technical support within a couple of days necessitated the establishment of a support organization. Organizing support became crucial when foreign contracts emerged, and Hollerith could choose between using locals and Americans living in those countries. He initiated no subsidiaries in the 1890s.

Although Hollerith returned empty-handed to the United States in May of 1889, seeds were sown that would be germinated by references in American periodicals of the punched-card processing of the United States census in 1890. The eventual outcome was an Austrian clone production and orders from Norway, Russia, and France.

The location of the Austrian clone production was the precision making shop of Theodor Heinrich Otto Hermann Schäffler in Vienna.[7] In 1890, Schäffler requested and received a license to produce Hollerith's punched-card system for processing the Austrian census of 1890.[8] Probably, Otto Schäffler had seen Hollerith's exhibit at the Paris exhibition in 1889.

Schäffler had been trained as a mechanic and studied precision making in Württemberg, Paris, and London. In 1865, he established his shop in Vienna and for the next twenty-five years, his business was based on implementing foreign inventions of telegraph and telephone equipment in Austria.

In 1890 Schäffler repeated this business strategy by adopting production and support of Hollerith's punched-card equipment. As he had done earlier, he cooperated closely with the state, this time the Austrian statistical bureau. Based on the contact with Hollerith, Schäffler acquired a contract for the new production in his workshop. Hollerith furnished the plans and shipped a tabulator to ease the technology transfer, while Schäffler's company would build the machines for the Austrian census. In Austria, the arrival of the American-produced tabulator was considered a violation of the Austrian patent law that required patented devices to be domestically produced.[9] The patent was annulled, but Hollerith traveled across the Atlantic to Vienna and made a deal with Schäffler who then built the twelve tabulators used for the Austrian census in 1890.[10]

Schäffler improved the tabulator significantly by adding plugboard programming.[11] At that time, Hollerith programmed his tabulators by screwing wires to contact points. The inspiration for Schäffler's refinement was close at hand. His main production was telephone equipment, and telephone operators established electrical connections by use of plugs on switchboards. Plugboard programming eased the tabulator adjustments for a new application. In 1895, Hollerith also introduced plugboard programming in the United States.[12] This shows how the one-way transfer of punched-card technology as early as the 1890s prompted new inventions by the receiver, thus transforming the transfer into an exchange of technology.

Schäffler's version of Hollerith's punched-card system processed the Austrian census in 1890. The Imperial and Royal Statistical Commission (Kaiserliche und königliche statistischen Zentralkommission) had been established in 1863 as a joint statistical office for the area governed by the Habsburg monarchy. Following the segregation of Hungary as a separate kingdom in 1867, a new statistical office was organized in Budapest, while the former joint office in Vienna remained the national statistical office for Austria. The first Austrian census had been held in 1869 and the next followed in 1880. Processing both censuses was decentralized and based on household lists, which limited the accuracy of the returns and resembled the situation in the United States prior to 1850. Corresponding to the development in the United States, punched cards were introduced in Austria in 1890 as an instrument to enable central processing of individual records in the census, and the processing was centralized in Vienna.

Like in the United States, the Austrian tabulators were rented out but, after the completion of the census, the Vienna statistical office kept a few tabulators for processing various smaller statistical assignments. However, this application of punched-card processing ended in 1896 when Schäffler sold his company, and its punched-card business was discontinued. The next Austrian census was in 1900 but, based on their experiences ten years earlier, the Austrian statistical office did not consider punched-card processing a sufficient advantage to cause them either to find a local maintenance successor to Schäffler's company or to acquire equipment from the United States. They only resumed punched-card processing for the census in 1910 when they rented improved equipment for general statistics processing from the newly established agency in Germany of the Tabulating Machine Company.

During the 1890s the national statistical offices in Norway, Russia, and France adopted punched-card processing. The Institut international de statistique was the international statistical organization at that time and held a conference in Vienna in 1891, which was attended by statisticians from Europe and North America. Among the participants were the heads of the national statistical offices in Norway, Russia, and France, who also attended a presentation of Schäffler's version of the Hollerith original punched-card machine from 1890; this became the starting point for their punched-card deliberations.[13] But Hollerith—not Schäffler—came to supply equipment to these three countries, and no other country in Europe acquired a punched-card installation before 1904.

In Norway, Anders Nicolai Kiær headed of the national statistical office (Det statistiske Centralbureau). In addition to the Vienna conference in 1891, he participated in the international statistical conference in Chicago two years later, which shared a venue with a great exhibition. At the exhibition, Kiær saw Hollerith's punched-card system, and Hollerith offered Kiær a tabulator free-of-charge for tests in Norway. The machine arrived in 1894—as did Hollerith to assemble it—and the statistical office contracted with a Norwegian engineering office for maintenance. By then the Norwegians had completed processing the 1890 census.[14] Although it is true that the Norwegians had introduced individual records on the schedule at that census, processing by hand was manageable as there were only 2 million Norwegians. As the result of a test, the Norwegian statistical office bought the tabulator at a price of $1,100 (NOK 4,000), which was a little more than a one-year rental in the United States.[15]

Subsequently, this original Hollerith tabulator was used to process the Norwegian census of 1900 and for several small projects over the following years. It only proved advantageous to use this tabulator for large projects. For small projects, planning for punched-card processing, punching the information, programming, and manual reading of counters consumed too much time, compared with manual processing using written index cards or the original statistical forms. To process the Norwegian census in 1910, the statistical office rented improved equipment for general statistics processing from the newly established agency in Germany of the Tabulating Machine Company.[16]

Russia held its first general census in 1897, which was late compared with other European countries. Previous censuses had been held in various parts of the empire, and their returns had been counted locally. For the 1897 census, central processing of the returned record on each person was the major reason for introducing punched-card-based processing.[17] The inspiration originated from the international statistical conference in Vienna in 1891, which had been attended by the director of the Russian Central Statistical Committee, Nicolas A. Troïnitsky. The Russian government explored both possible punched-card suppliers: Hollerith in the United States and Schäffler in Austria. As Hollerith had no Russian patents, Schäffler was free to supply a punched-card system to Russia. However, Schäffler's punched-card business ended in 1896, as mentioned earlier, and Hollerith was awarded the contract to process the Russian census the same year.[18]

Of the foreign projects emerging in the 1890s, only processing the Russian census was comparable to the United States census in 1890. Russia's territory was 2.32 times bigger than the United States, and Russia had 129 million inhabitants, 70 percent more than in the United States at the census seven years earlier. The size of the Russian contract compelled Hollerith to organize delivery and maintenance differently from what he did in Austria and Norway. The Vienna installation used local production and maintenance, and his small Norwegian business was based on production in the United States, Hollerith's personal assembly of the tabulator, and local maintenance. For the Russian census, Hollerith chose a third solution. He used machines produced in the United States that were assembled in St. Petersburg and maintained by Western Electric's agent, an expatriate national of the United States. Hollerith approached Western Electric, as they produced some of his punched-card machines.

The French statistical office (Statistique générale de la France) approached Hollerith in 1896 to acquire punched-card equipment to process some of the returned data in their census that year. This became the last contact from one of the European national statistical offices that had been represented at Schäffler's punched-card demonstration in Vienna in 1891. The French statistical office had been represented by its head, Victor de Turquan. He also attended the international statistical conference in Chicago two years later and probably saw Hollerith's punched-card machine at the great exhibition there.

The French request once more opened the issue of assembly and maintenance in Hollerith's foreign business. Hollerith received the French request at a time when he had no income and, thus, no money to travel to France. Two years earlier, a futile attempt had been made to form an English stock company based on British capital to sell and support Hollerith's equipment in the British Isles. Now, he decided to rely on an American company in a neighboring field that was established in the foreign country, as he had done for his Russian business. He chose to organize his French business through the Library Bureau in the United States, which had locations in Paris and London.

Founded in 1876, the Bureau was an offspring of the American Library Association. By 1896, the company supplied all kinds of office machinery and equipment. Its growth was based on the systematization wave, which gained momentum in the United States in the decades around 1900. Hollerith arranged for the Library Bureau to supply and support punched-card equipment abroad. On this basis, the Library Bureau contracted with the French national statistics office to supply punched-card equipment to process the statistics of residents' occupation in the French census of 1896. Maintenance in Paris was provided by French master of mechanical engineering Lucien March.[19] However, the Library Bureau did not acquire any additional punched-card customers abroad, and Hollerith terminated his contract with their company in 1899.

French census processing before 1896 had been carried out by local authorities. For the census in 1896, statistical authorities outside the French national statistical office recommended centralizing census processing based on punched-card processing to obtain more detailed statistics, like those in St. Petersburg, Vienna, and Washington, D.C. Most of the 1896 census processing was done as usual by local authorities, but processing occupation statistics was centralized in Paris.[20] Thus, also in France, the punched card became a tool to facilitate detailed and centrally processed statistics.

Central processing of occupation statistics went well, but Lucien March considered the initial transcription of data to punch cards unnecessary. He also found the Hollerith system expensive and difficult to maintain.[21] Therefore, March built an exclusively mechanical machine, the *classi-compteur*, which applied a different principle than punched cards. The *classi-compteur* was used to process the French census of 1901, which was the first census to be entirely processed in Paris and was used for successive French censuses until the 1930s.

Dynamics of Punched-Card Diffusion in the 1890s

The accounts of the application of punched cards for census processing in Austria, Norway, Russia, and France can be complemented by the refusal of Germany and Great Britain to apply punched cards for census processing before 1910. The inclusion of these two countries extends the basis of analysis to information on six European national cases out of the about twenty nations. The rejection of the technology in Germany and Great Britain came in spite of canvassing by Hollerith and his associates in both countries.

Germany was a federation under Prussian leadership that had first been established in 1871. Prussia had 24.7 million inhabitants in 1890, which constituted 59 percent of the total population of the federation, and Germany had twenty-five additional states.[22] The various German states had held censuses before unification, and census processing remained the responsibility of the various states until 1939; the census endeavor was thereby divided into smaller, more manageable tasks. Prussia had introduced individual records as a basis for census processing as early as 1874, and this approach was applied in several additional German states for the census in 1890.[23] The state statistical office in Prussia found the price of $1,000 (marks 4,182) to rent a tabulator for a year exorbitant, and the task of census processing had an additional social objective; a thousand people with disabilities and recipients of unemployment benefit were employed to assist in the processing.[24]

Similar to the situation in Germany, the censuses in Great Britain were conducted separately in Scotland, England, and Wales. England and Wales had 29 million inhabitants in 1891.[25] The country had had a census every ten years since 1801, all of which had been processed in London and based on individual records from 1841. Consequently by 1890, they had managed to process five censuses with individual records, without any problems that could justify introducing punched cards in 1891. Every British census was enacted by Parliament as a special law, but processing was done by the office of the Registrar General, whose main task was the national register of births, marriages, and deaths.[26] Though census processing was not a permanent assignment, a few managers remained from one census to the next and who, together with papers kept in the institution, constituted the organizational memory.[27]

Furthermore, after the initial punched-card projects in Europe had been completed, activities died out. After 1900, no new European census was processed by use of the first Hollerith equipment from 1890. This was not caused by the emergence of improved systems, but by the limited success of Hollerith's original punched-card system.

The poor outcome was caused by the Europeans' assessment of technology and price and by Hollerith's limited sales efforts. The diffusion of the technology in Europe was curtailed by the sizes of the individual nations. In 1890, only four European states had more than 20 million inhabitants: France (38.1 million), Italy (30.5), Prussia (30.0), England and Wales (29.0), and Austria (23.5),[28] but this did not prevent Norway (2.0 million) from acquiring a punched-card installation. In addition to size, three factors were decisive in the assessment of punched cards by the European census offices: the introduction of central processing of census returns, the use of the individual as census unit, and the fact that permanent institutions had already been established to house censuses. Austria, France, and Russia introduced punched cards to manage the introduction of central processing and of the individual as the census unit. Austria and France both abandoned punched cards once they had assisted the introduction of these two features. Russia's next census was not held until 1926. Austria and England saw no decisive reason to change their way of processing the census returns, as they managed several times the central processing of individual census records by use of simpler means.

In contrast to the United States, the basic distinction was that all six European states had permanent institutions to house their censuses, which (except for Russia) maintained the expertise of census processing between two censuses. No state statistical office used their first punched-card system to process more than one census. Therefore, the original Hollerith punched-card system from 1890 only got a chance in Europe as a means to ensure the transition to centralized processing of one record per individual.

Hollerith's limited sales efforts constitute the second factor in the dynamics of census processing in Europe in the years from 1889 to 1900. He relied on his personal efforts and his positive customer references.

Europe was far away from his company in Washington, D.C., and his main sales effort was to write letters with attached copies of articles praising the technology. These efforts were not sufficient to conquer the highly atomized European census processing market of the approximately twenty national statistical offices.

When foreign business emerged, Hollerith organized his business on an ad hoc basis, and he did not develop a more general business strategy encompassing an organization in Europe to attend to maintenance. Some kind of representation in several European countries was needed to attract customers, but the size and number of the national statistical offices hardly enabled them to provide a basis for a durable organization.

SIX

Punched Cards for General Statistics in Europe

Europe embraced Herman Hollerith's technology for processing general statistics as the technology became stabilized in the 1900s. This contrasted with the lukewarm reception of his first punched-card system in the 1890s. The positive reception was the result of improvements to the technology and the establishment of marketing and maintenance organizations in several countries in Europe.

Foreign marketing could be organized through either independent agents or subsidiary companies. The independents provided the cheaper of the two solutions, while the subsidiaries were easier to control. Recent studies of the internal relations in current multinationals found significant differences between the agendas in a multinational company's headquarters and in its foreign subsidiaries. In particular, the studies identified distinctions based on the subsidiary's national business network, its own technological development, and its relations to its national government.[1]

The Tabulating Machine Company made the inexpensive choice of establishing agencies in London (1902) and Berlin (1910). These agencies were intended as simple agents to distribute equipment produced in the United States. However, the diffusion of the technology proved to be more complex. People in different countries demanded machines with varying capabilities that required them to be adapted. Simultaneously, challengers also emerged in Europe, as the Powers Accounting Machine Company established agencies in Great Britain and Germany. The outcome of this was diverging developments in various European countries, shaped by differing national traditions in private business and public organizations. Particularly, the differences in patent legislation in the various countries played an important role.

Slow Start of Punched Cards in Britain

The British reaction to punched cards had been negative in the 1890s. The English census office had decided not to use punched cards to process census returns and an attempt failed to establish a British agency in 1895. However, while Herman Hollerith had selected the continental locations for his demonstrations in 1889, journalist Robert Percival Porter selected Great Britain. Porter was an English-born, naturalized United States citizen who had headed the office of United States' census in 1890.[2] In 1894, he arranged the first introduction of Hollerith's punched-card systems to the British statistical community through a lecture in the Royal Statistical Society.[3]

Porter moved back to his native England in 1901 and there established contact with Ralegh Phillpotts, a lawyer who was well connected both in politics and in the military. They agreed to establish a Hollerith agency, where Phillpotts would act as general manager and Porter would be chairman. The following year, they obtained an option from Hollerith to organize an agency for the British Empire that would import American-produced machines at cost plus an overhead of 10 percent. To do this, they needed to raise $98,000 (£20,000), of which one half would be paid to the American company for assignment of patents.

A historian observed that this amount was comparable to the total nominal value of the shares at the incorporation of the Tabulating Machine Company in Washington, D.C., in 1897. However, the core of the original arrangement for the British agency was that they would be allowed to take possession of American-produced machines. At any rate, over the next two years they only succeeded in raising one-tenth of the required total amount, and they therefore persuaded the Tabulating Machine Company in the United States to accept payment in installments. On this basis, the British agency was incorporated in 1904 as the Tabulator Limited with an authorized registered capital of $24,000 (£5,000).[4]

The new company experienced a prolonged start-up, partly due to initial problems with the machines' ability to handle the shillings and pence of sterling currency. The slow build-up of business was a heavy drain on the finances of the company. More shares were issued, and the agreement with the Tabulating Machine Company was renegotiated for the second time, as the British company could not honor the agreed once-and-for-all payment for the Hollerith patents of £10,000. This caused the American company, still controlled by Hollerith, to introduce tighter control over its first agency.

Although Hollerith originally had agreed to sell his machines outside the United States, the Tabulating Machine Company now retained ownership of the machines and charged the British agency a royalty of 25 percent of their revenues from sorter and tabulator rentals. Consequently, due to the inability of the burgeoning British company to raise the considerable amount required to obtain the full patent assignment, the status of the British agency was reduced from being on equal standing with the American company to a situation comparable to one of the American company's domestic customers. In 1907, the British agency was renamed the British Tabulating Machine Company (also known by its initials of BTM). Soon the agency's business grew, and it authorized an increase in the share capital by a factor of 10 to finance machine leases. But the agency once again faced difficulties in raising capital; in fact, the problem of financing machine rentals would remain through the interwar years.[5]

Shortly after the launch of the British Tabulating Machine Company in 1907, its number of customers grew; the company had only two customers in 1908, rising to about thirty by 1914 and to thirty-four in 1916.[6] This progress was reflected in the company finances, which only showed a profit after 1911. Three years later, the shareholders saw their first dividend. In spite of the profits that started in 1911, in 1913 a row erupted with the Tabulating Machine Company over delayed royalty payments. The American company threatened to terminate the supply of machines, and this drastic measure made the British Tabulating Machine Company pay their debt.[7]

The technical basis for the British business had been established in 1903 by hiring Everard Greene, a graduate in engineering from Cambridge University. As Greene's job was to facilitate the transfer of the punched-card technology, he spent the first year studying the production of punched-card machines at the Tabulating Machine Company and their applications in the United States.[8]

Although there had not been much difference between the processing of census statistics in the United States and the statistics applications envisaged for the planned agency in 1895, it proved a challenge to handle sterling currency—an important capability as business statistics was becoming the prime application. Greene had studied this application in the United

States, and the British company started to cultivate this market. The machines were designed to accommodate the dollar currency of one hundred cents in a dollar, but in Great Britain at that time there were twenty shillings in one pound, each worth twelve pennies. (In 1971, the currency was decimalized and one pound sterling became worth 100 new pence.)

The sterling currency problem only appears to have been realized during the first trials at the Woolwich Arsenal Ordnance Factory in London and at a plant of Vickers, Sons and Maxims in Sheffield, as both these customers needed to process a large number of calculations involving money. The first tabulator was adapted for sterling currency at the Tabulating Machine Company in Washington, D.C., in 1905, and it used four adding machine wheels to represent shillings and pence. Later, the adaptation of tabulators for sterling currency moved to the British Tabulating Machine Company that, in 1908, started to assemble the American-produced machines and to manufacture cards. The British Tabulating Machine Company devised a simpler system of representing shillings and pence that only used three adding machine wheels.[9]

Back in 1904, Greene had started to sell punched-card systems, and his first trials had been industrial and railway applications similar to those he had studied in the United States. Further, he approached the Registrar General who was in charge of the British census operations. The first two organizations to give the machines a trial were the Woolwich Arsenal and the Vickers, Sons and Maxims Sheffield plant. Both projects were to tabulate the distribution of costs and wages to each job at the works, and both trials were unsuccessful.

Nearly half a century later, Greene ascribed the rejection at Woolwich to Luddites among the staff: wires in the machines were disconnected repeatedly, and the staff whistled the funeral march when the rejected machines were carried through the office. However, the Woolwich Arsenal staff's negative attitude was justified by two shortcomings of the machines. The problem of dealing with sterling currency was only realized during the Woolwich and Vickers trials, and a mechanized sorter only became available in Britain later. In fact, Vickers accepted the machines once they had been adapted for sterling and revenues started to flow to the British Tabulating Machine Company. During the years prior to 1916, at least nine similar installations were established at other industrial producers, and two more were acquired to process sales statistics.[10]

The British Tabulating Machine Company's third trial was for operational statistics at the Lancashire and North Yorkshire Railway in 1905. The project was to calculate locomotive mileage and consumption of coal and oil, and the Railway approved the punched-card system. During the years prior to 1916, five similar installations were established at other railways. In addition, by 1916, the British Tabulating Machine Company had four insurance statistics customers, and their machines processed public statistics in two towns and produced operational statistics for two gas and electricity utilities.[11]

Greene's persistent approaches to the census authorities eventually paid off as, in 1908, he was asked to provide price estimates and trial machines. The Registrar General conceded that the advantage of punched cards over manual card systems was generally recognized, but he considered the price excessive and was worried about relying on a foreign company.[12] Census processing only required counting tabulators, but the Tabulating Machine Company in Washington, D.C., by then only produced the more costly adding tabulators as they had terminated their business with the Census Bureau in the United States. However, the British Tabulating Machine Company, so eager to land the census contract, decided to build their own counters to replace the adding units on an American tabulator.

The British Tabulating Machine Company's engineers built a new counter, but it did not perform satisfactorily during trials at the Census Office and the prototype was scrapped. However, the trials concluded that the most appropriate design would be a tabulator equipped with thirty-six counters, in contrast to the five adding units on a standard tabulator from the American company. In spite of the short time available until the census, an improved counter was built and eight census-tabulators processed the census in England and Wales in 1911.[13] Tabulators were also used to process the census in Scotland in 1911. Further, in 1913, the British Tabulating Machine Company negotiated the sale of census-tabulators to Deutsche Hollerith-Maschinen Gesellschaft mit beschränker Haftung (Dehomag) in Germany. However, this contact was discontinued at the outbreak of the First World War.[14]

A challenger emerged when James Powers went to Europe in late 1913 and demonstrated his prototype punched-card machines in Berlin, notably before he started to produce machines in the United States. While

Hollerith—twenty years before—had used census applications as the basis for his first attempt to sell punched-card machines abroad, Powers went directly for business applications. The American Powers Accounting Machine Company paid an individual from each of three British organizations to spend a few days working with Powers' machines in Berlin, and two of these individuals were impressed by the printing capability of the Powers tabulator. Therefore, two of the three organizations, namely His Majesty's Stationery Office and the Prudential Assurance Company, became Powers' first customers in Britain.[15]

A British Powers agency, owned by the American parent company, was established in 1915 to market American-produced punched-card machines. It was named the Accounting and Tabulating Machine Company of Great Britain Limited. The printing tabulator provided a good sales pitch and the new company's business grew fast. In late 1916, after only one year's operation the company had eleven customers, compared with the thirty-five customers of the British Tabulating Machine Company.[16]

The British Powers agency seems not to have encountered problems with the Tabulating Machine Company's patent in Britain—in contrast to the situation in the United States and Germany. Hollerith's patent on the automatic group control facility was an important patent. Filed in the United States in 1914, it caused problems for the Powers companies in the United States and Germany. Hollerith's equivalent patent in Britain was granted in 1918, one year after Powers' similar patent had been granted. These patents seem not to have generated any conflict in Britain until Hollerith's automatic group control patent was granted in the United States in 1931.[17]

Following this, IBM in New York tried to persuade the British Tabulating Machine Company to enforce Hollerith's British automatic group control patent on the British Powers company. However, this attempt at enforcement was discontinued in 1933, which a historian explained as being the result of a gentleman's agreement between the two British companies.[18] Gentlemanly behavior might well have been a reason for the absence of patent litigations on the punched-card trade in Britain, but the patent laws provide a more immediate explanation.

Herman Hollerith's automatic group control patent had been filed in the United States in 1914, but it was only granted in 1931 when it gave rise to IBM's enforcement attempt in Britain. As patents in the United

States were valid for seventeen years from the day they were granted, this patent lasted until 1948—that is, thirty-four years from the date the application was filed.[19] In contrast, the equivalent British patent was filed in 1917 and granted in 1918.[20] The British patent expired in 1931, as a British patent was valid for fourteen years from the date when the application was filed.[21]

British patent law resembled the patent laws in Austria, Belgium, and France with respect to protecting British production against imports. Production in Britain of the patented device was required within four years of filing the patent application.[22] This requirement explains why the British Tabulating Machine Company never seems to have tried to enforce Hollerith's sorter patent that had been filed in the United States back in 1901.[23]

It was first in 1921 that the British Tabulating Machine Company started to assemble machine components imported from the parent company in the United States, and only in 1924 did they commence assembly of the Tabulating Machine Company's first printing tabulator with automatic group control. This suggests that neither the sorter nor the automatic group control patent could have been enforced in Great Britain. Finally, the British patent laws facilitated enforced licensing of patents, which was encouraged as a private agreement between patent holder and licensee.[24] These clauses explain why the British punched-card trade experienced no patent litigation during the 1930s, in contrast to the situations in France, Germany, and the United States.

At the outbreak of the First World War, the British armed forces totaled 733,514; this was a relatively small number compared with the German and French armies. However, the British armed forces were rapidly expanded to 4,231,670 people in December 1916 and to 5,757,457 people in December 1917.[25] But establishing an army of more than five million people for the large-scale warfare of the First World War posed extensive demands on production and transportation. The rapidly expanding mobilization and warfare put great pressure on the British offices that were to process more transactions with fewer staff. British historian Martin Campbell-Kelly has investigated office mechanization at two big private and one big public organization, and his study reveals a rather limited level of mechanization in British offices.[26]

His Majesty's Stationery Office was responsible for printing and sale of government publications and, for several years, the Stationery Office had shown a keen interest in mechanization. Their records confirm the modest level of office mechanization in government in 1914.[27]

When war broke out, the Stationery Office had just concluded trials of the Powers punched-card machines, which resulted from an invitation to inspect his machines the previous year in Berlin. The Stationery Office was impressed with the possibilities offered by the printing Powers machines for their accounting and statistical tasks, in contrast to the nonprinting machines from the British Tabulating Machine Company. The war meant that many of the Stationery Office's experienced clerks were mobilized and, simultaneously, the office was confronted with an enormous increase in their workload because of processing supplies to the military. The Stationery Office discarded their system of accounting by hand and introduced Powers punched-card machines; the printed lists and totals enabled the office to retain their well-established system of lists and receipts for their audit.[28]

During the war, punched cards became an important tool in producing the statistics needed to monitor the war effort, resulting in rising revenues for the two British punched-card agencies. During the early part of the war, armaments policy was organized by free enterprise, meaning that the government contracted with companies for arms and munitions, and direct control was rare. However, this system did not supply the vast munitions needed for the scale of warfare. In 1915, a separate Ministry of Munitions was established which, gradually, incorporated the country's industrial capacity into the armament industry.[29] A statistics office in the Ministry of Munitions monitored this development and, to do this, they acquired a punched-card installation from the British Tabulating Machine Company.[30]

Punched-card installations were also acquired, for monitoring purposes, by the National Service Department, Ministry of Labour, the Admiralty (the ministry responsible for the Navy), the War Office (responsible for the Army) and, after the war, this effort was followed by several institutions dealing with demobilization and pensions.[31]

The Quick Success of Punched Cards in Germany

Herman Hollerith's agency in Germany was established in 1910 as part of a new strategy to establish agencies in Europe. Whereas the British agency had arisen unsolicited, now Hollerith contracted with American engineer Robert Neil Williams to form companies on the European continent to sell Hollerith's machines and manage his patents. Williams had engineering offices in Berlin and Paris, and he embarked on his assignment in Germany.

Williams approached Carl Duisberg to obtain money to establish a company. Duisberg was managing director of Farbenfabriken vorm. Friedrich Bayer and Co. (Bayer), which was the ninth biggest company in Germany.[32] Duisberg was contacted because he already had visited Hollerith's company in Washington, D.C., and had ordered a punched-card machine to process detailed statistics on Bayer's turnover figures. However, Duisberg refused to invest in a German agency, as he thought the German market for punched cards too small. In Duisberg's opinion, punched cards were only suitable for very big industries and, moreover, only a few of these were interested in systematic management.[33] However, Williams got in contact with Willy Heidinger, who was the director of a small company representing the American Elliott-Fisher mechanical office machines in Germany.

Unlike Duisberg, Heidinger saw market potential in establishing an agency in Germany of the Tabulating Machine Company. He raised the necessary funds, $71,000 (marks 300,000), and the German agency was established in 1910 as the Deutsche Hollerith Maschinen Gesellschaft mit beschränkter Haftung (the German Hollerith Machine Limited Liability Company, abbreviated to Dehomag); 91 percent of its capital was held by Heidinger and 9 percent was owned by Williams, who shared the position of CEO with Heidinger. Dehomag gained the right to market machines produced by the Tabulating Machine Company on the same conditions as in the contract with the British Tabulating Machine Company in 1908. Dehomag would pay royalties of 25 percent of the revenues from sorter and tabulator rentals and the Tabulating Machine Company would retain ownership of the machines.

Before the First World War, the royalties consumed 12 percent of Dehomag's total revenue, 48 percent of which came from sorter and tabulator rentals. The agreement of 1910 granted Dehomag the exclusive rights to sell and lease the American company's machines and cards in Germany, Switzerland, and Scandinavia. In addition, Dehomag could sell German-produced punched cards, but they were not allowed to manufacture the Hollerith machines.[34] To increase the Tabulating Machine Company's sales in Europe, in 1914 the American company expanded Dehomag's territory to include Austria-Hungary and the Balkan states.[35]

In addition to funds, Heidinger brought with him his business network from selling office machines. From this Dehomag soon built up a customer base, both statistical offices of the German Reich, the states (*Länder*), and industry, where punched-card processing of operational statistics became a big success. During the two first years, Dehomag obtained contracts for equipment to process the census in 1910 in four of the German states, which together comprised 16 percent of the population.[36] Further, Dehomag opened a new market segment of municipal statistics processing when the town governments in Berlin and Cologne placed orders for punched cards for their statistics processing.[37]

Big industries constituted another group of early customers. Williams' original approach to Duisberg of Bayer shows that Dehomag, from the outset, targeted this market. True, Duisberg refused to invest but, in the following year, his company acquired a punched-card installation for sales statistics in addition to their original application. Punched cards enabled more differentiated statistics than did other methods. Duisberg had reorganized Bayer in the period from 1891 to 1907, and a key element had been to introduce various kinds of statistics to monitor the activities in this large, diversified company. Originally, these business statistics had been compiled in a newly established statistics department using less sophisticated technology than punched cards.[38] Punched cards were subsequently introduced to enhance the production of extensive operational statistics. Similarly, punched cards were used to produce operational statistics in other big German companies, including heavy industries.[39]

Census statistics had been the initial field for punched-card application in the United States, then came railway and insurance statistics. In 1912, Dehomag launched a sales promotion periodical, the *Hollerith Mitteilungen* (*Hollerith Bulletin*). The first issue argued for the introduction of punched cards for railway statistics in Germany,[40] and the second issue argued for insurance statistics.[41] However, railway applications proved to be no immediate success. The state railways in Württemberg and in Prussia considered using punched cards but found that the machines were not suited to handling their freight accounting and statistics. The reason for this was most probably the lack of a printing tabulator in Hollerith's system. The same applied to trials by several other customers: The Hapag shipping company in Hamburg tried punched cards for bookkeeping tasks and the Berlin Post Office tested it for the calculation of interest, although, in the end both decided not to adopt punched cards for these tasks.[42]

Dehomag also tried, without success, to attract the *Grossbanken* (large banks) by making a proposal to audit their vast securities portfolio by use of punched cards.[43] In the banking field, too, the absence of a printing capability seems to have been the reason for the lack of interest as banks became punched-card customers in the 1920s when Dehomag started to supply printing tabulators.[44]

Back in 1913, James Powers demonstrated his tabulating system in Berlin, an event attended by Heidinger and Williams of Dehomag. Powers' printing tabulator and his horizontal sorter offered major advantages over the equipment from the Tabulating Machine Company. The American Powers company proposed the founding of a common German company to market machines from both producers in the United States. This would cause no technical problems, as both lines of machines were based on the same 45-column punched-card and, in fact, several installations in Europe in the 1920s used a combination of machines from both producers.[45] The American Powers company offered to invest in Dehomag to acquire half of the shared capital. Heidinger and Williams could not agree on this proposal. Heidinger wanted to remain exclusively with the Tabulating Machine Company marketing only their machines, while Williams wished to merge the two agencies to market both lines of equipment. Unable to resolve their disagreement, they went their separate ways. Williams established a German Powers agency, Deutschen Gesellschaft für Addier- und Sortiermaschinen mit beschränkter Haftung (German Company for Adding and Sorting Machines with limited liability) in Berlin and became Powers' first manager for Europe. Williams and Thomas J. Felder provided the required capital of $24,000 (marks 100,000) to establish the new agency. Like Williams, Felder was a citizen of the United States living in Paris.[46]

Dehomag considered Powers' printing tabulator a major threat to their business—and for good reason.[47] For its part, Powers' weak point was the dependence on basic Hollerith patents with valid equivalents in Germany. In the United States, the Powers company would enter a license agreement with the Hollerith company in 1914 that required them to pay about 20 percent of gross revenues to the competing company. For Powers, a common Hollerith-Powers company in Germany was a way to avoid having to pay royalties in Germany. Heidinger responded to the creation of the German Powers company by filing a patent infringement suit in 1914. This prevented the Powers company from doing business until the German High Court, in 1916, ordered a compulsory license of 5 percent on machine sales and rentals. Two years of waiting ensured Powers a considerably better arrangement in Germany than in the United States.[48] However, only few months would pass before the United States entered the First World War in April 1917, providing Powers with a narrow window of opportunity.

Before the High Court verdict was issued, the German Powers agency had established a test punched-card installation at a major pipe producer, Mannesmannröhrenwerke, in Düsseldorf. Mannesmannröhrenwerke used punched cards in their central bookkeeping department to monitor their current account bookkeeping at the end of every month. For this task, a printing tabulator was essential. The pending patent case prevented the arrangement from becoming commercial, but the German Powers company supplied the installation and cards free of charge, seizing the opportunity to establish a showcase at a well-known company.[49] Lack of income as a result of the patent litigation drew heavily on the company's basic capital, and American financier John Isaac Waterbury bought all shares of the agency in 1914, just as he had acquired the American Powers company the year before.

After the German High Court verdict in 1916, the Powers agency started to receive revenues from its machines at the Mannesmannröhrenwerke, but the United States' entry into the war in 1917 prevented the agency from getting additional machines and caused it to become subject to German custodianship.[50] (Williams was a citizen of the United States, but he had left Germany before the United States entered the war.) In the same year, all Powers' equipment in Germany was requisitioned to produce war-related statistics, which also provided revenues. However, the cost of running the agency consumed both revenues and capital, and the agency vanished.[51] It was not until in 1923 that a new German agency was established and Powers' business really began in Germany.

At the start of the First World War, the Allies imposed a blockade on the Central Powers, which Germany came to manage for four years. In order to manage the economy under the blockade, a comprehensive system was established for sharing out provisions and raw materials, organized within the Ministry for War and through a large number of special war companies, like the company for the supply of bread grain (Reichsgetreidestelle, Geschäftsabteilung Gesellschaft mbH) and the company supplying sauerkraut (Kriegsgesellschaft für Sauerkraut mbH).[52] However, the various companies managed their limited fields without very much overall coordination, in marked contrast to the United States where the economy was controlled through fewer organizations.

At the national company for the supply of bread grain, a small Dehomag punched-card installation produced a variety of statistics on the sale and consumption of bread grain and flour which they called control bookkeeping (*Kontrollbuchhaltung*). It was based on copies of all the receipts when bread grain and flour was traded and enabled the monitoring of the supply of bread grain, the production of flour, and the use of flour in bakeries and in retail outlets.[53] Like in many big German companies, this process was organized through extensive operational statistics and bookkeeping; since manpower was in short supply due to the general mobilization, the process relied heavily on machines. Both Dehomag and Powers equipment was used for this purpose.

Expansion of business was difficult during the First World War, as all American supplies ceased in 1917 following the United States' entry into the war. But by then Dehomag had a substantial number of punched-card machines in Germany, their revenues rose, and the company increased dividend payments from 4 percent in 1915 to 8 percent in 1916 and to 10 percent in 1917.[54] The American entry into the war caused Dehomag to become subject to custodianship, as some of the shares were held by an American citizen, but IBM later acknowledged that the custodianship did not curtail Dehomag's business.[55]

Much effort was put into keeping all the machines running, and, in 1918, Dehomag established a small factory in the town of Villingen in the Black Forest to rebuild old machines and produce spare parts. It was a major limitation in the contract with the Tabulating Machine Company that Dehomag had no authorization to produce American machines. Dehomag tried, in vain, to obtain a German government order to break this clause.[56] A government order would have relieved the company of the responsibility of breaking their contract with the parent company. Heidinger made the best of the situation during the war to improve his financial position and his standing in relation to the American parent company. With this end, Dehomag established its own technological basis through building its own punched-card machines. This attempt was based on construction work by engineer Heinrich Tolle, hired in 1916, who had experience with calculating machines and punched-card sorter development.[57] Tolle's development work aimed at building independent punched-card machines, although this goal was not reached by the end of the war.[58] In the years just after the war, the company finalized the design of two complete punched-card machines: a sorter and a punch. The greatest achievement was a number printing unit that could upgrade the nonprinting tabulators already available from the Tabulating Machine Company as these enabled the company to win customers requiring printing capability.[59] However, these inventions never came into production, and Dehomag started to receive American-produced number printing tabulators in 1923.[60]

Germany proper was untouched by acts of war as hostilities ceased in November 1918, but industry was worn down by four years of war production. Further, five years of crisis followed caused by harsh peace terms with huge reparations and the new German government's lack of ability to control the economy. The outcome was runaway inflation and soaring exchange rates, which created a catastrophic business climate.

Under these circumstances, Dehomag's success during the war turned into a liability, as a large part of their revenues had come from tabulator and sorter leasing. As these machines remained the property of the Tabulating Machine Company, Dehomag was required to pay royalties that had to be paid in dollars. During the war, Dehomag was unable to pay their royalties due to currency restrictions, which was serious as the amount in dollars grew in step with the falling exchange rate of the German mark, and additional royalties accrued after the war.

Dehomag's position became critical in the years immediately after the war. Willy Heidinger tried to postpone payment until the mark regained its strength to save his Dehomag stocks. But the mark continued to fall and, in 1922, the debt rendered Dehomag insolvent.[61] At the same time, the Tabulating Machine Company was pressuring to receive their money and, in 1922, Thomas J. Watson arrived in Germany to settle the matter.

The Tabulating Machine Company planned to take advantage of the opportunity to replace the foreign-owned agency with its own subsidiary.

However, Dehomag's success was a strong argument against breaking all ties and starting from scratch in Germany, as IBM later did in Norway in 1935 where a subsidiary succeeded an unsuccessful agency.[62] The 1922 settlement made Dehomag a subsidiary of the Tabulating Machine Company, which acquired 90 percent of the shares, while Heidinger kept the remainder. This ensured that he stayed in the company where he kept de facto control as long as his success lasted.[63]

Assigning operational autonomy to the nationals leading the subsidiaries emerged as a key element in IBM strategy and was found in several European subsidiaries and agencies, in sharp contrast to the limited delegation in IBM in the United States. At the same time, however, Watson closely controlled the Dehomag board.[64] This strategy ensured Heidinger remained with Dehomag with his business network and abilities to produce profits which, in turn, yielded revenues to the American company. But Heidinger's attempt during the First World War to use the German government against the American company indicated that his loyalty could cause problems if conflicts again should arise between Germany and the United States. Probably Watson never learned about Heidinger's double-dealing, but Heidinger resented Watson's capture of Dehomag. The elimination of Austria and Hungary (which were separated after the war), as well as Switzerland, Scandinavia, and the Balkan states from the company's business territory represented major reduction in Dehomag's scope of business. In accordance with the American company's new business strategy, Dehomag was reduced to serving the German market.[65]

The Late Start of Punched Cards in France

Except for Austria, France was the first of the major European countries to introduce punched cards for processing census statistics in 1896, and the last of these countries to introduce them for general statistics in 1921. France shared Austria's somewhat negative assessment of Herman Hollerith's first punched-card system created to process census returns in the 1890s. However, while the Austrian census office refrained from applying punched cards for processing the next census, their French counterpart commissioned the building in France of an alternative device to facilitate processing their census returns.

French engineer Lucien March was in charge of the first punched-card application in France that was used to process employment statistics in the census in 1896. The punched-card equipment leased from Hollerith was able to carry out this task, but March was not sufficiently satisfied with it to use the system again. He found the initial transcription of data onto punched cards unnecessary, the system too expensive, and maintenance costs substantial. His assessment of the technical aspects of Hollerith's simple electromechanical device was based on his training as a mechanical engineer. Subsequently, March built the *classi-compteur* to facilitate processing population census returns. The *classi-compteur* applied a different principle to Hollerith's machine and was exclusively mechanical. This equipment was used to process the French census in 1901.

Lucien March was trained as an engineer at the prestigious École Polytechnique (Technical University) in Paris that gave its students access to senior positions in the civil service.[66] The curriculum at École Polytechnique placed great emphasis on theoretical subjects, such as mathematics and physics.[67] March followed the French scientific engineering tradition by contributing to a large number of publications. He also contributed to the development of the science of statistics, national statistical education, and international collaborations in statistics.[68]

March's *classi-compteur* had a full keyboard of sixty keys, each connected to a counter that consisted of four ten-digit printing wheels placed in the rows on the lid behind the keyboard, enabling the counting of up to 9,999 entries in each of its sixty positions.[69] When the *classi-compteur* was used to compile a table, each of its keys corresponded to one entry in the table. Processing was accomplished by the operator keying in the relevant information from a form. During this process the key or keys pressed stayed locked in a lower position, and the operator could correct an error by first unlocking the key or keys and then entering the information correctly. When all the information from the form was entered, the operator pulled a handle that made the counters connected to the pressed keys advance one unit, and the operator proceeded to the next form.

When the operator had completed entering the information for a table, she placed a sheet of carbon paper above the counters and tilted the movable frame that held a series of rollers between which paper ran from a large roll. This left an impression of the printing wheels on the white paper. The rollers moved the paper along to a new zone of blank paper, and the

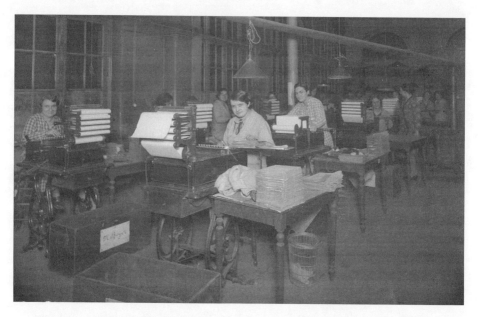

Women working on *classi-compteurs* in the French national statistics office, c. 1930. (Reference 9183-1, Agence photographique Roger-Violet, Paris, France)

printed area was torn off. The subsequent aggregation was accomplished manually or by use of an adding machine. March's *classi-compteur* was a mechanical engineer's response to Hollerith's simple, electromechanical design, possibly inspired by an Italian design from 1881 that was never applied in actual census work.[70]

It was simpler to organize the compilation of statistics by use of a *classi-compteur* than by using punched cards. On a *classi-compteur,* one person could carry out the whole task of processing the data, one table at a time, using one technical device—but the device had four limitations. First, producing the tables required onerous physical handling of the original forms, which damaged the forms themselves. Second, the sixty keys severely limited the size of a table that could be compiled in one operation. Bigger tables were broken down into part-tables that were later merged manually, for example, the table of age, sex, and marital status for the French census of 1911, which held 1,150 entries for each county (*département*) as opposed to 166 entries in the similar manually processed tables from the census in 1891.[71] Third, it was not possible to verify the compilation of a table except by reprocessing all data. Fourth, operating a *classi-compteur* was hard work, as the activation of the counters, the printing operation, and the rolling of the paper was done manually. Later the physical strain was relieved by the introduction of electrically powered handling of these operations, a solution resembling the U.S. Census Bureau's improvement, for the census in 1910, of the reading mechanism on the Hollerith tabulators from 1890.[72]

As with punched cards, the *classi-compteur*'s main advantage was to facilitate dividing the work of statistical processing among female machine operators with modest training, while the mainly male statisticians edited and interpreted the results for various publications. The French national statistics bureau (Statistique générale de la France) bought *classi-compteurs* to process the census in 1901.[73] In 1900 the first model of the *classi-compteur* was completed, 157 machines had been build by 1912, and it was still being produced in 1930.

The *classi-compteur* seems to have been marketed exclusively for processing population censuses. It was applied at the Belgian national statistics office in 1912, at the Dutch national statistics office in the 1920s, and the producer tried, in vain, to sell the machine to the English census office in 1912.[74] At the time of its invention in the late 1890s, the *classi-compteur* was comparable as a census processing tool with Hollerith's punched-card system from 1890. Punched cards only became a superior tool to compile statistics when the Tabulating Machine Company introduced the subsequent improvements: sorting cards by machine, automatic group control, and number printing.

In France, *classi-compteurs* were used to process the censuses between 1901 and the 1930s. During this period the French national statistics bureau was in a weak position compared with other national statistics offices. While other national statistics offices grew due to the importance ascribed to their role by politicians and civil servants, the French national statistics bureau enjoyed limited political support. The bureau used twenty-five *classi-compteurs* but did not receive any appropriation for a punched-card installation. In 1921, they were presented with a British or United States Powers sorter and several punches that were used to process statistics relating to the movement of the population.[75]

Any lack of progress by the French national statistics bureau was not caused by passivity in the statistics community. In 1920, a professional committee proposed improvements at the bureau, which included the acquisition of a punched-card installation. This wish was repeated several times during the following twenty years. Further, three statisticians connected to the French national statistics bureau established the Institute of Statistics at the Université de Paris in 1922 to improve training in statistics in France. The reasons for the lack of modernization are, thus, to be found either in the management of the French national statistics bureau or in its supplier of assignments, the French state. To some extent, the lack of development of statistics production at the French national statistics bureau between 1900 and the 1920s was a manifestation of the weak French state.

France during the seventeenth and eighteenth centuries had experienced strong state intervention in the economy, but the intervention of the state was rather limited during most of the French Third Republic (1870–1940). At that time, the French economy was weak, the country had perpetual trade deficits, and French industry lost market shares abroad.[76]

The First World War exposed the weaknesses of the French government. Like other belligerent nations, France failed both to anticipate a protracted war and to make adequate preparations for munitions production. During the hostilities, neither the civilian authorities nor the military high command played a prominent role in managing munitions production. In 1916, the government entrusted the federation of employers in the iron and steel industry (Comité des Forges) with the control of the import of pig iron and steel, and only in July 1918 did the government establish centralized control over pig iron production in France.

The war also put pressure on the supply of food. France had been a net exporter of food before the war but after it started French agricultural production declined drastically, due in part to mobilization and, further, to Germany's occupation of northern and eastern parts of the country. As a result, France became a major importer of food, leading inevitably to price increases. However, the government only introduced price control on grain in October 1915, a measure that was subsequently extended to other basic foods. State monopolies that controlled the entire supply chain from producer to consumer were established for sugar (1916) and grain (1917). But a ration card for bread was not introduced until the summer of 1918.[77]

Punched-card technology was not used in France for monitoring munitions production or food supply during the First World War: none of the introduced regulations demanded extensive statistics. For example, the import and domestic production of iron and steel could be monitored more easily in other ways, and ration cards were a simple means to distribute the supply of a commodity to the entire population.

It was first in the 1920s that the French state started to become more interventionist to improve the country's economic performance. From 1923, the government started to extend the telephone network to rural areas and to bring electricity to the countryside as well. From the mid 1920s, investments followed in social housing, transport, and the education system. In addition, the government encouraged mergers between French companies from 1928 to 1932.[78]

It would be tempting to assume that the weak governments in the Third Republic were the reason for the slow progress at the national statistics bureau. This assumption is substantiated by the absence of any major statistical study of French industrial development between 1860 and 1931.[79] However, the management at the national statistics bureau did not take advantage of the situation when the government intervention increased. The French censuses in 1931 and 1936 could have served as an opportunity to extend staff and to improve the equipment used to process the returns. After all, by then the *classi-compteurs* had been in use for thirty years. Moreover, one government punched-card installation was established in 1926 and another in 1927. Responsibility for the slow progress at the French national statistics bureau should primarily be attributed to the management of the bureau.

In contrast to Germany, Great Britain, and the United States, France had neither significant punched-card applications at the national statistics bureau nor punched-card processing of business statistics during the first two decades of the twentieth century. The first commercial punched-card application in France was established in 1921. The late introduction of this technology can be explained in two ways. First, by the assumption that punched-card processing at the national statistics institutions in other countries served as a basis for the diffusion of the use of punched cards beyond that of population statistics.

The second explanation is based on the combined observation of the late introduction of punched cards for processing business statistics and of the late establishment of large integrated companies in France and French firms' hesitancy to diversify into new products or processes. The former supply explanation accounts for the late establishment of a subsidiary of the Tabulating Machine Company in France. However, the late start of punched cards can also be explained by the late emergence of demand for equipment to compile business statistics, as the diffusion of punched cards in France coincided with the breakthrough of industrial rationalization. Further, the demand explanation reaches beyond the emergence of punched cards to give reasons for the subsequent use of punched cards for business statistics and bookkeeping in France.

Since 1906, a few companies, like tire producer Michelin and car producer Renault, had introduced time and motion studies of their working routines and processes as a step toward improving productivity, along the lines suggested by the American engineer Frederick Windslow Taylor. But it was only after 1920 that many other companies joined them. Similarly, Michelin and Renault were among the pioneers introducing punched cards to process their operational statistics in the early 1920s. In the 1920s, the rationalization in France was not confined to the factory floor. Also, offices were rationalized, which included the adoption of new accounting methods.[80]

The Tabulating Machine Company established a French subsidiary in 1920, Société internationale de machines commerciales (SIMC, translated as the International Business Machines Company). It is true that the Time Recording Company, another of the companies in the Computing Tabulating Recording Company, had established its own subsidiary in 1914, but it did not do any punched-card business. SIMC marketed punched-card equipment, while the other subsidiary sold clocks and time recorders. The two companies merged in 1935 to form a new company named Compagnie électro-comptable (CEC, or Electric Bookkeeping Machine Company).[81]

The Tabulating Machine Company's subsidiary in France won its first punched-card customer in 1921 at the French subsidiary of SKF (Svenska Kullager Fabrik, the Swedish ball-bearing producer), which leased the punched-card equipment to generate their statistics. Over the next few years, the number of customers grew to twelve in 1923, to twenty in 1925, and to thirty in 1927 with a total of fifty-six tabulators and fifty-one sorters, a growth also reflected in the French IBM's companies' aggregate net earnings.

Several large French companies followed the lead of the French subsidiary of SKF during the 1920s. First, there was a group of industrial producers: Renault (1922), Thompson-Houston (1923), Compagnie Continentale des Compteurs (1923), Roger Gallet (1924), Citroën (1925), and Société Kodak-Pathe (1925). Second, SIMC attracted a railway company as a customer in 1923 and two additional companies in 1925.[82] Public sector diffusion of punched cards started later in France than in the private business sector. Only in 1926 did the Caisse des Dépôts et Consignations[83] acquire the first punched-card installation in the public sector, and four more public institutions introduced the use of punched cards over the next six years, including three ministries.[84] All seem to have used their punched-card installations to produce various statistics and sorted lists, for example, as a basis for monitoring and controlling payments.

As early as 1922 progress was so promising that the subsidiaries of the Computing Tabulating Recording Company (the outcome of a merger in 1911 that included the Tabulating Machine Company) bought premises at Vincennes, near Paris, for a machine shop which they opened the following year. The aim was to be able to assemble machine parts and components imported from the United States, as the import of components rather than complete machines was subject to more flexible customs regulations. The intention could also have been to secure French patent protection, which required the patented device to be produced in France within two years of the filing of the patent application.[85]

This explains why the French IBM companies never took legal proceedings against Powers in France for infringement of the patent—rather like the situation in Great Britain. The French application for Hollerith's important automatic group control patent had been filed six years before the Vincennes machine shop was established, meaning that this patent was not valid in France.[86]

France became a battleground between the American Powers company and the British Powers company. In 1918, the British company Morland and Impey acquired the right to market the American Powers machines in France. Morland and Impey was the British importer of the Kalamazoo loose-leaf ledger system from the United States, and they had built up an extensive Kalamazoo business in France. Furthermore, Morland and Impey enjoyed good relations with the British Powers company and managed to obtain an additional agency for the British Powers company's machines in France.

As all Powers machines during the 1920s were based on Hollerith's numerical standard punched card from 1907, it was no problem to use a combination of machines from the two Powers producers. The Powers business in France prospered, and it was incorporated as the Société anonyme des machines à statistique (SAMAS, translated as Statistics Machines Company Limited) in 1922. However, the American Powers company cancelled their French agency the following year and established a French subsidiary company, while SAMAS continued to market the British Powers machines. The outcome was a lawsuit filed in France by the American company against SAMAS for patent infringement.[87] The American company lost the case, as neither of the companies had any machine production in France that invalidated the French patents.

The link between SAMAS and the British Powers company was strengthened in 1929 when the British Powers company bought a controlling interest in SAMAS. In 1936, the American Powers company merged their businesses in France with SAMAS, so that it again became an agency for both the American and British Powers companies. This arrangement only lasted until 1939 when the American Powers company sold their French interests to the British Powers company.[88]

SAMAS seems to have done well during the 1920s, and in 1927 the company had thirty customers—the same number as IBM had in France.[89] In 1931, the SAMAS customer base included the Finance and Naval Ministries, several banks, four railway companies, and many insurance companies.[90] This information apparently conflicts with information from the French IBM company, which claims that during the 1920s the same railway companies and the Ministry of Finance used their equipment. There could be several reasons for this: First, each institution could have had two separate installations. Second, the conflicting information could have been caused by change in supplier. Third, they could have had combined installations, as both suppliers used the same 45-column punched card.

Dynamics of Technology Transfer and Adoption

The extensive demand in Great Britain, Germany, and France for punched cards to process general statistics after 1904 contrasted sharply with the lukewarm reception to Hollerith's first punched-card system in the 1890s.

The lack of demand in the 1890s was caused by the exclusive focus on processing census statistics. This was an area in which Europe had efficient organizational structures—in contrast to the United States. Further, the situation was exacerbated by the absence of organizations for selling and maintaining punched-card equipment in Europe.

Within this short time-frame the situation had changed significantly. Hollerith's punched-card technology had been improved in the United States to facilitate the processing of general statistics, thus providing access to the much larger market for operational statistics in private companies and public organizations. In addition, organizations for selling and maintaining the installations were established in Great Britain (1904), Germany (1910), France (1920), and several other countries in Europe in the 1920s.

The rate of diffusion varied in Great Britain, Germany, and France. The technology spread faster in Germany than in Great Britain, and France saw the lowest rate of the three countries. The difference between Britain and Germany is clearly reflected in the slower increase in revenues for the British Tabulating Machine Company than for Dehomag. A historian explained this performance as being the result of inadequate leadership of the British Tabulating Machine Company.[91] This was endorsed by Dehomag's rapid establishment of an efficient sales organization within the first few years of its existence, including the publication of a regular sales bulletin. The British Tabulating Machine Company only established a nationwide sales organization in the early 1920s and first started to publish a regular sales bulletin in 1936.

This explanation through leadership in the European subsidiaries is complemented by an explanation based on different demand. In the 1910s and 1920s processing operational statistics became the prime application field of punched cards in all four countries. Alfred D. Chandler compared the forms in big industrial companies in the United States, Great Britain, and Germany.[92] He documented the importance of the development of hierarchical organizational forms for the development of big industrial enterprises. He found the United States to have the most complex hierarchies, closely followed by Germany, while Great Britain had smaller hierarchies in their industrial enterprises. This observation is supported by the differences in the speed of punched-card diffusion in the three countries in the 1910s and the 1920s.

Chandler's analysis did not include France. However, French economist Maurice Lévy-Leboyer compared French companies with businesses in Britain, Germany, and the United States. He found that French companies lacked industrial integration, so that the size of the firm, measured by total assets, was smaller than in other big industrial economies.[93] This offers a further explanation for the lower use of punched cards in France in processing operational statistics. However, it cannot explain the total absence of punched-card processed operational statistics in France until 1921. There were companies in France that were rationalizing, diversifying, and introducing new organizational forms before the First World War. In Germany, director Carl Duisberg of Bayer ordered a punched-card installation from the United States before the Dehomag agency was established to process business statistics. A French punched-card installation for business statistics in the 1910s would only have required one comparable enterprising manager—and several such individuals became crucial in the subsequent history of the development of punched cards in France.

It is also essential to look at adaptations, as the later transfer of punched-card technology can be distinguished from the previous phase by the significant modifications to the technology and new facilities. Otto Schäffler in Vienna added plugboard programming to his version of Herman Hollerith's first punched-card system from 1890. However, the reason for this modification is not documented and could simply result from his experience of telephone equipment rather than being a response to market demand.

The problem of computing sterling currency appeared in the very first business applications in Great Britain. This problem was limited to Britain as this was the only industrialized country using a nondecimal currency. But it exemplifies the problem of tackling national distinctions. A later example was the issue of how to cope with national characters, like ö, œ, ø, and ñ, on alphanumeric punched-card machines and computers. Resolving such issues was essential in diffusing first punched-card and later computer technology.

The ability to print processed numbers and results was considered essential by several punched-card users in Germany and Great Britain and became a crucial competitive advantage for the Powers company in Great Britain. The German agency of the Tabulating Machine Company designed a number printing unit for the machines imported from the United States.

A prototype was completed in 1921 but was never produced, as American-produced numeric printing tabulators started to arrive in Germany.

There does not seem to be any record to substantiate that this German and British demand for technical improvement influenced the American Tabulating Machine Company's introduction of a number printing punched-card machine. But this became the German company's first experience in developing its own technology, which was continued in the interwar years when Dehomag developed a punched-card technology that was discernibly different from that of IBM in the United States. European users influenced the direction of punched-card technology and its applications in different ways than users and producers in the United States.

SEVEN

Different Roads to European Punched-Card Bookkeeping

Mechanization of bookkeeping in Western Europe was similar to expansions of offices in the United States between the late nineteenth century and the Second World War. This mechanization enabled the Tabulating Machine Company and the Powers company to extend their positions in the United States to Great Britain, Germany, and France. However, their control was not tight enough to prevent discernible differences from surfacing among punched-card-based bookkeeping systems in the various countries in Europe and in the United States in the interwar years.

British punched-card applications were shaped by the emergence of the earliest letter printing tabulator, which was marketed by the British Powers company in 1921. Having this capability failed to become a major asset for the company, which only developed an alphanumeric system after the Second World War. Instead, the British Powers company focused on developing and marketing cheap punched-card machines using small nonstandard cards, a strategy that proved rather successful.

In Germany, the Powers agency marketed the alphabet printing tabulator, which had been developed by the British Powers company, but its success was limited. The German demand at that time called for numeric calculation capability—while the need for alphanumeric punched-card systems only emerged during the Second World War. In contrast, French punched-card applications were already developing alphanumeric printing systems in the 1930s; these subsequently became crucial for establishing a national register between 1940 and 1944. These bookkeeping systems were shaped through a combination of varying demand in the different countries, different national business conditions, and different relations to IBM and to the Powers company in the United States.

Great Britain: Hesitant Transition to Punched-Card Bookkeeping

The first development of punched cards for bookkeeping in Great Britain originated in the sphere of social legislation. The welfare reforms of the Liberal governments from 1906 to 1914 extended the range of government intervention, although the significance of much of the legislation was in its pioneering nature rather than in the level of expenditure involved. Three of the most important innovations were old age pensions, health insurance, and unemployment insurance. These innovations were derived from two basic principles. The old age pension program created in 1908 was based on local organization of public expenditure, which was simple to administer.[1] In contrast, the compulsory national health insurance and unemployment insurance from 1911 were financed through contributions related to the individual employee. Virtually all employees were covered and thereby entitled to sickness benefits and assistance with medical treatment. The national health insurance program relied on weekly contributions from all employees and their employers, supplemented by state contributions. This program required much more administration than one based exclusively on public funding. The collection of contributions and payment of benefits were administered by "government-approved societies," mostly consisting of trade unions and commercial insurance companies, which individual employees could opt to join.[2]

The government-approved commercial insurance companies were all industrial life insurance companies. This market was dominated by the Prudential Assurance Company in London, which held several million policies. In 1911, the only aids the company had to administer their policies were simple key office machines and address plates that were used to address notices and receipts.[3] The new large-scale project of health insurance was expected to be substantial, and in 1913 Prudential established a completely new system to administer it.

Prudential's chief executive officer, Joseph Burns, had attended James Powers' demonstration of his machines in Berlin in 1913. He shared the Stationery Office's enthusiasm for Powers' printing tabulator, and Prudential acquired a Powers installation in 1914 to compile statistics. Initially, Prudential used punched cards to compile the obligatory statistics required by national health insurance and, in 1919, punched cards were also used to produce statistics of the company's life insurance policies. Two years later, punched cards were successfully applied in the annual valuation of the company's insurance policies.[4]

Meanwhile, Prudential used typewriters to produce the lists of policies for the frequent collection of industrial insurance premiums. In 1914, Frank P. Symmons of Prudential suggested to the British Powers agency that they could ease production these lists by using punched cards. This would require letter representation and a tabulator that could print the names of policyholders in addition to policy numbers and amounts to be collected. Charles Foster, an engineer working at the Powers agency, accepted the challenge and designed an alphabet printing unit for the standard numerical American produced tabulators. The alphabet printing unit was completed in 1916 and could print a reduced alphabet of twenty-three letters but not digits. The remaining three letters of the English alphabet were provided through double use of three letters.[5] This alphabet printing unit could print the names of policyholders for internal use in the company, but it was not able to print full addresses, as these required a combination of letters and digits. This limited the alphabet printing unit's usefulness in policy administration, but the unit was used to print specifications in bookkeeping projects elsewhere.

The British Powers company was a then simple agency, owned by the American company and selling and maintaining exclusively American-produced machines. Therefore, the alphabet printing unit would either have had to have been produced in the United States or the British company would have had to establish production of machines. As the American Powers company had no interest in an alphabet printing unit, the British decided to establish their own production.[6]

At the same time, the First World War was raging, making it precarious for the Prudential Assurance Company to rely on the continued import of office machine equipment, particularly from the Powers company, which was experiencing financial problems. Moreover, transatlantic transport was unreliable due to the First World War, which complicated acquisitions of new machines and spare parts from the factory in the United States. Therefore, Prudential decided that they needed to control the supply of punched-card equipment before they could extend their own use of punched cards.[7]

For these reasons, in 1918, the Prudential Assurance Company of

London acquired the British manufacturing and sales rights for the machines from the American Powers company. They paid $90,000 (£20,000) for the rights to manufacture and sell Powers equipment throughout the British Empire, and they would neither have to pay royalties nor be bound by any pricing restrictions. Therefore, despite the hefty price, this arrangement was highly preferable to the 1908 agreement between the British Tabulating Machine Company and the Tabulating Machine Company in the United States.

The new British Powers company became closely tied to Prudential. It acquired a board of high-ranking Prudential officers, which lasted until 1945 when the Vickers company gained a substantial shareholding. In 1919, Frank P. Symmons became general manager of the British Powers company, a post he retained until 1945. He came from a position at Prudential and he continued his career in that company, becoming its chief executive officer in 1925.[8]

From the outset, the new British Powers company chose to use domestic machine production. Initially, they attempted to subcontract manufacturing, but probably due to the lack of experience in the industrial manufacture of office machinery in England, they were not able to make a satisfactory arrangement. Instead they built a factory of their own in Croydon, in 1920. This facility gradually took up full production of all punched-card machines, starting with the sterling currency attachment and the alphabetical printing unit in 1921 for the imported tabulators. The tabulator was the most complex punched-card machine and was the last to go into production, in about 1924.[9] This made the British Powers company entirely independent of punched-card machines produced by the Powers company in the United States. The two companies remained in contact and exchanged patents and designs, but even so two distinct lines of punched-card machines evolved.

The American Powers company accepted the independence of the British company. However it twice tried to improve its position in the 1920s. First, immediately after the American Powers company had been reconstructed in 1922, the new management tried to renegotiate the territories specified in the agreement from 1918, but the British Powers company saw no reason to make any changes. As the British company by this time was approaching self-sufficiency in machine production, the American Powers company instructed its suppliers of printing ribbon and card stock to stop supplying the British company. The ribbons were easily replaced, but the British company was only able to find inferior card stocks. Despite this, they managed to run the machines with these stocks until relations with the American company improved.

This story was repeated when Remington Rand acquired the American Powers company in 1927. This time, the British company managed to get a British paper producer to produce cards of suitable quality. At the same time, Remington Rand tried to improve its position by selling its equipment within the British Empire, and the company began marketing in India, in breach of the 1918 agreement. The British Powers company threatened legal action, and Remington Rand withdrew.[10] The two foiled attempts by the American Powers company to strengthen its position in relation to its former affiliate demonstrate the strong position attained by the British Powers company through the acquisition of the patent rights as opposed to a license.

However, though the alphabet printing unit had been a major reason for Prudential to invest in the British Powers company, it proved only a limited success in the company's policy administration. In the 1920s, Prudential mechanized the administration of their "ordinary" life insurance policies by using a combination of Powers punched cards and address plates. "Ordinary" policies were for middle- and upper-class people for considerable insurance sums and in which premiums were collected between one and four times a year. The policy administration used a punched card containing all numerical information for every policy, which was used for internal operations in Prudential. In addition, an address plate for every policy existed, containing the policy number, the holder's name and address, the insurance company agent, the total amount insured and the premium to be paid.

To ease identification in the administration, the policy punched card showed an imprint of the corresponding address plate. The address plate was utilized to generate renewal notices, receipts, and letters to the policyholder, while the punched card was used to value the company's policy holding, compute bonuses, and print the lists needed to control the collection of premiums.[11] It is noteworthy that the British Powers company's reduced alphabet unit was neither applied nor developed for this application and that the dual system of punched cards and address plates remained in place until after the Second World War.[12]

The alphabet printing unit was used in business other than insurance companies in the 1930s to specify entries on invoices or ledger pages, while addressing was accomplished by the use of a separate address plate system. For example, a customer number ensured the customer received the correct invoice, as it was printed twice on the invoice, both from a punched card and from an address plate.[13]

In 1925 the British Powers company realized that their sales strategy, which was based on contacts in a few insurance companies, neither attracted many insurance companies nor new customers outside the field of insurance. They decided to improve the current sales and extend their scope to include both commercial and local public organizations. To this end, they outsourced their sales operations to Morland and Impey, who sold American Kalamazoo loose-leaf binders in Britain and who had successfully conducted negotiations with the American Powers company in 1918 on behalf of British Powers. Morland and Impey's sales people had contacts to many bookkeeping departments and they drew on these when they started to sell Powers' punched-card machines.[14]

Improved numerical capability of the British Powers company's machines was essential if they were to attract customers from commercial and local public organizations and so, in 1926, a separate development department was established.[15] Their initial ambition was to introduce the use of punched cards for various bookkeeping and accounting purposes, including costing and sales analysis. Punched cards were used in factories to compile payrolls and carry out wage analysis, to perform stock-taking, to produce purchase journals, and to monitor waste material.[16] In other words, all applications requiring calculations or lists of separate letters and numbers printed on rolls of paper that the current Powers machines could produce.

To enhance their competitive edge over the machines from the British Tabulating Machine Company, the British Powers company improved the numerical capability of their machines in two significant ways: They introduced Y-wiring in their connection boxes, and they integrated automatic group control in their tabulator. These improvements ran parallel to the simultaneous machine development in the American Powers company; in fact, the British company's introduction of automatic group control seems to have been based on the American company's implementation of this feature.

Y-wiring in connection boxes was introduced in 1926 and facilitated more complex calculations. Connection boxes were interchangeable units composed of a rigid frame into which rows of wires were assembled that transmitted the movement from the hole sensed in the card through to the unit performing the calculation and to the printing units on the tabulator. Y-wiring enabled information to be transmitted from a perforation on a card to two different units on the tabulator, which was of great importance in preparing invoices and statements for which debit and credit values needed to be recorded in separate calculating units, because the tabulators at that time were unable to perform subtractions. In comparison, competing machines from the Tabulating Machine Company in the United States had, from the outset, the ability to transmit information from a perforation on a card to two or more calculation units because the information was transmitted electrically.

Automatic group control was introduced by the British Powers company in 1927, the same year as the American Powers company. Automatic group control ended the need for the time-consuming insertion and removal of total cards, which informed the tabulator where it was to print the total of certain amounts. The automatic group control mechanism allowed the tabulator to pick up the change of designation on the cards without action by the operator, to record the total at the end of each card group (for example a customer number), and to proceed to the next group of cards without interruption. The first automatic group control mechanism was subsequently improved so that it could control computing of both subtotals and grand totals.[17]

Tabulators with automatic group control had first been marketed in the United States in 1921 by the Tabulating Machine Company and were introduced on the British market in 1927 by the British Tabulating Machine Company. The basic Hollerith and Powers patents for this feature had been granted in 1916 and 1917 in Great Britain, where, in contrast to the situation in Germany, this did not give rise to litigation.

In 1929 the first paper-feeding system for the British Powers company's machines was created, which allowed better control over printing. However, the tabulator still used rolls of blank paper and separate items, like invoices, were torn off the roll. The heading of an invoice was rubber stamped on the paper. Paper control was improved during the 1930s to enable the use of preprinted forms, although the addressing operation still relied on separate address plate systems.[18]

The strategy from 1925 of outsourcing sales was not a success. Much later, L. E. Brougham of the company's engineering staff described Morland and Impey's sales efforts as a "flop."[19] Selling loose-leaf binders was different from selling punched-card machines, and so special punched-card salesmen were introduced. It should be remembered, however, that the British Powers company's machines only started to become well suited to bookkeeping applications through the improvements introduced between 1927 and 1929.

In 1929 a dedicated sales company, the Powers-Samas Accounting Machines Limited, was established owned jointly by the British Powers manufacturing company and Morland and Impey. SAMAS, the abbreviation of Société anonyme des machines à statistique (Statistics Machines Company Limited), was the agency for the British Powers machines in France that Morland and Impey had established earlier. The British Powers company bought a controlling interest in SAMAS in 1929. Powers-Samas became the brand-name of Powers' machines in Britain, even though the Powers-Samas company was much smaller than the manufacturing company. Integration of the manufacturing and sales companies was achieved through product planning and development committees, an organizational structure that worked well. However, the two-company structure was confusing for outsiders, and after the Second World War the sales company was absorbed as a sales division of the manufacturing company.[20]

To improve sales, an American, Harold R. Russell, was hired in 1929 as the first general manager of Powers-Samas. Russell had worked in office machinery sales since 1910. He joined the American Powers company in 1919 and became its general sales manager after the company was reorganized in 1923. He returned to the United States in 1931 to become general manager of the Powers Division of Remington Rand. Russell's appointment indicated the desire of British Powers-Samas to professionalize the company's sales efforts. This he did, reorganizing sales entirely. The new organization was centered in London and had district offices in Birmingham, Manchester, Leeds, Glasgow, and Dublin.

Russell introduced selling on commission, he organized the sales force in such a way to ensure frequent calls to customers, and he advertised Powers-Samas machines in accountancy journals. In fact, prior to his appointment, there does not appear to have been any advertisements for punched-card machines in British journals. Finally, a company magazine,

Powers-Samas Punch, was introduced. (It was renamed *Powers-Samas Magazine* in 1935.)

So far, competition between the various companies manufacturing and selling punched-card applications had been based on the practical capabilities of the machines to perform certain administrative tasks. However, IBM fundamentally changed the entire basis for competition in this field in 1928 when they introduced an 80-column card that was intended as a new proprietary standard to replace the existing industry standard of the 45-column card. There were essentially two ways to counter the new card: either to introduce another extended card or to diversify by offering cheaper punched-card systems with smaller cards.[21]

Remington Rand, the American Powers company, chose the option of an extended card which was created through a redefinition of the old standard 45-column card. They divided this card into two decks, which yielded ninety columns. The British Powers company could thus attain an extended card by adopting Remington Rand's card and machines, or by following IBM's lead and squeezing the forty-five columns on the existing card, or, finally, they could introduce a longer card. British Powers considered squeezing sixty columns onto the existing 45-column card, which should have been possible using the existing technology.[22] Instead, however, they decided to diversify by offering a cheaper punched-card system with smaller cards to attract customers who could not afford to use a 45-column punched-card system.

The cheaper punched-card system was launched in 1932 as Powers-Four. It used smaller punched cards of 2 by $4^{11}/_{16}$ inches (5.1 × 11.9 cm) with 26 columns each of 11 punching positions, one less than on the 45-column and 80-column cards. The new card was only 39 percent of the size of a standard 45-column card but had 53 percent of its punching positions.[23] Further, it had the same column spacing as the 45-column card, which minimized the amount of machine development required. Both cards and machines cost only about half the price of the 45-column cards and machines.

The lower price of the machines was achieved by a reduced number of units and a simpler design. Powers-Four offered four units, instead of up to seven units on the 45-column card machines. The small card made the machines physically smaller, they had fewer components and operated at a lower speed. Initially a range of only three machines was built: a hand punch, a sorter, and a numerical tabulator.[24]

The success of Powers-Four led the company to broaden its scope through additional machines and an alphanumeric capability. Between 1932 and 1935 the company added an automatic key punch, an implementation of automatic group control, and an alphanumeric tabulator with a reduced alphabet, and subsequently a full range of ancillary machines followed. The performance of the alphanumeric tabulator was constrained by the card only containing eleven punching positions. It had only twenty-nine characters, ten digits and nineteen letters, and each remaining letter was substituted by a digit or another letter.

This tabulator had fewer letters than the alphabetic unit for the 45-column tabulator, but it was alphanumeric, which meant it could print either a letter or a digit in all its printing positions. The alphabet capability was improved in 1935 to encompass twenty-one letters in addition to the ten digits.[25] Powers-Four's alphanumeric capability was marketed specifically for entries on invoices or ledger pages, just like the alphabet printing unit for the 45-column cards; a further similarity was that Powers-Four recommended a separate address plate system for the addressing operation.[26]

The success of Powers-Four for the company was not unqualified, however. Powers-Four was launched in 1932 when British Powers was still suffering from the effects of the 1929 economic crisis. The company only managed to raise factory output in 1933, and these are the only figures handed down as a record of the turnover. In 1933, doubts arose in the company surrounding the choice of marketing small cards. It was feared that Powers-Four would have a negative impact on the sales of 45-column equipment, which was already under pressure from the competitor's 80-column card equipment. Some at Powers-Samas were even afraid that Powers-Four would become the main product of the company. The higher sales volume of the small machines could only partly offset their lower profitability. It was therefore decided to condense the columns on the 45-column card to restore the competitive position of the company's big machines.[27] The columns on a punched card could be condensed without infringing IBM's 80-column card patent, as this only covered rectangular or oblong perforations.[28] Smaller, round perforations worked well for the Powers company's pin-sensing method.

The process of condensing the columns of the British Powers company's cards was achieved in two phases. First, in 1936, they put sixty-five columns on the original 45-column card, which was accomplished by squeezing the columns together and using smaller perforations. Second, the card was made to contain eighty columns in 1954, twenty-six years later than IBM's card. The number of columns on the Powers-Four card was similarly increased in two phases, from twenty-six to thirty-six columns in 1936, and then to forty columns in 1950.[29]

There was an important distinction between IBM's conversion from forty-five to eighty columns and British Powers' change from forty-five to sixty-five to eighty columns. While the IBM machines for the new card could be plugged to handle the old cards, Powers' mechanical technology was less flexible. British Powers' customers had to replace all their old machines each time the card changed. To ease the transitions, both Powers and IBM supplied machines that transferred information from the old to the new punched-card formats.

In the mid-1930s, the British Powers company was not only working toward increased representation on their cards, but they also designed a punched-card system, launched in 1936, that was the smallest and cheapest punched-card system on the market. It used a 21-column card, 2 by 2¾ inches (5.1 × 7.0 cm) with eleven punching positions. The tabulator was and remained a standard numerical machine with one calculating unit that operated at a modest speed. Powers-One was originally developed for the Cooperative Wholesale Societies to compute dividends for their members, with the expectation that the system would spread to other organizations that were unable to justify the greater cost of punched-card equipment using larger cards.[30]

The new machines improved British Powers' competitive position in relation to the British Tabulating Machine Company, and the two smaller sets of punched-card equipment facilitated cheaper, though simpler, punched-card processing. The enhanced competitiveness was reflected in the company's rising factory output until 1937, but output was lower in 1938 and 1939. Output figures are not available from the British Tabulating Machine Company, but the company had steadily growing assets from 1932, including 1938 and 1939, which indicate improved performance. In 1938 and 1939 British Powers produced approximately the same number of Powers-Four and standard punched-card size machines, while the

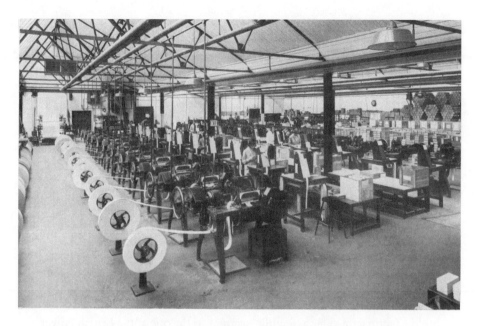

Punched-card production at the Powers-Samas Company in London, 1937. Production of the cards themselves contributed up to a third of the punched-card producers' revenues. (*Powers-Samas Magazine* October 1937, 7)

number of Powers-One machines was approximately half as large, which shows the limited success of Powers-One. The falling machine production indicated a company in crisis, while the distribution of the company's production between the three different machine lines showed that the forebodings of some employees' in 1933 had been realized. Measured in number of machines, the company had primarily become a producer of small-size machines; any advantages gained from this did not, however, offset the disadvantages facing the Powers-Samas' 65-column card in the competition with the 80-column punched cards from British Tabulating Machine Company.

Problems of raising capital had been a major reason for the sluggish start of the British Tabulating Machine Company (BTM) from 1902 and until 1908, and the company had financial problems with the Tabulating Machine Company in the United States until the early 1920s. In 1912, the company fell behind in its royalty payments, and the American parent company pressured for payments. This was repeated in 1916 when the

British company raised the question of whether they were to pay royalty on the part of the revenues that they used to pay the company tax, which had been imposed on British industry due to the First World War; according to the 1908 agreement, the 25 percent royalty was to be based on the gross revenue.

At that time, the British company tried to get the American company to lower the royalty percentage and take the royalty debt as shares. However, the Americans were unwilling to accept shares instead of royalty payments, as they were only interested in a controlling interest in the British agency. Only first in 1919 could the British Tabulating Machine Company sell additional stocks, which, finally, enabled the agency to pay off the debt to the American company.[31]

The British Tabulating Machine Company had established a workshop in London back in 1912 to assemble machines out of parts supplied from the United States. The growth in business during the First World War turned this facility into a bottleneck, and the company decided to build a factory for their assembly—similar to the one at the British Powers company. The British Tabulating Machine Company built a factory in Letchworth, near London, which opened in 1921. During the 1920s, their main job was to assemble pieces imported from the mother company in the United States. By and large they sold machines identical to IBM's but for adaptations to sterling currency.[32] However, the introduction of products in Britain lagged behind a few years, for example the number-printing tabulator was introduced in Britain in 1924, compared with 1921 in the United States.[33]

Assembly in Britain made the British Tabulating Machine Company's operations less transparent for the Tabulating Machine Company in New York. In 1922, the American company questioned the British company's calculation of royalties, as goods supplied by the American company did not agree with the royalties remitted. The American company demanded detailed royalty statements and threatened to take over the British company by buying up shares, a threat which was credible in light of the takeover of its German affiliate the previous year. Dishonest accounting would entitle the American company to void the contract, which would force the British Tabulating Machine Company out of business, as it only assembled machines, except for the attachment designed to handle sterling currency.

A report was prepared by auditors from the American company who found a deficit of $183,000 (£40,000), 28 percent of the British company's assets in 1922. During the subsequent negotiations, the British Tabulating Machine Company raised the problem of the fluctuating prices based on currency exchange rates, which was a contrast to the fixed rate that had been the basis for the agreement between the two companies since 1908. In the agreement reached in 1923, the British Tabulating Machine Company only had to pay 10 percent of their deficits, and they were granted the freedom to decide on the cost of rentals. Further, they no longer had to pay royalties on the devices and machines which they designed and produced themselves.[34] At that time, this arrangement was only relevant for the tabulator modification to handle sterling currency, but it made machine development advantageous for the British company, just as it was for the German subsidiary of the American company, Deutsche Hollerith Maschinen Gesellschaft mit beschränker Haftung, or Dehomag.

To exploit this enhanced freedom, the British Tabulating Machine Company formed a development department in 1923. Its first accomplishment, in 1926, was a "pence translator." The new numeric printing tabulator also required reasonable printing of amounts in sterling currency, which required an improvement of the existing modification to handle sterling.[35] Apart from converting machines for sterling, very little technical development took place at the British Tabulating Machine Company during the 1920s, and the important improvements, such as numeric printing and subtracting tabulators, came from the United States. The subtracting tabulator greatly widened the company's scope for application in bookkeeping.[36] However, as late as 1924, the company only envisioned the application of punched cards for statistics processing.[37]

Originally, the British Tabulating Machine Company's sales operations were exclusively based on the managers' contacts and word of mouth, like Hollerith's original sales promotion in the United States. In 1922 the British Tabulating Machine Company improved their sales promotion. They held the first training course for service engineers that year, and they established a branch office in Birmingham, followed by offices in Manchester (1923), Glasgow (1924), and Leeds (1930).[38] In 1936 the company started to publish a regular sales bulletin, *The Tabulator*.

The company's sales organization evolved more gradually and systematically than that of the British Powers company. A historian explained this by observing that several of the company's directors were familiar with selling capital goods when they joined the company.[39] Additional inspiration came from their frequent contact with IBM's efficient sales organization.

During the economic crisis that started in 1929, protective barriers sprang up in Britain as they did in France, Germany, and most other countries. In 1932 the government introduced general tariffs, which for punched-card machines amounted to 25 percent of their value and 10 percent for punched cards. As British Powers produced almost all the machines themselves, they were less affected than the British Tabulating Machine Company, whose business was based on machines produced in the United States. Although the British Tabulating Machine Company assembled about 70 percent of the machines, which reduced their duties, duties caused an 8 percent rise in costs, giving British Powers a competitive advantage.[40]

This advantage was furthered by the growing national hostility to non-British manufacturing expressed in the "Buy British" campaign of populist newspapers like *The Daily Mail* beginning in 1930. British Powers chose to capitalize on this trend and advertised that their machines were completely manufactured with British material, with skilled British labor and were backed entirely by British capital. This put pressure on the British Tabulating Machine Company as an importer of machines produced in the United States. Their brilliant counteraction was to appoint to their board of directors Conservative politician Leo S. Amery, who was a leader of the "Buy British" campaign.[41] Thus, in spite of the new tariffs and the "Buy British" campaign, the British Tabulating Machine Company experienced steadily growing assets from 1932.

Perhaps ironically, these impediments probably contributed to the British Tabulating Machine Company embarking on its first major building of a new subtracting tabulator, as the contemporary restrictions in Germany encouraged the German subsidiary to build its own tabulator. In Britain, the company's stated objective was to reduce their royalty payments. Since the late 1920s, the development department had established the basis for this endeavor through inventions by Harold Hall Keen. Their new tabulator from 1933 was based on 80-column punched cards and it could subtract and print numbers. After some initial problems, this machine became the British Tabulating Machine Company's standard tabulator, and they no longer marketed the IBM-designed tabulators.[42]

Although this new tabulator was exclusively numeric and intended for statistical tasks, both IBM in the United States and the competing British Powers company saw bookkeeping as their main field of application at this time. To accommodate bookkeeping, an improved paper-control device was designed and produced for the new tabulator to facilitate printing on forms, and an alphanumeric version of the tabulator was produced in 1935. This British-designed machine was supplemented in 1939 by a multiplying punch that could handle sterling currency. The British Tabulating Machine Company also produced its own hand punch, hand verifier, and sorter. However, the other new machines, such as smart punches, reproducing punches, and collators were all IBM products.[43]

As stated earlier, the British Tabulating Machine Company's main objective in developing their own tabulators had been to reduce royalty payments. Their claims were based on three of Harold Hall Keens' patents, but IBM found that they all incorporated inventions that they already had sole ownership of. The outcome was a "proportionate" reduction in the royalties for the Keen claims IBM found to be novel.[44]

On the British market, the promotional newsletter from the British Tabulating Machine Company indicated that they envisioned their tabulator's alphanumeric capability nearly exclusively to be used to write specifications, for example, on invoices.[45] However, punched cards were used for addressing in a few of the reviewed applications, while the alternatives were to use typewriters or address plates.[46] This observation that specifications were the main use of the British Tabulating Machine Company's alphanumeric capability can also explain why the company in 1936 introduced smaller cards and machines to compete with the small card machines from the British Powers company. Both companies seem to have thought that they would attain more business in Britain using these cheaper formats than using alphanumeric applications of their bigger machines, a fact that is confirmed by the record of punched-card applications in British insurance companies mentioned earlier.

The British Tabulating Machine Company introduced a half-size 38-column card measuring $3^{11}/_{16}$ inches by $3\frac{1}{4}$ inches (8.3 × 9.4 cm). The shorter card enabled them to halve many of the internal components of the machine, while at the same time still be able to apply many parts from the 80-column card machines. Subsequently, the British Tabulating Machine Company introduced similar 24-column and 60-column cards, but none of these short-card formats ever caught on.[47]

In contrast, the British Tabulating Machine Company was unparalleled when it came to large-card machines, which Powers could not rival. In fact, the markets of the two companies were starting to diverge. The British Tabulating Machine Company went for the high-end market for their 80-column equipment, while Powers focused on the cheaper small-card machines.

Germany: Numeric Punched-Card Bookkeeping

The Tabulating Machine Company's agency, Dehomag, successfully introduced punched cards in Germany in 1910. The use of punched cards grew rapidly for public and business statistics in the few years up until the First World War, and Dehomag's business soared during the war.

The German postwar crisis resulted in a runaway inflation that was halted by both the introduction of the new mark in November 1923 and also the first mutually agreed arrangement for Germany's payment of reparations the following year. Following this, Germany experienced substantial economic growth until the world economic crisis in 1929. This growth caused the relative cost of labor to rise and a major wave of rationalization swept German industry.[48]

In the postwar economic crisis lasting to 1923, many punched-card users gave up their rented equipment. But once the German mark stabilized, punched-card applications once again grew rapidly both in private companies and public organizations. First, punched cards regained their function of compiling public statistics and operational statistics in big companies. Then punched card-based operational statistics spread to the public sector and the cards started to be used for bookkeeping tasks. In the public sector, traditional statistics punched-card applications returned and new ones emerged between 1923 and 1927. The German National Statistics Department (Statistisches Reichsamt) in Berlin resumed punched-card processing of their foreign trade statistics in 1923, and early the next year the National Bank (Reichsbank) used punched cards to improve their surveillance of the foreign debt and assets, crucial to stabilize the new mark after 1923. The use of punched cards facilitated more detailed and quickly available statistics than did traditional methods.[49]

In addition, public institutions started applying punched cards to processing operational statistics, as had been the case in German industry since 1911. The German national railroads and the city of Frankfurt am Main exemplify the dynamics of this application. The German Reich was a federation of many states and in many respects had a decentralized structure. The organization of the railroads illustrates this. They were not a nationwide company, but they were organized through eight regional railroad systems operated by different states.[50] By the end of the First World War, the Versailles treaty caused large railroad sections to be surrendered, and the regional railroad systems merged, forming the German National Railroad (Deutsche Reichsbahn) in 1920.[51]

According to the 1924 agreement between Germany and the Allies on reparations, the national railroads had to provide a third of the reparations through the amortization of nearly half the national railroad's estimated total assets.[52] Therefore, reparations became a decisive motivation for extensive rationalization during the following several years to achieve the advantages of scale in the amalgamated company. To this end, an important tool was extensive punched-card-based operational statistics. Further, within a few years, the operational statistics applications were extended to monitor line construction and train building to minimize stocks and costs. Finally, this was supplemented in 1928 with the introduction of punched-card-based bookkeeping for line construction expenses.[53]

Similarly, the city of Frankfurt am Main acquired punched cards in 1927 to process their municipal statistics. Then in 1929, they started to apply punched cards for expense control, which facilitated daily surveys of payments for various purposes by the use of numeric lists of expenses.[54] Punched-card-aided bookkeeping was starting to emerge.

Energy supply, like income tax and real estate tax, brought a municipality in direct contact with a large part of the population. Municipal energy suppliers introduced punched cards to make various lists on consumption and to monitor the expenses of the utilities, for example, in Dresden (1925), Berlin (1929), and Frankfurt am Main (1930). While these applications only required addition, punched-card-based calculation of the amounts to be paid using the meter readings was a more complex task, the solution to which only emerged in 1935.[55]

German industry provided a parallel story. They too experienced a rapid expansion of operational statistics in the late 1920s, after which punched card-aided bookkeeping emerged. Between 1910 and 1912 the chemical and electrical industries were the first to introduce punched cards, which they used for various kinds of operational statistics. Three big ironworks introduced punched cards for operational statistics in 1920, three additional steelworks followed in 1926, and the Friedrich Krupp concern the following year.[56]

The extension of the use of punched cards in big companies to bookkeeping started in the late 1920s and can be exemplified by the development at the Siemens concern. After two years of preparations, Siemens established two punched-card installations, one in Berlin with Powers equipment, and the other in Nuremberg using Dehomag equipment.[57] The Dehomag installation in the Siemens factory in Nuremberg was established in 1928 for internal accounting of materials in two production units. In the early 1930s, the application was extended to wage administration and wage statistics and, by 1937, an exclusively numeric punched-card register of workers, including their marital status and number of children, which influenced the calculation of wages, was operationalized. Now all Dehomag punched-card handling was performed in a centralized office, which, in 1942, had seven members of management and 158 additional employees.[58]

The Siemens factory in Berlin introduced Powers punched-card machines in 1927 for wage administration, including wage calculations, the handling of deductions like health insurance, and the printing of records of payment of wages and health insurance contributions. In 1928 tasks at this Siemens factory were extended to encompass operational statistics, similar to the Nuremberg factory.[59]

In a textbook on punched cards from 1929, Robert Feindler argued for the use of punched cards for wage calculation and wage control. However, the lack of multiplication capability in the punched-card machines impeded wage calculations being done with punched cards using the time rates and the number of hours. Further, the absence of subtraction prevented punched card-based calculation of the payable wage by deductions from the gross wage. Therefore, Feindler proposed that each worker's deductions be added separately and to subtract the total amount from the gross wage either manually or by the use of a key calculation machine. Feindler suggested lists of wage components and deductions as output.[60]

The German punched-card companies, Dehomag and the German

Powers company, pursued different roads to bookkeeping. Following the stabilization of the mark, economic growth in Germany was reflected in Dehomag's rising turnover, which already in 1925 in dollars was 4.7 times that in the peak year of 1918–1919. This growth accelerated during the rationalization wave in the second half of the 1920s and became the basis for renewed machine development and production consented to by IBM as majority shareholder. The company also established branches in major German cities: Frankfurt am Main (1925), Hamburg (1925), Stuttgart (1926), Düsseldorf (1926), Dresden (1927), Leipzig (1929), and Munich (1930).[61] Dehomag, which already in 1924 had bought a machine works factory in Sindelfingen in the industrialized Stuttgart area, moved machine repair and spare part production from Villingen.[62]

In the ten years following 1925, Dehomag's main development endeavor was to improve the calculating capacity of punched-card machines, which reveals their perception of the role of punched cards in bookkeeping. First, they improved balancing and later features for the calculation of interests. The two main figures in this endeavor were Hermann Adalbert Weinlich, the company's chief engineer since its inception in 1910, and Ulrich Kölm, an engineer who joined the company in 1925.[63] They built numerical punched-card machines with calculation capabilities that were unsurpassed elsewhere. The prime users of this shaping of punched-card equipment seem to have been the banks.

Dehomag had seen the potential of the bank market as early as 1914, particularly at the large and internationally recognized *Grossbanken*, which they approached. Dehomag envisioned punched cards being applied to deposit administration, transaction of payments, credit administration, and possibly current accounts (*Kontokorrent*).[64] However, these applications were initially held back by their tabulator's lack of a printing capability but eventually emerged when improved tabulators included this feature.

In 1923, Dehomag started to import number printing tabulators from the United States and subsequently secured several banks as customers, who primarily used their installations for current account administration. Short-term loans through overdrafts on current accounts were a prime way to supply working capital for German industry, and many German banks had a large number of current accounts. Further, many current accounts had frequent and large entries, which made it crucial to establish effective monitoring of transactions to ensure that each costumer and bank would be able to meet their outstanding debts.

Using a separate punched-card for each transaction, the numeric printing and adding tabulators of the mid-1920s facilitated printing of sorted and summed up lists of the activities on every current account, which was cumbersome and time-consuming to compile by other means. However, the absence of subtracting tabulators hampered the appeal of punched cards for current account administration, as balancing was crucial due to frequent mixed debit and credit movements. Further, calculating interest on current accounts was complex due to the frequent activities on these accounts, but punched-card processing of this task would require a punched-card multiplier. Consequently, punched-card applications for current account administration called for developing features for subtraction and multiplication.[65] Current account administrations that included the use of punched cards were established at Vaterländische Bank in Berlin in 1924, at Provinzialbank Pommern in 1925, and at Dresdner Bank in the late 1920s.[66]

The then-available account applications showed the potential of three possible punched-card machine improvements, namely subtraction, interest calculation, and addressing statements to the customers. Due to the absence of subtraction, the calculation of balances was awkward. Credits and debits were recorded in separate fields on the punched cards and balances were established by use of complementary figures, making it difficult to decide whether a balance was credit or debit. The punched-card companies subsequently addressed these shortcomings, but Dehomag distinguished itself by being the company that most stubbornly concentrated on improving the calculation capabilities, while disregarding alphabetic features.

In 1926, the engineers at Dehomag solved the balancing problem by modifying an IBM tabulator produced in the United States to provide this capability, and in 1928 IBM started to produce a tabulator that included subtraction capabilities. However, this modification came with the major disadvantage that negative numbers needed to be punched as complement figures, which was demanding for the operator and bred errors.[67] IBM marketed a tabulator with a similar subtraction capability in 1928 and a full subtracting tabulator in 1933.[68]

The success in 1926 of Dehomag's development team in modifying an IBM tabulator to perform subtractions became the basis for their subsequently building tabulators with high computing capability and forms control system, which was crucial for advanced printing on forms.[69] In 1933, they completed an improved tabulator that did not require negative figures to be punched as complement figures. This meant that all numbers could be punched regularly; negative signs were indicated by using perforations in row 11. IBM marketed similar improvements the same year in the United States, fundamentally facilitating the use of punched cards for bookkeeping applications involving addition and subtraction. Dehomag produced 250 copies of these tabulators from 1933 to 1936, until the D11 tabulator replaced them.[70]

Following this, the tabulator was improved to also multiply. In 1934, Dehomag completed a version that could multiply by up to three digits, which, for example, could calculate interest on current accounts in a bank.[71] By the following year, this line of development was completed by Dehomag's D11 numeric tabulator, which could multiply, divide, and perform complex calculations. This tabulator was produced in Germany from 1935 to 1960, with a total of 1,120 machines supplied by 1943.[72] The absence of letter printing during this same period is striking. Until the outbreak of the Second World War, little interest in letter printing using punched cards appeared in Germany.

The German Powers agency had the disadvantage in developing its road to bookkeeping that its initial start-up had been foiled by Dehomag's patent infringement suit in 1914, the First World War, and the subsequent runaway inflation. Thus, the Powers agency reestablished itself from scratch in 1923 in Berlin. Jan Büchter headed the agency, and it retained the former name, Deutsche Gesellschaft für Addier- und Sortiermaschinen mit beschränkter Haftung (German Company for Adding and Sorting Machines with limited liability). Their business, based on machines imported from the United States and a card printing shop, marked the first permanent presence of the American Powers company on the European continent. They made the German agency the center of their continental business.[73]

Dehomag had established a robust first-mover position in Germany because of their business since 1910 and because the Powers agency had been prevented from operating in the country between 1914 and 1923. Jan Büchter now countered his inferior position through an offensive strategy in marketing and machine development that illuminated his visions for new punched-card application fields. Büchter's marketing strategy focused on the *Grossbanken* and insurance companies. The main technical asset the Powers agency had compared with the IBM machines was the British Powers company's tabulator with a reduced alphabet of twenty-three letters.

The German Powers agency, which implemented the alphabet reduction differently from the British company, had no feature for four national German characters (ä, ö, ü, and ß).[74] However, in spite of this shortcoming, the German Powers agency argued in 1926 that punched cards be applied to wage administration systems that encompassed printing lists with the worker names for internal use in the company. Printing worker names for the workers' use required a more complete alphabet.[75] Moreover, the German Powers company tried to improve its position through machine development.[76] These actions indicated an independent position for the German Powers agency during the American parent company's weakness in the 1920s.

Remington Rand's acquisition of the Powers company in the United States in 1927 caused changes in Germany. Remington Rand maintained a strategy of conducting marketing abroad through its subsidiary companies, which distinguished it from its punched-card predecessor. Therefore, they took over the German agency and established a child company, Powers Gesellschaft mit beschränkter Haftung (Powers company limited) with new management.[77] During the next several years, the German Powers company marketed machines produced in the United States and printed punched cards in Berlin, while terminating their own machine development efforts. As in the United States, the German Powers company introduced the double-deck 90-column punched card in 1929, and began marketing the new Remington Rand range of machines (Model 2) intended for the new punched card.[78]

It is difficult to assess the German Powers company in the 1920s, as no turnover figures survived from this period. Company revenues in 1940 only amounted to 13 percent of Dehomag's revenues that year, implying a modest-sized turnover.[79] However, Dehomag launched a second patent infringement suit against the German Powers company in 1924, indicating that they saw the German Powers company as a significant opponent. This patent infringement suit was in reference to Herman Hollerith's and James Powers' automatic group control patents. Both patents were filed and granted in Germany, Powers' patent in 1921 and Hollerith's patent in 1924, but Dehomag contested the validity of Powers' patent in German courts. This dispute was only solved in 1929 when the Supreme Court ruled that Powers' patent was valid.[80]

During the world economic crisis starting in 1929, the German right-wing governments of the early 1930s chose to concentrate on the balance of payments by introducing government import controls. This contributed to a fall in imported industrial goods, in particular, with imports between 1929 and 1932 being reduced in value by 45 percent.[81] When the Nazis assumed power in January 1933, they promised full employment through the creation of new jobs and advocated for "self-sufficiency" and "independence from the World Economy," to be implemented by strengthening the previous governments' import restrictions, thus causing further reductions in the import of manufactured goods.[82]

These Nazi objectives made Dehomag vulnerable. Though the introduction of punched cards was frequently justified because it saved on manpower, nearly all of the company's machines were imported, and the company was 90 percent foreign owned. However, Dehomag's substantial machine development, which culminated in 1935 with the D11 numeric tabulator, improved the company's position both in Germany and simultaneously in relation to IBM in the United States.

In the summer of 1933, Dehomag knew its work building machines would come to fruition, requiring it to have manufacturing capacity. The company bought a site in Berlin and built a new factory, which it opened in 1934 with the explicit consent of IBM. From 1935, this factory housed the highly successful D11 tabulator production, which became the major reason for Dehomag's fast growing turnover during the second half of the 1930s. Also, the number of employees increased from 462 in 1933, to 1,100 three years later.[83] This accomplishment enabled Dehomag to avoid conflict with the National Ministry for Trade and Industry (Reichswirtschaftsministerium), which controlled foreign trade, whereas Dehomag's competitor, a Powers' subsidiary, had to commence production in Germany using American-designed machines to avoid importing complete machines.

Further, Dehomag's management took the opportunity at the opening of their new Berlin factory in 1934, where there were representatives from the government and the Nazi party present, to voice their allegiance and

Cabling of D11 tabulators in the factory of IBM's German subsidiary, Dehomag, in Berlin, 1935. This tabulator shows the size and technical complexity of large punched-card machines in the 1930s. (*Festschrift zur 25-Jahresfeier der Deutschen Hollerith-Maschinen Gesellschaft.* Berlin: Dehomag, 1935, 45)

enhance their connection to the new regime as well as stress the company's "Germanness." Director Willy Heidinger told the audience that, "We have firm trust in our physician and will follow his orders blindly, because we know that he will lead our nation towards a great future. Hail to our German people and their Führer."[84] In 1935, the company further emphasized its Germanness in a publication commemorating Dehomag's twenty-fifth anniversary.[85] The German origin of punched cards was asserted because of Herman Hollerith's being the son of German immigrants, his first name (mis-)spelled with a double n, according to German tradition. He was referred to as a German-American, though he was born in the United States and had never lived in Europe. Finally, the company drew attention to its own development and production work. However, development and production did not carry enough status within the company for an engineer to gain a position among the executive managers before

1945. The delegation of authority to national leaders within IBM eased the requisite adaptation during the chauvinistic mood of the 1930s.

Dehomag's second director, Hermann Rottke, also used the factory opening in 1934, attended by the Nazi party and Deutsche Arbeitsfront representatives, to claim that Dehomag punched-card systems created jobs, as people with Dehomag punched-card training easily found work in spite of the high unemployment nationwide. But, he avoided any discussion about whether punched-card machines saved on the number of jobs or not.[86]

The various sectors of German industry experienced different ramifications of the self-sufficiency and rearmament policy. While, for example, the production of shoes for the civilian population declined and caused shortages, industries of military importance, like the Siemens concern, experienced rapid growth.[87] Dehomag experienced even bigger growth than Siemens. Between 1936 and 1942, in addition to a growth in turnover 19 percent larger than Siemens', Dehomag experienced equivalent relative increases in its workforce. The company also increased its number of branches, spreading its distributor net across Germany. They opened branches in Bielefeld (1935), Breslau (1935), Nuremberg (1935), Saarbrücken (1935), Königsberg (1937), Bremen (1938), Dortmunt (1938), Essen (1938), Hannover (1938), Karlsruhe (1938), Cologne (1938), and Magdeburg (1938).[88]

Between 1933 and 1940, Dehomag's high level of profit caused IBM yearly problems, because the repatriation of funds was not possible, and because Willy Heidinger, who was the sole German shareholder, wanted his dividends. Declaring the profit was delayed, and then, with the exception of payment to Heidinger, they were either reinvested or invested in property.[89] Punched cards became increasingly important for the arms buildup and later for warfare.

Hitler's assumption of power in 1933 signaled the start of the establishment of a totalitarian state. Punched-card-based control became a major sales argument for Dehomag, who supplied equipment for statistical monitoring of the members of the Nazi party, SA, SS, Hitler Jugend, and Bund Deutscher Mädeln.[90] These applications were remarkable for their limited ambitions, compared with the simultaneous introduction in France and the United States of punched cards for managing records.

The German Powers company was much more vulnerable than

Dehomag to the import restrictions of the German autarchy, as it was exclusively owned by Remington Rand in New York and was a much smaller company. Like Dehomag, the German Powers company voiced its support for the new age by offering their expertise for the construction of a scientific national planning system, but even that could not eliminate the company's problems.[91]

Machine development in Germany had been terminated in 1927 when Remington Rand took over the agency, leaving the German Powers company two options: to import machines produced in the United States or to establish production in Germany of machines designed in the United States. The first option had already been selected before the economic crisis started, but tight German currency restrictions from the early 1930s curtailed this possibility, and production started in Germany of punches designed in the United States but not of sorters or tabulator. This eased the German subsidiary's relations with the authorities for a while, and the company earned money in the rearmament. Profits, however, were invested in bonds or equities in Siemens and Halske, IG Farben, and other German companies because the authorities did not allow them to be exported.[92]

The choice of Siemens and Halske was hardly a coincidence, as they had used Powers' punched cards for several years and were substantially developing punched-card equipment.[93] Siemens and Halske included the low-voltage part of the Siemens group and produced telegraph, telex, and telephone equipment. It was a proliferating, innovative company with technology similar to Powers' punched-card technology, especially Siemens' alphanumeric telegraph transmitters. Siemens and Halske, who had acquired a Powers' installation for wage administration in 1927, soon began developing their own punched-card machine.[94] By 1934, they had designed their own punched-card code, a sorter, and a card reader. The Siemens card had ninety columns organized in two decks of four rows and was inspired by the Powers double-deck card even though that card had five rows in each deck.[95]

Further, Siemens utilized their expertise in the transmission of digital information. In 1934 they had completed a link between a key subtracting and multiplying machine and a punch that facilitated the punching onto cards without extra action, with information that had been keyed in and processed using an ordinary bookkeeping system. This feature enhanced the modest calculation capability of a standard tabulator and resembled a similar Dehomag construction from 1929. In addition, Siemens had built a machine that transferred information from punched strip tapes for telegraph transmitters to punched cards.

If Siemens wanted to market this technology, they had the choice of either cooperating with one of the existing punched-card producers or producing and marketing their own machines, as they had done for many other products when they diversified in the 1920s. Siemens possessed several punched-card patents and vast technological abilities, but they lacked key patents, like the automatic group control patent Dehomag and Powers had in Germany. In addition, the economic crisis caused falling revenues for Siemens from 1930 to 1933.[96]

In 1934 Siemens chose to establish a partnership with the German Powers company. Powers acquired the right to the use of the Siemens' innovations, and Siemens would produce the machines they designed. However, the agreement exclusively concerned the machines designed by Siemens, while Powers would continue to import sorters and tabulators from the United States.[97] For Powers, the alliance would prevent a new punched-card competitor from emerging, and Siemens was an attractive, prestigious partner. The alliance, which would ease the concerns of the National Ministry for Trade and Industry regarding imports, could not solve the problem that Powers continued to rely on complete tabulators imported from the United States.

Two Siemens punched-card machines appeared on the market in 1935, which consisted of a sorter that could handle regular 45-column cards, as well as double-deck Siemens and Powers cards, and a link between a key calculating machine and a punch.[98] The innovation of a converter to change from punch strips to punched cards ran into trouble when no eager first customer appeared, grounding development in 1938. It was only revived due to the subsequent emergence of an eager customer, the army's punched-card service (later called Maschinelle Berichtswesen).[99]

In 1934 the Siemens-Powers partnership envisioned two contradictory developments of the punched card. They planned to develop the Siemens punched-card either into a triple-deck 135-column card or to introduce alphabet representation. Siemens' four-row number representation provided space for a third deck on the card, but four rows only allowed fifteen characters, whereas exclusive letter representation required at least twenty-six characters (five rows) and alphanumeric representation needed at least thirty-six characters (six rows).[100] These plans were probably based on Siemens' extensive expertise in alphanumeric telegraphic transmitter systems with paper strips.

Neither of these options was pursued by the German Powers company. They did not even adopt the Remington Rand 90-column alphanumeric system that was introduced in the United States in 1938. The German Powers company only marketed the old reduced alphabet printer for 45-column cards, designed by the British Powers company, which had been on the market since the 1920s.[101] The German Powers company's focus on bookkeeping remained the calculation capability of their punched-card systems, while letters were only used as an auxiliary for specifications.

During the second half of the 1930s, the German Powers company contracted with Siemens to produce punches and verifiers based on designs from the United States.[102] Remington Rand, which only allowed Siemens to produce minor machines like punches and verifiers, wanted their sorters and tabulators to be imported from the United States. Siemens produced a sorter of their own design, but the production of tabulators became a problem. The Ministry for Trade and Industry pressured the German Powers company to produce all their machines in Germany. In 1935, Powers was permitted to import tabulators for another three years, while preparing for production in Germany. However, import was blocked in 1937 as the Ministry for Trade and Industry realized that Powers did not plan to move their tabulator production to Germany, forcing Powers to start their own tabulator production in Berlin. The American company would not allow Siemens to produce their machines, as they feared Siemens would use their expertise to enter the punched-card market when the Siemens-Powers partnership expired in 1949. Siemens, for their part, was eager to adopt tabulator production.[103]

France: Alphanumeric Accent in Punched-Card Bookkeeping

The first bookkeeping application of punched cards in France was security administration at the Bank of Alsace and Lorraine (Banque d'Alsace et de Lorraine) in Strasbourg in 1930.[104] It was derived from equipment from the American Powers company, and the tasks involved monitoring holdings and frequently producing account statements for account holders with letter specifications of the various securities. Letter specifications in these reports were produced by the reduced alphabet printing unit from the American Powers company. Customer account statements were addressed by use of address plates.

By 1933, the Bank of Alsace and Lorraine had established an additional Powers punched-card installation for current account administration. As soon as the bank was informed of a transaction, it was recorded on a key bookkeeping machine that produced advice for the customer and carbon copies for use in the bank. On this basis, information of the transaction was punched onto a card, which was used daily to produce various sorted and consolidated lists for internal use in the bank. Some of these lists were sent as statements to the account holders. A special subtraction unit on the tabulator consolidated the entries for each department or account. Similar bookkeeping applications using punched cards were established at several French banks in the 1930s, making French banks a noticeable market segment.[105]

Another distinction of the punched-card trade in France was the emergence of the state as a user that contributed to shaping punched-card systems for bookkeeping in France. The public sector diffusion of punched cards started later than it did in the private business sector. The first installation in the public sector was only established in 1926, which grew to six installations in 1932.[106] At that point, most punched-card machines were still operating in businesses, but in 1936, 60 percent of punched-card machines operated in the national administration.[107] The introduction of punched cards in the national government was part of the process to reform government from the ground up, while attempts for a general reform failed.

The Third French Republic (1870–1940) suffered from basic constitutional weaknesses. It had a two-chamber parliament that faced severe problems due to the numerous, frequently changing parties that rendered it virtually inoperable. The French president, who was elected by parliament, held little power, the government's position was seriously weakened by the many unstable parties, and by the fact that it was not empowered to respond to a vote of censure by dissolving parliament. These constitutional inefficiencies reinforced the volatile political development in France during the 1930s and contributed to the failure of a general reform. The outcome was the disintegration of the Third French Republic which, in 1940, transformed into the autocratic Vichy state.[108]

The state's majority share of the punched-card business in the 1930s distinguished France from Germany and Great Britain, with the French state gaining a significant role in shaping punched-card systems. Simultaneously, the emergence of the Compagnie des machines Bull in France, independent of the IBM and Powers companies, made the competition in France unique in the 1930s.

The installations at the Bank of Alsace and Lorraine were modeled on American Powers equipment, which was marketed separately from British Powers equipment in France starting in the late 1920s. Furthermore, these installations were associated with plans to establish a production of punched-card equipment from the American Powers company in Hagenau, Alsace, an idea supported by two *polytechniciens*, Georges Vieillard and Elie Doury, engineers trained at the distinguished École polytechnique (Technical University) in Paris.[109] However, this plan was not implemented, and Vieillard and Doury moved to assist in establishing the Bull company in France, while the American Powers company chose to merge their business into SAMAS (Société anonymes des machines à statistique, Statistics Machines Company limited), the French subsidiary of the British Powers company.

SAMAS, which imported foreign-produced machines, had neither production nor machine development in France, making the company vulnerable when foreign trade became complicated after the breakdown of the international gold standard in 1931.

The state came to control imports and, simultaneously, the French state became the most important punched-card customer. National preferences grew in public acquisitions and became evident during the Pierre Laval government from 1935 to 1936.[110] Furthermore, the French government insisted on Frenchifying foreign-owned businesses in France, which meant having French ownership of a majority of the stocks, and a majority of French nationals on the board. To accommodate this request, Frenchmen took over a majority of the shares of SAMAS. To accomplish this, all the British directors on the board resigned and French nationals were appointed in their places. However, the British Powers company retained the right to dismiss the directors, which implied they had not relinquished control of the French company.[111]

SAMAS established their own production facilities to assemble foreign-produced parts and started machine development. However, they ran into financial problems due to the costs of production and development combined with the falling value of the French franc, which declined by 63 percent compared with sterling between 1935 and 1937, and by 61 percent compared with the dollar. By the end of 1937, this had caused SAMAS severe deficits, and the British Powers company and Remington Rand began negotiating a reconstruction. The outcome was that the British Powers company paid for the reconstruction of SAMAS in 1939, while Remington Rand terminated their business in France, selling their French interests to the British Powers company in 1939.[112]

IBM companies in France prospered in the 1920s under the rationalization of businesses. Their total profits rose nearly every year from 1922 to 1931. Net profits are the only figures available for the IBM companies in France, but the impression of progress is upheld by a rising number of customers. IBM in France had thirty customers in 1927, which grew to seventy-six customers in 1935 and to 102 customers in 1938. In addition, the company established branch offices in Lille (1923), Lyon (1924), Marseille (1928), Bordeaux (1932), Nantes (1935), Alger (1936), Strasbourg (1938), Casablanca (1938), and Toulouse (1940).[113]

In 1935, the French IBM companies were also pressured by the French government to Frenchify, just as SAMAS had. To accommodate this request, IBM's total business was united in a single company in 1935 with the lengthy name Société française Hollerith, machines comptable et enregistreuse (French company for bookkeeping and registering Hollerith machines), which the following year was simplified to Compagnie électro-comptable de France (CEC, electric bookkeeping machine company of France). Although the economic control of the company in France remained with IBM in the United States, Frenchmen took over the board in France. Prior to that, there had only been one Frenchman on the boards of the two IBM companies in France. Now two-thirds of the members were French.[114]

National development and production became crucial components of IBM's strategy in France after the IBM company got a board with a French majority. However, this strategy had been initiated much earlier. As early as 1924, IBM had acquired a factory proper at their Vincennes machine shop to establish production.[115] In addition, they had started machine development in France in the late 1920s that seems to have focused on improving machines designed by IBM in the United States.[116] The Vincennes workshop finished a verifier in 1931, which they subsequently produced. Further, they improved the calculating capability of an IBM tabulator in 1934 to meet the special needs of French railroads, and they finished an improved paper feed for tabulators in 1935.[117]

In 1935 the French IBM company concluded an agreement with the parent company in the United States to assemble a printing numeric tabulator and a sorter, in addition to the verifier they had produced since 1931. This decision was the result of severe problems with importing completed machines from the United States. Customs duty was substantial and customs officers were zealous. By 1937 most machines were produced in France, including IBM's alphanumeric tabulators.[118]

The falling and low total net profits of IBM's French subsidiaries between 1933 and 1940 indicated poor business for IBM in France, but these numbers were influenced by the prices in the trade between IBM and its subsidiaries as well as the fluctuating value of the franc compared with the dollar. Moreover, the dynamic actions of French subsidiaries indicated a different and more positive development. The competing Bull company did receive preferential treatment as a genuine French company. In 1935, the French army acquired Bull equipment to monitor army transport expenses, in spite of the recommendation from the office processing those expenses to purchase IBM machines. They considered the IBM machines technically superior.[119] However, Bull remained a small company in the 1930s compared with the French IBM company, and the French IBM company had a dynamic development of punched-card applications in the French army, showing the importance of the army in the transition from statistics to bookkeeping applications in France and of IBM's role in that transition.

René Carmille was the key person in the French army's use of punched cards in the 1930s. He was trained in engineering at the elite École polytechnique in Paris and was employed in the French army after graduation. In 1924, he was assigned to the Corps du contrôle de l'administration de l'Armée de terre, where he rose to become the head in 1936.[120] This corps was a tool for controlling the French army's expenses and for promoting efficient administration.[121]

Back in 1931 or 1932, René Carmille became interested in using punched cards while studying ways to control costs in the army's artillery workshops. As a *polytechnicien*—a graduate from École polytechnique—he was influenced by the stories of rationalization propagated by his fellow engineers stemming from their key roles in the rationalization of the industrial sector dating back to the 1920s. Still, the Bull company had not yet emerged on the French market, and in 1932, IBM machines were chosen for use in cost control at the national weapons factory in Puteaux, near Paris. Management control was to be improved by the use of frequent and regular budgetary statements.

Key numeric information on invoices received or wages to be paid for work to produce ammunition and weapons was punched on cards as these invoices or wage accounts were approved for payment. This produced statistics on expenditures before they were met, facilitating a more detailed breakdown of costs according to the various departments or projects than could be provided by the records on payments. During the 1930s, the Puteaux factory maintained its IBM installation while it was expanded to include additional applications and machines.[122]

Between 1934 and 1937, the army established similar punched-card installations for army operational statistics at the army's explosives establishment in Sevran-Livry near Paris and in four other locations. In contrast to the Puteaux installation, the army's punched-card installations were all modeled on equipment from the French Bull company. However, national preferences in public acquisitions were not mandatory. In 1935, René Carmille started to extend the scope of punched-card applications in Sevran-Livry. Carmille intended to move beyond operational statistics to include bookkeeping of the weekly payroll. Since the start of the first army installation at Puteaux in 1932, punched-card producers had improved the capability of their machines. Most important, the machines had become alphanumeric—they now included a full alphabet and printed combinations of numbers and letters.

The new application at Sevran-Livry was based on the IBM punched-card standard and required some IBM machines to complement the original Bull installation.[123] The Sevran-Livry application involved a bookkeeping system that allowed machines to add together various wage components, to subtract any deductions, and, of particular interest to the army, to print lists with names and various figures, as well as a receipt with the wage earner's name and the amount to be paid.[124] The IBM machines were selected, in spite of being American, because it would have taken an additional year before Bull could supply a subtracting tabulator.[125]

In 1936, Carmille described in a book his conception of punched-card applications as demonstrated at the Sevran-Livry works. This conception encompassed general statistics and bookkeeping, and used letters to improve the readability of various items, for example, for goods listed on inventories or the names of employees and wages owed to them. By and large, until this time punched cards had remained a processing tool and the cards were discarded when processing ended. However, in the Sevran-Livry wage administration application, each worker had a basic punched-card that was kept and used each time wages were paid. Each worker's basic card held his name (up to thirteen letters), wage number, numeric wage, and the tax and insurance deduction information used to calculate the wage to be paid. The basic cards comprised a register, though this only constituted an auxiliary function to the system, which each week consumed four additional cards per worker that were subsequently discarded.[126]

The full alphabet found on the IBM tabulators was crucial to print receipts with the wage earners' names, which distinguished this from similar applications in Germany. In 1935, the French army also began developing the idea of a register with alphanumeric information for use in conscription and mobilization. A substantial register for this purpose would be an attractive and prestigious application that all three punched-card companies on the French market tried to attain. However, the order for this application was only decided after the fall of France to the German invasion in 1940 and was awarded to the Bull company.

The original Bull punched-card machines were designed in Norway by engineer Fredrik Rosing Bull and built by a local precision maker. Bull died in 1924, but his machine building was continued by engineer Knut Andreas Knutsen in Norway. In 1927 and 1930, Swiss-Belgian interests acquired the Norwegian patents and expertise, which only encompassed a punch, a sorter, a nonprinting tabulator, and the design for a number-printing tabulator.[127] Building of these machines relocated to Paris in 1931, where a new company, Egli-Bull, was established for this purpose. The Egli name came from the majority share holder, the H. W. Egli adding machine producer in Zürich. Production was established in an existing workshop in Paris, which changed from manufacturing American office machines to producing Bull machines. The new company had a share capital of $140,000 (3.6 million francs). However only one-third was paid, which provided very low funds for starting production of the machines built in Norway and for designing new machines, as the remainder was bonus shares awarded for the patent rights and the workshop facilities.[128]

The new company did not have an easy start. The transfer of technology from Norway to France proved complicated, demand was curtailed by the economic crises, and the new company had to design and produce additional punched-card machines to become a full competitor to IBM and SAMAS. In spite of these problems, the Bull company during its first year succeeded in building and supplying their first major punched-card machine to the French Ministry of Labor. However, this accomplishment exhausted the company's economic resources and they ended up with a significant deficit. By the end of 1931, the company needed about $94,000 (2.4 million francs) in additional share capital to alleviate their current economic problems and to complete building a competitive set of punched-card machines; in dollars this amount was 2.4 times the originally paid share capital.

To aggravate matters, the Swiss majority shareholder, H. W. Egli adding machine producer in Zürich, now needed to sell some of its shares due to falling demand caused by the world economic crisis.[129] The need for additional capital to the small Egli-Bull company and financing for further development of punched-card machines could be solved either through a takeover by a foreign producer or through the acquisition of French capital combined with further machine development within the Bull company. Both possibilities emerged successively in 1931.

Back in April 1931, Remington Rand had approached the main shareholder in the burgeoning Egli-Bull company, the H. W. Egli Company in Zürich, to buy the rights to Bull machines outside France. Remington Rand had acquired the only competitor to IBM in the United States in 1927, and they now needed to improve the performance of their machines. In the summer of 1931, directors from Remington Rand visited the workshops in Paris.[130] During these negotiations, which lasted until December 1931, Remington Rand became aware that Egli-Bull was seeking capital, and they offered to invest $230,000 (5.9 million francs), implying that Remington Rand would take over the company. In addition, Egli-Bull would be allowed access to Remington Rand's technology, though the advantages of this would be curtailed due to their different basic technologies.[131]

As the negotiations between Egli-Bull and Remington Rand were approaching an agreement, Georges Vieillard and Elie Doury emerged with an alternative. For several years, they had been working on a failed project to establish the production of Remington Rand punched-card machines in France. Now, they offered to promptly raise the needed capital in France for the Bull company.[132]

While Remington Rand's offer was backed by a big company, Vieillard and Doury rapidly needed to raise the required capital from a group of punched-card users they were trying to organize. The most urgent concern was to raise $24,000 (francs 600,000) to acquire shares from the Swiss H. W. Egli company in order to relieve that company's financial situation. In early December 1931, Georges Vieillard established a consortium of punched-card users in France who were willing to provide the amount required to buy the H. W. Egli shares in Switzerland and finance the much-needed expansion of Egli-Bull's capital. The consortium consisted of various industrial producers, railroads, and insurance companies. Prominent among them was the Société des Papeteries Aussedat, a manufacturer of paper and cardboard, which later became a large punched-card producer.

The consortium of users offered the Egli-Bull company a second choice, and it chose this French option. In April 1932, the capital increase was achieved, allowing the French to attain a majority of the company's share capital, and the Egli company in Zürich received the funds they needed.[133] However, the French rescue meant no technological alliance with Remington Rand as required by that firm if the French were to become the majority shareholder. Therefore, the Paris company had to accomplish the much-needed development of their machines alone.

Again during 1932, development costs exceeded revenues, which once more were relieved by a substantial investment by the consortium of punched-card users in France in early 1933. The same year, the company was renamed Compagnie des Machines Bull (CMB).[134]

The increases in share capital in 1932 and 1933 reflected the young company's problems in producing reliable machines and in developing a full line of punched-card machines to compete with IBM. Only in 1935 did Bull accomplish producing a basic line of punched-card machines, but the company's financial situation was aggravated as a large part of its trade was leasing machines, which required more capital than selling. The outcome resulted in deficits throughout 1935, making the need to raise additional capital urgent. Simultaneously, the company was bolstered by a growing number of customers and increasing orders.[135]

Meanwhile, the Bull company had for several years been working to alleviate their problems and build up exports. From 1932 to 1934, they had tried in vain to find a partner in Great Britain, Italy, Czechoslovakia, the Soviet Union, and the United States. In all of these negotiations, it was a condition that two-thirds of the share capital should remain with French and Belgian shareholders.[136] In 1935, Bull's financial problems grew serious, and the company initially tried to obtain aid from the French state to support research and development, but the state hesitated. Then, Bull turned to find a partner in the United States, where they approached both Remington Rand and IBM. The chairman of IBM, Thomas J. Watson, came to Paris where he met with Jacques Cailles and Georges Vieillard of Bull's board of directors and offered them an agreement of cooperation on the condition that IBM took over the Bull company. The discussions were tough, and many years later, Georges Vieillard remembered Watson to have said, "Mr. Frenchmen, you got up too late to conduct this trade." The French answer should have been, "Mr. Watson, the sun shines on all of us, we will not do anything against IBM, but France has the right to have a punched-card industry. That is exactly the objective we pursue and which fills us with enthusiasm. We will stay independent and do everything to remain so."[137] IBM did not manage to buy the Bull company, which would continue to navigate alone.

As a consequence, Bull had to rely on French capital supplied by the Cailles family, who owned the Aussedat paper mills. The family had contributed to raising additional capital for the Bull company in 1932 and 1933, and Jacques Cailles was their representative on Bull's board of directors. The Cailles family now became the majority shareholder through additional investment. There can be two reasons for this investment, to safeguard the Bull company in French hands or to consolidate the Aussedat company through a forward integration.

The reason for safeguarding the company in French hands is evident in the previously mentioned story regarding Watson's negotiations in Paris, and this reason explains why Frenchmen invested in the Bull company, though it does not explain how the Cailles family was selected to accomplish this task. This can perhaps be explained through a strategy by the Aussedat paper works for forward integration. By 1935, they were providing Bull with punched cards, a business relation dating back to the early days of the Bull company in France. Then Bull approached Aussedat to acquire punched cards, but Aussedat had difficulties manufacturing cardboard of the same quality as the paper mills in the United States. To improve the quality of their punched cards, Aussedat obtained a license from the Racquette River Paper Company in the United States, who was a manufacturer with a reputation for punched-card production. Since that time, Bull had been providing ever-increasing prospects for the Aussedat paper mills. Now the threat of a takeover by IBM worried Aussedat, as IBM required their costumers to buy punched cards from them.[138]

After Bull had received a third increase in share capital, the company faired well during the remaining part of the 1930s. From 1935, the company managed to produce well-functioning machines, and by 1942, it accomplished becoming full competitor to IBM when it built the last of several required punched-card machines. From 1936, the French Bull company attained a surplus. While the increases in share capital in 1932, 1933 and 1935 had been caused by operating deficits, from 1936 to 1939, the surpluses provided the basis for the company's expansion.

This is a case that appears to confirm the success of a company in France during the 1930s, that is, once national capital was attained. However, this story does not explain how the asset of national capital provided business, a feat that was accomplished by people in the Bull company via three interrelated social networks that also provided capital for the company's infusions of capital in 1932, 1933 and 1935: the Association of French Industrial Producers (Confédération générale de la production française), a network of (former) commissioned officers in the French army, and a network of *polytechniciens*.

First, the Bull company was admitted to membership of the Association of French Industrial Producers, distinguishing them from their foreign-based competitors, the IBM subsidiary and SAMAS.[139] Notably, this took place as early as December 1931 at time when the majority of the company's capital was foreign.[140] This admittance was most likely facilitated through the good offices of Lieutenant-Colonel Emile Rimailho, who was a member of the board of directors between 1931 and 1937 and its chairman from 1931 to 1934. Rimailho was a distinguished French artillery officer who had contributed to the design of a gun and, subsequently, had moved into industry, where he became a specialist in scientific management—hence his interest in office mechanization.[141]

Rimailho illustrates the company's early close relations to a network of commissioned officers in the army and to French industry. Close relations to French industry were also mediated through two *polytechniciens*, Georges Vieillard and Elie Doury, who suddenly turned up in late 1931. Using their extensive network of *polytechniciens*, they accomplished within a short period the establishment of a consortium of punched-card users in France, which acted as the network that raised capital for the company in 1932, 1933, and 1935. Vieillard was appointed managing director of Bull in 1933 to secure his continued services for the company.[142]

Jacques Cailles, who was the second commissioned officer who became important for Bull, graduated from the Military Engineering University (École spéciale militaire) of Saint Cyr. He succeeded Rimailho in 1934 as chairman of Bull's board of directors and represented the Cailles family, who had contributed to the raising of additional capital for the company in 1932 and 1933 and became majority shareholders in 1935.

Although René Carmille, the third key military officer in the story of Bull's success in France, was employed outside the company, he became a core supporter, which was managed by his fellow *polytechnicien* Georges Vieillard. In 1935, Carmille advised the government to make a substantial direct investment in the company, which was suffering from financial problems arising from their difficulties in producing reliable machines.[143] Carmille's investment proposal was not accepted, but he subsequently provided advice that was favorable for the Bull company. Carmille resolved in 1935 to acquire Bull equipment to control army transport expenditures—in spite of a recommendation from the responsible office to purchase IBM machines.[144] Further, Carmille selected Bull equipment for the tests for a conscription and mobilization system in 1935, which failed due to malfunctioning machines, and as the basis for the national register that started in 1940.

These examples illustrate the initial foundation of Bull's success in the 1930s. The effective establishment of business networks provided access to a dynamic set of customers, and the establishment of a basis of French capital for the company was but one component in this process. Similarly, in Germany, Dehomag's well-established and extensive business network was a crucial reason that it avoided the brunt of autarchy until 1940, in spite of the fact that the company was 90 percent owned by IBM in the United States. In contrast, SAMAS and IBM in France focused on becoming Frenchified and IBM lost steam, which can be explained due to a weaker business network. The second foundation of Bull's success in the 1930s was the company's ability to develop and produce reliable machines with the features required in France, which again resembled Dehomag's experience.

The young Bull company had serious problems until 1935 in producing reliable punched-card machines. In addition, they needed to extensively develop the machines to catch up with the competitors. This was the basic reason for the company's need to raise additional capital in 1932, 1933, and 1935, a situation that was aggravated by the company's original low-paid capital and the leasing of 37 percent of its tabulators.[145]

It proved difficult to move the production of machines from Oslo and Zürich and to establish an industrial production in Paris, and many of Bull's machines frequently made errors. The company only managed to produce reliable machines in 1935.[146]

The Bull company's development of punched-card machinery in the 1930s was closely linked to its attempt to produce equipment that equaled the technical capabilities of that of IBM and SAMAS, which extended the scope of Bull's strategy from statistics processing to encompass bookkeeping applications. In this process, Bull distinguished itself by shaping its own path to bookkeeping with punched cards. In addition, though the Bull machines were modeled on the same basic electromechanical technology as IBM's, the Bull people chose separate designs.

Bull started as a producer of statistics-processing equipment in France.[147] Their first production of equipment derived from earlier machine building in Norway and Switzerland of a nonprinting tabulator, a horizontal sorter, and a simple manual punch. In comparison, its competitors had produced numeric printing tabulators for several years. Demand for numeric printing capability had been mentioned by the people from the Egli company during negotiations with Norwegian engineer Knut Andreas Knutsen in Oslo in 1927.

In 1924, Knutsen had succeeded Bull as the leader of the development of the machines.[148] This demand caused Knutsen to design a numeric wheel printer, an accomplishment that proved to be a major reason why the French Bull company was established and why Knutsen was hired to manage Bull's development activities. In late 1930, Knutsen hired Norwegian engineer Anders Eirikson Vethe to innovate Knutsen's patented design for a printing tabulator for production. In September 1931, the first numeric tabulator-printer was supplied to compile social statistics at the Ministry of Labor. It used printing wheels and printed at a speed outdistancing its competitors' tabulators. The reliability of this tabulator was subsequently improved.[149]

The Bull company had the choice of focusing on the development of several features for their equipment for the extension of the scope of punched-card applications to encompass bookkeeping, alphanumeric capability, an extension of the 45-column card, and improved calculation capacity. First, influenced by considerations by insurance companies to introduce punched cards for insurance policy administration, they focused on alphanumeric punch card printing.[150] However, by 1934 they had only gained two French insurance companies in addition to Rentenanstalt Zürich (today, Swiss Life), which had used Bull equipment since 1926.[151]

The problems involved in implementing alphanumeric systems were so great that all producers initially introduced systems that were not yet completely alphanumeric, for example, the British Powers company's reduced alphabet unit from 1921 could not print numbers, and IBM's first alphanumeric system with a reduced alphabet, introduced in 1931. Knut Andreas Knutsen initially built a tabulator with alphanumeric representation in a third of its printing positions and exclusively numeric representation in the remaining two-thirds in 1932. However, it had different representations of numbers in the two sections, reducing flexibility. At the same time, this tabulator was based on printing wheels instead of type bars, which proved to be an excellent basis for an extension from numeric to alphanumeric representation.[152]

The alphanumeric printing tabulator was particularly suited to print addresses that required a combination of letters and numbers, possibly indicating that insurance policy administration was the objective of this development. IBM's research department in New York also developed a wheel printing tabulator, but their work was discontinued as a result of

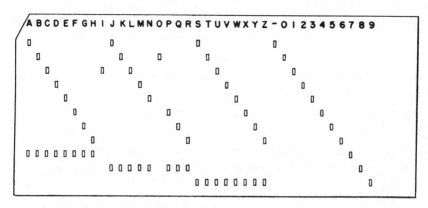

The Bull company's alphanumeric code, 1935. The uppermost row (12) was used for qualitative indications, for example, a negative sign. (K.A. Knutsen, [U.S.] Patent 2,175,530, filed in France in 1935 and issued in the U.S. in 1939)

Knutsen's patents.[153] Eventually, in 1949, at which point Knutsen's patents had expired, IBM marketed a wheel printing tabulator with a printing mechanism similar to that on the Bull tabulator.[154]

After completing the tabulator with reduced alphanumeric capability, Knutsen finished an alphanumeric design in 1934 that used the old digit representation standard for combinations with letters that allowed calculation on all numbers punched on to cards. This representation was distinct from IBM's and from the subsequent alphanumeric systems designed by the American and British Powers companies, making alphanumeric punched-card systems proprietary, distinguishing them from the exclusively numeric systems that was an industry standard. A tabulator based on Knutsen's improved design was produced in 1935 and continued to be manufactured without major modifications until 1968.[155]

Extending the old standard 45-column card was the second basic feature of the Bull punched-card system to be improved. After 1928, IBM used their new 8-column card and began supplying alphanumeric machines in 1933. From 1931 the Bull company prepared their machines to be built for 80-column cards.[156] However, because of Bull's electric reading of cards, this implied adopting rectangular or oblong perforations that would constitute an infringement on IBM's French patent on this card design.[157] Bull first applied 80-column cards for the machines used by the French army to test a new conscription administration system in 1935, which failed because of unreliable machines. For this application, the extension from forty-five to eighty columns facilitated additional information about the conscript to be stored on the card, avoiding the use of a second card for every conscript that would have made processing more complex.

The French army required 80-column cards for this application, as the extra capacity was needed and as IBM had alphanumeric machines for 80-column cards. Subsequently, the French army made 80-column cards a prerequisite for orders, and the Bull company marketed the 80-column machines in 1938.[158] This caused IBM to instigate legal proceedings. Bull lost this lawsuit in 1941 in the lower courts, but they later won the appeal in 1947.[159] Only during the appeal was it disclosed that this patent had been dissolved in the United States in 1933, as the design had been disclosed in the United States prior to the filing of the patent application.[160] Thus, IBM's suit had been based on a dissolved patent.

It is remarkable that a subtracting tabulator first reached the market only in 1936, subsequent to the alphanumeric facility and the initial introduction of the 80-column card. One of the partners of the burgeoning Bull company was the Egli company in Zürich, which produced key subtracting machines. The 80-column card was introduced before the subtracting tabulator for reasons of demand, but the late introduction of a subtracting tabulator was the result of problems the people developing the Bull machines had in completing a reliable design. Bull's development people planned a subtracting tabulator as early as 1931, and they started to build a prototype in early 1934.[161] However, it took another two years before the subtracting tabulator, which was based on the designs of the Toledo key adding machine, to reach the market.[162]

The Bull company had a development department, which from the outset was managed by Knut Andreas Knutsen. In 1936 he had six engineers and three assistants, and he proved apt at attracting able engineers to accomplish the substantial development of the Bull machines in the 1930s. He hired Anders Eirikson Vethe to implement his basic design for a printing tabulator in 1930. Georges Ziguelde was hired from the French IBM company and André Ziguelde from the key office machine producer, Olivetti, in 1935, to develop a subtracting tabulator; they subsequently designed the punched-card multiplier, launched in 1938. Roger Clouet was hired and designed a tabulator with improved programming facilities that was marketed in 1939.[163]

Dynamics of Distinctions in Europe

Punched-card based bookkeeping systems were shaped in the years between the two world wars in Great Britain, Germany, and France with discernible differences. The earliest letter printing tabulator was developed and marketed by the British Powers company in 1921. However, this feature did not become a major advantage for the company, which only developed an alphanumeric system after the Second World War. Instead, the British Powers company focused on developing and marketing of cheap punched-card systems based on small nonstandard cards, which had substantial success.

In Germany, the Powers agency marketed the letter printing tabulator developed by the British Powers company, but they also experienced limited success. The German demand focused on numeric calculation capability, demand for alphanumeric punched-card systems only emerging during the Second World War. In contrast, the French punched-card applications developed an emphasis on alphanumeric printing systems in the 1930s, which Bull responded to with machine building, while the French IBM subsidiary marketed machines designed in the United States.

These varying systems for bookkeeping were shaped through a combination of different interpretations of national demand and in response to IBM and the Powers company (which joined with Remington Rand in 1927) in the United States. Like in the United States, most of the development of punched-card systems in European companies originated in actual and imagined applications. An example of actual applications was in Britain, when the Powers company based their building of the first alphabetic punched-card system on specifications from the Prudential Assurance Company in the years around 1920. An example of imagined applications was the Powers company's envisioning the potential of their small size punched-card equipment for medium-size companies.

In Germany, the market account administration in banking in the 1920s provided the focus for Dehomag's development of a numeric interpretation of punched-card bookkeeping. In France, the emphasis on alphanumeric features at Bull originated with the Swiss producer of office machines, Egli, which was instrumental in the transfer of the original Norwegian technology to France. However, alphanumeric features only became prominent at Bull when the French army started to demand alphanumeric punched-card equipment in 1935. Subsequently, the French army became the prime user of the Bull company, which shaped the technology of their alphanumeric features and the conversion of the company's basic punched-card from forty-five to eighty columns. In contrast, IBM's French subsidiary did not contribute significantly to the shaping of the technology in France. Instead, they concentrated on producing equipment designed by IBM in the United States, which satisfied the French government by saving foreign currency.

Through the development of these different punched-card systems, European punched-card producers contributed significantly to shaping punched-card technology. The most noticeable examples were the British Powers company making the first letter printing tabulator, Dehomag producing punched-card machines with advanced calculation facilities, and the Bull company making alphanumeric tabulators that used printing wheels. After the Second World War, alphanumeric became a standard capability in punched-card systems, Dehomag's D11 tabulator was marketed by IBM in the United States, and IBM started to produce a tabulator with printing wheels once the relevant Bull company patents had expired.

The variations served to enhance the independence of the punched-card companies in Europe in relation to the parent companies in the United States. The relations between subsidiaries and parent companies had two interrelated dimensions, legal connections and a degree of operational freedom. The Tabulating Machines and Powers agencies in Great Britain and Germany were established by 1914. Originally, they were domestically owned, which provided great operational freedom. In contrast, the subsidiaries in France were only established in the early 1920s and were from the outset completely owned by the parent companies, which provided less operational freedom. Reduced operational freedom was also caused by the Tabulating Machine Company's exploitation of the German mark crisis of the early 1920s by taking over their German agency, Dehomag. However, Dehomag's successful founder, Willy Heidinger, remained and managed the company with substantial operational freedom. Assigning operational autonomy to nationals leading the subsidiaries in Europe was in contrast to the limited delegation within the company in the United States.

Heidinger's subsequent success and autonomy in relation to IBM was closely linked to Dehomag's development and production of punched-card machines in Germany, independently of IBM. Similarly, the degree of operational freedom in the two British punched-card companies was not based upon their legal independence of the parent companies in the United States, but on their success in business as well as the development and production of punched-card equipment. Powers-Samas had more independent development and autonomy than did the British Tabulating Machine Company. National development of punched-card technology became a means to gain autonomy.

While the various agencies, child companies, and the Bull company in France had different relations to IBM and the Powers company in the United States, they interacted on their national markets. The high degree of competition in Germany and France seems to have focused the national emphasis in the two countries, with computation in Germany and with alphanumeric systems in France. In contrast, the low competition British market segmented into two, punched-card systems based on standard punched cards from the British Tabulating Machines Company like those in the United States and cheaper systems based on small nonstandard punched cards.

Finally, shaping diverse punched-card equipment and applications in the various countries in Europe was enhanced by the reduction of international trade after the breakdown of the international currency standard in 1931. British development was least affected, as both the British Tabulating Machine Company and the British Powers company exclusively used British capital. Only the British Tabulating Machine Company relied to a large extent on foreign design, but they do not appear to have been much effected by the "Buy British" campaign in the 1930s, as they had steadily growing assets in sterling after 1925.

In contrast, the punched-card business was very much affected by the Frenchifying that occurred in France in the mid 1930s and the German autarchy from 1932. These measures did not eliminate the foreign control of the Powers and IBM companies in either country, but in Germany they contributed to the development of a separate version of punched-card technology with heavy emphasis on computation capability.

EIGHT

Keeping Tabs on Society with Punched Cards

The Social Security Act of 1935 was a core part of President Franklin D. Roosevelt's New Deal policy to raise the United States out of the economic crisis that had started in 1929.[1] The act authorized a national system of contributory old age benefits for people who turned 65 in 1942 and later a state administered, federally supervised unemployment insurance system and a program to induce nationwide financial aid and to provide aid for the needy, the aged, the blind, and those children deprived of parental support.[2] It also promoted national financial assistance for improved public health services by expanding research, by helping states improve their health staffs, by giving special aid to disadvantaged children and children with disabilities, and by renewing promotional health work for mothers and infants. Finally, the act established a three-member Social Security Board to manage the program.[3]

Compared with other Western societies, however, the United States was tardy in introducing compulsory national programs for old age pensions and unemployment compensation. The United States had extensive examples from abroad in the form of old age pension programs in Germany, Great Britain, and France to draw on in creating their system.

For the first time, the Social Security Act conferred a general federal commitment in social matters in the United States but only help to blind people and orphans was to come from the transfer of public revenues. The programs for unemployment insurance and old age pensions were intended to pay their own way. To take care of each of the millions of participants in the old age pension program, however, the federal government organized and funded an extensive central bureaucracy more ambitious than any other undertaken in the United States or abroad. The national government wanted to care directly for the individual.

The objective of a just relationship between contributions and pensions was shaped by the politicians and civil servants who formed the Social Security Act of 1935. They knew of the old age pension programs in Germany, Great Britain, and France and of policy administration in insurance companies. A bureaucratic program designed to execute the intentions of the law was only developed subsequent to its enactment. It was only at this point that the punched-card method was introduced, and the bureaucracy became based on a vast register of IBM punched cards of all wage holders who had contributed to an old age pension.

Previously, punched cards by and large had remained a processing tool and were discarded when the processing was completed. Exceptions were punched-card systems for insurance policy administration. For every premium collection, one card per policy of the same set with the basic information was used in issuing invoices. Administering the Social Security program introduced for the first time a similar register on a national scale, giving it unprecedented size and scope. In turn, the size of the program provided the essential impetus for further development of the established closure of punched cards for bookkeeping. This further development opened up the industry to considering changes to their model, reversing their usual inclination to be conservative about changes to an established closure.

Furthermore, this large, register-based national system for administering old age pension savings in the United States became the first of many large registers used to mobilize all people. Large registers became an essential tool in modern societies. France established a comparably large register between 1940 and 1944, while in Germany, the government tried in vain from 1943 to 1944 to establish a large register of people. In the Second World War, additional large registers in the United States became a means of enhancing warfare capabilities, and subsequent large registers were used to manage a more advanced society.

Franklin D. Roosevelt's Social Security Program

Franklin D. Roosevelt promised "social justice through social action" in his presidential campaign in 1932. After his election, the most pressing problem for his new administration was providing relief for destitute families, and this was done through relief and support measures for industry and agriculture. In the summer of 1934, Roosevelt revived his ideas concerning general social legislation, establishing a committee to study economic insecurity in the United States and European experiences with social security. Their report became the basis for the bill on social security, which was adopted by Congress in August 1935.

The contributory old age pension of the Social Security Act from 1935 was compulsory and encompassed all wage holders in the United States with several notable exceptions, including, agricultural labor, domestic service, sailors, and public employees.[4] In 1937, the scheme was comprised of 32.9 million people and grew to 46.4 million people in 1945.[5] Employers were obliged to report frequently to the Office of the Commissioner of Internal Revenue, the federal tax collecting authority, any wages they paid and to pay 1 percent of the wages as new tax, alluded to as pension contributions, plus the same amount as an excise tax on employers.[6]

The information about wages paid to individuals was forwarded to the Social Security Board, which was created as a federal organization to run the Social Security program. The Board was responsible for monitoring each individual's accumulated earnings, the basis for his or her pension. A small percentage of the accumulated earnings were then paid monthly from the age of 65 until death, and Social Security also paid widow benefits.[7] The taxes paid under this program were accumulated as a special fund in the Treasury, which was also responsible for paying pensions that were determined by the administration of the Social Security program.

The Social Security Act did not specify how monitoring wages reported for individual employees should take place, but the operation was to start in 1937. The president's committee in 1934 and its staff, who prepared the bill, had studied contributory old age pension programs in Germany, Great Britain, and France and seemed certain that a system could be devised to monitor the wages paid to tens of millions of individuals over many years.[8] This became one of the first great challenges of the Social Security Board.[9]

During its first year, the Social Security Board shaped the general design of the system for monitoring the wages earned, which included a way to identify the records of each of the millions of workers covered by the Social Security Act. The act mentioned several traditional means of reporting wages earned, indicating that manually processing the reports was envisioned.[10] While this system was being designed, punched cards emerged as a means of processing wage reports, becoming preferred over more traditional means.

The establishment of a system for monitoring wages needed a way to identify people. Simple considerations showed the importance of issuing identifiers and not simply of recording wages by name and identifying like-named individuals by their dates of birth, mothers' maiden names, or other similar information. A telephone book from Washington, D.C. (with about 500,000 inhabitants), revealed thirty-three individuals named John Smith and eighteen named Mary Jones, demonstrating how difficult, time-consuming, and inexact differentiating among so many individuals would be without some type of numeric or alphanumeric identifier.

In November 1935, the newly appointed Social Security Board tentatively decided that the identifier should have eight characters, three letters and five digits. This decision sparked two controversies. First, the Social Security Board found that several federal agencies used their own exclusive numeric identification schemes, and a wish emerged to define the Social Security identification in a way that would facilitate an extension to universal registration. Second, they learned that these agencies had rejected the use of alphanumeric systems largely because only IBM and Remington Rand produced machines that could work with such a system. Moreover, the government had filed an antitrust suit against these companies in 1932 that was only resolved by the United States Supreme Court in April 1936.[11] Thus the Social Security Board's tentative decision for an alphanumeric indicator was not only inconsistent with the exclusive use of identifiers government-wide, it also clashed with Justice Department strategy.

These considerations made the Social Security Board realize that only an all-numeric identifier would be feasible, reducing the focus to how many digits a Social Security number should have and what they would represent. The Social Security Board decided in June 1936 to identify each individual wage holder with a number composed of nine digits divided into three sections. The first three digits specified the geographical area where the individual lived when applying for a Social Security number, the next two digits indicated the group, and the remaining four digits constituted an individual serial number. The two-digit group number was originally planned to facilitate the location of the individual location within the wider area specified by the first three digits. However, the two-digit group number lost its meaning in the initial number assigning process in 1937, and it would, anyway, have lost meaning as people moved.[12]

A similar numerical identifier for employers was also originally designed to facilitate localization. This nine-digit number also had three groupings. The first grouping of two digits indicated the Internal Revenue District to which employers were to report the wages they had paid and through which they would pay the excise tax levied upon them as an employer as well as the income tax deducted from employee salaries.[13] The second group of two figures embodied the industrial code of the employer's business, and the final group of five digits was a serial number. Also the design of this number was subsequently simplified, as the industrial code was eliminated and added to the serial number, which now consists of seven digits.[14]

The initial task of enumeration was performed by post offices throughout the country, as the Post Office Department was an already established government organization that maintained offices in a majority of communities, while the Social Security Board had not yet established a network of offices. In the first step of this process in November 1936, each local post office identified every employer in its area, providing each of them with an application for an employer identification number. The applications, which included a question concerning the number of employees each employer had, were to be returned within a week. Although there was no legal compulsion for the employers to cooperate, nearly all complied with the request, indicating a high degree of public acceptance of the social security program. Next, the post offices provided employers with sufficient application forms for Social Security numbers for their employees. The completed applications were then sent to the Social Security Board's office in Baltimore, which assigned account numbers.

The Board originally anticipated an initial registration of some 26 million employees, and expected an increase of about 5 million during the first year of operation and 2 million each year thereafter, until an average load of 35 to 40 million active numbers was established. However, this projection turned out to be too low. In less than a month after the application forms had been distributed by the post offices, more than 22 million completed applications had been received. Some 30 million numbers had already been assigned in July 1937 when the Board assumed responsibility from the Post Office Department for the entire operation of issuing Social Security numbers.[15]

Monitoring every employee's wages, an enormous undertaking, was the basic task of the Social Security Act's old age pension program. Since wage reports were to start arriving near the middle of 1937, most of the decisions surrounding this issue were being made simultaneously with those concerning employee registration.

Comparable old age pension programs had already been established in Great Britain, France, and Germany, and the United States Social Security Board included these experiences in their considerations. All these programs, which accounted for people's varying wages, used written files and key office machines. In addition, they were all based on decentralized administration in a large number of organizations, but the total number of people encompassed by the schemes in Great Britain and France were comparable to the projected number of people in the United States. In Great Britain a total of 19.1 million people were insured in 1932, which was along similar lines to that planned in the United States.[16] In 1939 in France, a total of 21,000 organizations administered the pension insurance of 9.2 million people.[17]

The old age program for social security in the United States distinguished itself from all of the others by being administered by a centralized bureaucracy, which complicated the task. The key problem was to frequently monitor the income of tens of millions of people. In the summer of 1936, the Social Security Board considered two basic methods for obtaining information on peoples' earnings: payroll reports and stamp passbooks. The payroll report method would rely on quarterly or semiannual statements of each individual's earnings. Another method would entail stamp passbooks that indicated the earnings of individual employees and were to be sent to the Social Security Board once or twice a year. The stamp passbook method was simpler, particularly for small employers, who relied on handwritten records in ledgers. However, Social Security Board discussions from the summer of 1936 eliminated the stamp passbook option.[18]

Based on the choice of the payroll report method and the decisions concerning identifiers for wage holders and employers, the Social Security Board asked ninety equipment companies to submit bids or proposals for establishing and maintaining the record-keeping system. Eight of these companies submitted proposals that were comprehensive enough to be considered by the Social Security Board. Five of these were soon eliminated, leaving the Burroughs Adding Machine Company, IBM, and the Monroe Calculating Machine Company. Their proposals were comprehensive and explicit enough regarding operation plans and the breakdown of operating costs to make detailed comparisons possible. Because the differences between the operating costs in the three proposals were small, cost was quickly eliminated as a determining factor.

As a result, the choice came down to the method proposed rather than which equipment company. The Burroughs and Monroe proposals were based on key-bookkeeping machines and involved extensive manual handling of records. In contrast, IBM proposed a punched-card method. The committee that the Social Security Board had convened to study the proposals chose the most technically sophisticated option, concluding that the application of the highly mechanized IBM system was the best approach at the outset. This system was superior in terms of adaptability to future change, both concerning procedures and workload volume. Moreover, it reduced the human element, which the committee believed should be eliminated as much as possible. The committee recommended acceptance of the IBM proposal, to which the Board concurred.[19]

Next, the Social Security Board acquired space for a centralized office in Baltimore and began establishing wage records for all the coming beneficiaries of Social Security old age pensions. This operation began just after the Social Security numbers had been assigned after applications were received from wage holders. Based on the information on the application forms, punched cards were produced, and registers and ledgers were established.

A master punched card, produced for every employee, was used to maintain the wage record. Each master card contained the employee's Social Security number, first ten letters of the first name, first three letters or initial of middle name, first twelve letters of the last name, a three-digit phonetic code, date of birth (but not year), sex, race (eight categories), and date of issuance of the Social Security number. The Social Security number and name were mechanically printed on the top of the card by a punched-card typewriter (interpreter). Compared with the subsequent French national register, it is worth noting that employee addresses, while not punched, were kept in the written case record.

The phonetic code was applied to eliminate confusion and errors rising from variations in the spelling of a registrant's name on different records associated with his account, for example, on wage reports filed by different employers. The phonetic code represented the last name by a letter and a three-digit number. Though costly to implement, the Equitable Life Assurance Society in New York had claimed that the code had eliminated the misspelling of names in their handling of policies.[20]

Next, punched cards were used to prepare the visible index of registrants, an index made up of legible strips that came in large perforated sheets. Each strip provided for one line of print and was used for a single Social Security number. Since the strips appeared primarily in the same order as the phonetic code, all the names that were phonetically similar, but that had dissimilar spellings were brought together in the visible index. Examination of this index therefore revealed a surprising number of variations in the spelling of many common names. In spite of the complications caused by having literally hundreds of thousands of individuals with the same surname and a total of about 32 million names in the file in 1937, it was possible for the clerks familiar with the visible index to quickly find any name and its corresponding Social Security number.

The master punched cards were also used to produce a numerical register of accounts by listing the name and date of birth of each registrant in numerical order. This register, a printed list of the information from the sorted cards, was kept in loose-leaf books and was applied to locate available numbers for assigning additional Social Security numbers. And, finally, the master cards were used to head up the ledger sheets on which wage records were posted. This was accomplished on fanfold paper by a tabulator, which was subsequently separated into sheets. The master cards and ledger sheets were then filed separately in numerical order by regions. The same procedure was followed for subsequent applications for Social Security numbers.

Established registers were used to record wages paid from the beginning of 1937. Employers reported the wages paid quarterly to the Bureau of Internal Revenue, now known as the Internal Revenue Service;[21] the Bureau then verified the reported information for consistency and forwarded it to the Social Security administration. They processed employer wage reports by preparing a punched card for each wage earned, allowing wage reports to be mechanically processed. They also balanced the cards

Clerks filing Social Security punched-card records in Baltimore, late 1930s. (Social Security Administration, Baltimore, Maryland)

punched with the wage report information, verified the identity of each registrant for whom wages were reported, and posted the amounts reported to the correct ledger sheets so that records of individual earnings were available to determine the amount of benefits due.

As the ledger sheets of the registrants were filed in the same order as the Social Security numbers, employee earnings cards also were sorted in the same order before the mechanical posting of wages was carried out expeditiously. This system brought all the earnings cards for an individual's account together in instances where more than one employer had reported wages during the period. The earnings cards were then mechanically compared by use of a collator punched-card machine with the corresponding employee master cards. This step was taken to verify the identity of the individual for whom wages were reported and to segregate the cards of individuals who did not yet have an employee master card. If the name on an earnings card did not agree with the name on the master card bearing the same account number, the account was removed from the ledger section, the various files were searched to establish identity, and the account was posted at a later time.

The earnings cards in the ledger section were listed mechanically, the total number of cards and the amount of earnings recorded. This list of accounts indicated which individual ledger sheets were to be taken from the file for posting. The earnings were then mechanically posted on the individual ledger sheets by accounting machines which "read" the amounts punched on the cards. All records were then returned to the files.

The Social Security Board's choice in 1936 of a punched-card system to monitor the wages of tens of millions of people had a variety of implications. It provided extensive punched-card business, which shaped technology and the industry. It also provided the national government with a tool for direct access to individual citizens.

The monitoring of wages could have been accomplished by the standard IBM equipment available. However, two new items were introduced to ease the operation: a collator—a nonstandard punched-card machine—and a new posting attachment to the tabulator. People at IBM had worked to design a collator punched-card machine since the early 1930s to facilitate frequent file mergers, for example, permanent employee master cards, and weekly wage cards in a wage administration. However, the development of the machine was only completed in 1937, when it went into production as a response to the demand from the Social Security administration.

The collator merged two piles of punched cards that had been sorted in the same succession, for example, when the quarterly earnings cards were compared with the employee master cards to verify the identity of the wage earner. A standard sorter could accomplish this operation, but extensive and time-consuming sorting would be required. The core problem was how to handle unmatched cards and cards that had been misplaced. IBM built a collator that could reject unmatched cards. Misplaced cards could still derail the process, and merging punched cards remained a major problem. This problem was finally solved by the transition to magnetic storage combined with extensive sorting, which had not been feasible for punched cards.[22]

The new posting attachment to the tabulator was a technically sophisticated solution to the problem of adding information onto the wage earner's ledger sheet regarding a record of aggregate wages after the beginning of 1937. The solution was chosen from IBM's overall proposal in 1936, which suggested either strip posting or direct posting. Strip posting entailed the tabulator printing information to an adhesive strip that clerks subsequently affixed to the proper ledger sheets. The strips were heat laminated onto the ledger sheets using an iron.

To provide direct posting onto the ledger sheets, it was necessary for IBM to equip its regular tabulator with a new posting device. At first, the strip method appeared to be better, as it was less costly. However, there was no way to judge the life of such strips, or exactly how long they would remain firmly adhered to the ledger sheets. And when a strip fell off, an individual's record would be ruined. IBM, in contrast, was able to satisfactorily demonstrate that it could supply an appropriate device to print the posting directly onto the ledger sheets. The Board chose this solution.[23]

Notably, IBM's huge punched-card contract for monitoring wages was not contested by Remington Rand, who did not submit a comprehensive proposal for monitoring wages to the Social Security Board in 1936. The Social Security contract, which enhanced IBM's position as the leading punched-card producer in the United States, was probably a major reason why Remington Rand, in 1938, finally launched their Model 3 line of alphanumeric punched-card machines, which went into production and reached the market over the following five years.

The register for monitoring wages in administering Social Security was a noted success for Franklin D. Roosevelt's contested New Deal policy, and it opened up the possibility of using large, machine-readable registers. They also became an important tool for warfare during the Second World War. Many other administrations on the federal, state, and local levels developed an interest in establishing a Social Security number as an identifier for a large number of people in the United States and its application in the machine-readable register. However, to safeguard the confidentiality of the information in the register, the Social Security Board decided in 1937 to require that all records must be confidential and used only for administering the program.[24] In this way, they established a high standard early, distinguishing them and their successors from the people who built comparable registers in France and Germany between 1940 and 1944.

The Vichy Mobilization Register

In June 1940, France capitulated after five weeks of fighting against the German invasion. The country was divided into two parts, one was occupied by Germany and the other was under rule by a French government in Vichy. Armistice conditions were harsh, allowing the French government an army of a mere 100,000 men, which could not threaten Germany.

During the 1930s, René Carmille, who held a key administrative position in the French army, had tried in vain to improve the administration of conscription and mobilization via the use of punched cards or a combination of punched cards and address plates. Shortly after the fall of France, Carmille suggested establishing a mobilization register using punched cards to prepare an army with an amount of strength not permitted by the armistice. Over the following two years, he implemented a register that allowed for the conscription of 300,000 men. The scope of the register, however, went beyond this to become a national register enabling the state to monitor its subjects, for better and for worse.

The mobilization register was completed in the spring of 1942, but the Germans took over the unoccupied part of France in November 1942, removing the possibility of mobilization and raising the risk of the register's being detected by the Germans. Consequently, Carmille destroyed the mobilization register and related archives but continued to build up the national register. However, the autocratic French government in Vichy provided a far from ideal setting for exploring the concept of keeping a national register, its history illustrating the somber implications of big, machine-readable registers. The register was used to keep tabs on the population by improving both the control of identification cards issued and the distribution of ration cards. Since Jews were registered separately because of the anti-Semitic inclination of the Vichy regime, punched cards may have been used to locate Jews for deportation.

The civil administration of the French national register was established from scratch in 1940, which resembled the development of the punched-card register that the administration for the United States Social Security went through from 1935 to 1937.[25] However, René Carmille had promoted office mechanization based on punched cards for use in the French army since the early 1930s, starting with operational statistics and later including bookkeeping of the weekly payment of wages for workers at an armament factory.

Compared with these tasks, managing a vast modern army of conscripts was a huge administrative undertaking that was crucial for France's military performance and was accomplished by simple manual methods. The first step involved in conscription administration was to call the conscripts up and provide their basic training. Then, they were transferred to the reserves, and the military had to keep track of the conscripts to prepare them for exercises and for any future mobilization. The administrative task was complicated, as the conscripts remained in the reserves for sixteen years after their basic training. The size of this task is apparent from the fact that at the outbreak of the Second World War in September 1939, France called up 4.7 million men.[26]

Back in 1933, when France had adopted a new mobilization plan, the conscription and mobilization service began studying ways to mechanize this gigantic bookkeeping operation, taking into account where people currently lived, rather than their last contact address. The military wanted to improve the distribution of personnel to the various units, and they wanted to be able to print the diverse mobilization documents quickly. This assignment required writing or printing various kinds of information on each soldier, including name, address, profession, and unit.[27]

It was ascertained that the mobilization system could be improved, either through a system that used punched cards or a combination of punched cards and address plates. An exclusive punched-card system would have required an alphanumeric capability and a punched card with more than the standard forty-five columns to allow bigger records units.

Various companies in France had the ability to provide different solutions to the government's needs. After 1928, IBM was able to use their new 80-column card, and by 1933 they could supply alphanumeric machines. The French Bull company built their first alphanumeric tabulator the following year and were prepared to adopt the IBM 80-column card even though this constituted an infringement of IBM's French patent.[28] SAMAS marketed 65- or 130-column numeric punched cards and a combined system of punched cards and address plates.

For what were most likely patriotic reasons, only Bull was asked to provide machines for testing at the Versailles draft administration office. In the summer of 1935, the tests failed because of deficient machines. The tests, using one 80-column card per soldier, showed that addressing mail, for example, for mobilization orders for individual reservists was not possible.[29] However, this planned system introduced punched cards as a full register tool in the French army. Unlike the army wage administration system, the punched-card register comprised the core of the register. Bull was so eager to gain the draft administration order that they introduced IBM's 80-column card for this application.

After Bull's failure in 1935 concerning the army conscription administration test, the administration began to develop a conscription and mobilization system at their regional office in Rouen in Northern France, using a combination of SAMAS punched-card equipment and address plates. The zinc address plate was a well-established technology, widely used by big insurance companies. The drawbacks were that handling the zinc plates was heavy work and that the sorting was done by hand—for these reasons, an additional punched card was used for each conscript. These cards were used to produce statistics on the conscripts and to provide lists of the zinc plates to be selected. During the initial tests at an army recruiting office, a soldier was caught and convicted of industrial espionage for the Bull company against SAMAS. This incident probably contributed to the choice of the combination of the SAMAS punched-card equipment and address plates in 1936 as the basis for the conscription and personnel administration system.[30]

For the next two years, the army built up a system for the conscription administration in Rouen using SAMAS and addressograph equipment. The 130-column SAMAS punched card allowed a system that used only one punched-card per person in addition to an address plate. The punched card was used to assign the reservists to various groups for administrative and statistical purposes. The address plate printing machine enabled the operators to print selected information on the ten or so different forms issued during basic military training as well as for subsequent exercises and for an ultimate mobilization. A seven-digit number identified each person, his punched card, and his address plate.[31] The large storage capacity of an address plate changed the originally planned, but rejected, Bull system where it was necessary to store all information on one punched card holding eighty characters. At that time, the register of conscripts and army personnel held much more information, including the full name and postal address.

In spite of its potential, by the outbreak of the Second World War in September 1939, this system had not advanced beyond the regional conscription administration office in Rouen. One reason for this was a dispute over the choice of a French address plate supplier instead of the American Addressograph company, which had supplied the equipment for developing the system.[32] Clearly the project was considered low priority if the government allowed this controversy to impede it. The planned improvement of the conscript administration was not considered essential, as the French governments were blind to the growing power of the German armed forces after the revengeful Nazi government came to power in 1933. This blindness was closely linked to the French governments' conviction that their army was the best in the world, a victor from the Great War. In addition, the planned improvement of the conscription administration was impeded by the general inability of French society to implement reforms in the 1930s.[33] These factors explain both the lack of improvement in the French conscription and mobilization service during the 1930s and what followed during the early 1940s.

The foundation to French conscription vanished on 22 June 1940, when France capitulated to the German invasion. The country was divided into two parts: the northern and western three-fifths of the country were occupied by Germany, while the southern, nonoccupied two-fifths of the country came under a French government in Vichy. The latter was headed by the aging Marshal Philippe Pétain, known for his outstanding service during the First World War. The armistice conditions were harsh. Pétain's government was only to govern two-fifths of the country, while he formally remained in charge of much more land occupied by Germany, including Paris. The new French government was allowed an army of a mere 100,000 men, which could not threaten Germany, and the 1.8 million French prisoners of war would remain in German captivity to be exploited by them in later bargaining situations.[34]

The last parliament elected under the constitution of the unstable Third French Republic elected Pétain head of state and conferred on him great power, which he exercised to transform what remained of the democratic republic into the autocratic l'État français (the French state). This state was based on the virtues of work, family, and obedience to authority and had transparent fascist connotations. A flagrant example was the exclusion of Jews starting in late 1940 from all but low-ranking public jobs in both the occupied and the nonoccupied parts of France. This anti-Jewish policy represented a distinct break from the secular and egalitarian policies of the Third Republic.

Public employees had grown disenchanted during the 1930s. They had experienced considerable instability, right-wing governments, the People's Front government from 1936 to 1937, last-minute avoidance of a civil war, and extensive government by decree. Now, public employees agreed to serve the Vichy state, but, subsequently, their loyalty became strained as the German occupation was extended to include the whole of France.[35]

During the 1930s, René Carmille had experienced the success of his operational statistics assignments, but attempts by Carmille and his colleagues to improve the conscription and mobilization administration through the use of punched cards or a combination of punched cards and address plates had come to nothing. In August 1940, he took the major initiative of proposing to the Vichy government the establishment of a punched-card register of all inhabitants in France to be used as a permanent census tool. His avowed intention was to improve census processing at the ailing French Statistics Department (Statistique général de la France, established 1833), which for many years had needed basic improvement, an example being that they still processed the population censuses without punched cards.[36]

A register of all inhabitants in France would remove the work of having to collect and punch fresh material for each census, although it would be a major and costly task to keep the huge punched-card file up to date. However, a general punched-card register could make the individual vulnerable because of the confidential information provided. It would, for example, be much easier to locate people with special, recorded characteristics, than would otherwise have been the case by reading every returned census forms. This new option could give rise to requests from various authorities leading to the misuse of the information supplied through census returns. The register would supply the authoritarian regime with a tool to monitor its citizens.

However, the true purpose of the register, which at this point was to create a mobilization register, is evident in that the Vichy Minister of War was the recipient of Carmille's proposal. Two months after the armistice, the French army started to prepare a secret mobilization of 200,000 men, in addition to the army of 100,000 men permitted by the terms of the armistice. The implicit goal was, therefore, an uprising against the German oppressor. Such an uprising might not have been realistically possible, but the ability to mobilize a decent army was critical if the credibility of the regime among the commissioned officers was to be maintained. According to this interpretation, Carmille contributed significantly to the stabilization of Vichy France.

French economist Alfred Sauvy claimed in his memoirs from 1972 that the mobilization register was a smokescreen, allowing Carmille to achieve a national register of people in France.[37] In his interpretation, Carmille personified the ultimate technocrat, exploiting a national catastrophe not only to implement the punched-card-based conscription and mobilization register, which had failed in 1935, but also to establish a central register to monitor and control every inhabitant of France. The material recovered does not provide adequate proof of the correct interpretation, but the German occupation of Vichy in November 1942 made the national register the sole objective.

Whatever Carmille's original purpose, the Vichy government decided to establish the national punched-card register, which required vast administration. A new entity, the Demographic Service (Service de la démographie), was established in 1940 in Lyon (in Vichy France). Carmille donned civilian clothes to head the new institution, which the following year was merged with Statistique générale de la France in Paris (in the occupied parts of France) to form the Service national des statistiques (SNS, National Statistics Service), though they remained two separate geographical entities, in Paris and in Lyon, until after the end of the Second World War.[38]

The size of the register project is made apparent from the employment in the summer of 1941 of 1,968 people who operated 233 punches, 22 sorters, 14 tabulators, and 7 reproducers.[39] The project used the 80-column punched card and the Bull alphanumeric standard, which was incompatible with the IBM standard.[40] The SNS used both Bull and IBM equipment, but they only applied IBM equipment to process numeric information.[41] Bull shared the IBM numeric punched-card code that had been introduced as early as in the 1890s.

Carmille developed the national register of Vichy France and Algiers on the basis of a written file on each individual and two punched cards.

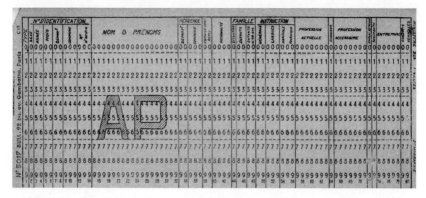

The French national register between 1941 and 1944 had two basic punched cards. This is the index card with the information needed for sorting and other functions. AP (recensement activité professionnelle, or census of occupations) shows that information on this card would have been derived from the returns from the census of professions in 1941. This card only allowed a name of up to eighteen characters. A member of the Jewish "race" was indicated in this card by specific perforations in the nationalities field. The second punched card held the address. (Box 55,359, Centre des archives économiques et financières, Savigny-le-Temple, France. Punch code in Jean-Pierre Azéma, Raymond Lévy-Bruhl, and Béatrice Touchelay, *Mission d'analyse historique sur le système de statistique français de 1940 à 1945*. Paris: INSEE, 1998, 19–20)

The written file was designed to include a description of the person's life from birth to death, including four photographs taken at different ages. It had all kinds of information—date and place of birth, parents, marriage(s), divorce(s), children, date of death, nationality, addresses—but nothing concerning religion. The first of the punched cards held encoded information for compiling various kinds of census statistics. The second card held the name and the current address to be used for addressing letters. Each person's written file and punched cards were identified through a thirteen-digit "national identification number" punched into the cards, which today is the French social security number.[42]

The register was divided into seventeen separate regional registers. Each person's file was kept in the SNS center of her or his residential region. When a person moved, he or she was required by law to report to the SNS.[43] This was crucial to updating the register, which in itself became a major task.

The information kept in the national register went far beyond that of population statistics or the requirements for a conscription and mobilization register. The register was used to strengthen government control over the inhabitants of France. The introduction in 1940 of the "national identification card" had the same purpose and remained in use after the Second World War. The identification card became an SNS responsibility, and the card was distributed displaying the thirteen-digit national identification number, starting in 1943.[44] The national register had become a central tool in monitoring the individual.

From the outset, the national register was based on information collected from soldiers following their demobilization in the summer of 1940. In June 1941, this information was complemented by a census of professions in Vichy France and Algiers of everyone between the ages of thirteen and sixty-five. This excluded processing information on men too young or too old to serve in the military, but it included women. This limited census allowed SNS to systematically update the addresses of male citizens who were vital for a mobilization. Furthermore, the register, for the first time, received information on women other than their change of addresses. From 1941 to 1942, most of the work at SNS concentrated on processing census information on men, while processing information on women had a lower priority.[45]

By early 1942, Carmille managed—without German detection—to establish a secret file of 300,000 males for mobilization in Vichy France (the 100,000 men allowed by the armistice agreement, plus an additional 200,000 men) that was stored in a separate location from the general files. This register was subsequently updated and, if used, would have allowed a mobilization by letter within thirty-six hours. During the spring of 1942, the mobilization was tested through a successful paper exercise that produced tabulator lists. Furthermore, in 1942, Carmille started to extend his service to the German-occupied regions of France, a demonstration of Vichy's authority in that part of the country.[46]

But the carefully prepared mobilization never materialized. On 8 November 1942 the Allied Forces landed in Algiers. Three days later, the Germans occupied Vichy France, which both removed the possibility of mobilization and raised the risk of detection of the mobilization register by the Germans. In consequence, Carmille destroyed the mobilization register and related archives.[47] If this register had been the only objective of Carmille's register project, he should have destroyed the unfinished national register as well; but this did not happen.

Failure to do this was probably the basis for Alfred Sauvy's claim that the national register from the outset formed the core of Carmille's register project with the mobilization register serving only as a pretext. But the register was a different entity in 1942 than when it had been proposed two years earlier. An organization had been established that provided register options and employment, thus carrying with it a significant momentum. Therefore, Carmille continued to build up a national register comprised of all the inhabitants in France, which became the key register for all national identity numbers and was used to monitor the issuing of ration cards. Carmille worked to consolidate his punched-card enterprise and to establish a corps to administer his register as a complement to the existing civil service corps. In 1942, he began to compile a punched-card register of French companies and institutions. In 1943, he established a spot-test service and a school of punched-card statistics applications. After the war, his organization became the core of the current French national statistics department.[48]

Keeping the Germans out had been a fundamental reason to establish Vichy France, and its occupation made many public employees reconsider their position. Before the occupation of Vichy, Carmille, as a loyal Vichy public servant, had insisted that there should not be any relation between his punched-card service and the resistance movement. But the Vichy regime strained the loyalty of its public servants and officers. Following the German occupation of Vichy, Carmille rebelled and joined the resistance movement while also working to improve a key Vichy monitoring tool. Arrested by the Gestapo in 1944, Carmille died in the Dachau concentration camp in early 1945.[49]

The French Third Republic had been secular, and, accordingly, its census statistics did not contain any information about religion.[50] After the formation of the Vichy state, the rights of Jews in France vanished. The German occupiers took the initiative, but the Vichy interventions went beyond the German requests. On 27 September 1940 the Germans ordered the segregation of Jews in the occupied areas, which caused the Vichy regime to adopt a law to the same end six days later, a law valid not only for the occupied zone but also in Vichy, the nonoccupied zone. As a direct result of this law, all Jews were swept from spheres of public influence. It was part of the Vichy struggle to exercise sovereignty over the whole of France, but the regime was infected by anti-Semitic prejudice, as had already been made evident from its encroachments on Jewish rights during the summer of 1940.[51]

In March 1941, Vichy honed its anti-Semitic policies by establishing a General Commissariat for Jewish Issues (Commissariat général aux questions juives), run by Commissioner Xavier Vallat. Back in September 1940, the German occupation force had ordered a census of Jews in the occupied parts of the country to map their number and distribution. In April 1941, Vallat ordered a census of Jews to be carried out by the French police and the municipalities in the nonoccupied zone to be processed by Carmille's SNS.[52] In both the German census and this separate Vichy census only Jews were requested to report, and no mechanism existed to identify those who evaded.[53] The two separate censuses enabled a detailed mapping of Jews in France, though the definitions of "Jews" used by the two censuses diverged.

The general census on professions in Vichy France in June 1941 differed from its predecessors by asking whether each inhabitant was of the "Jewish race."[54] This offered an additional mechanism to identify Jews in Vichy France. They were identified on the punched card for this census by specific perforations in the "nationality" field, thus segregating Jews as aliens. The cards on Jews were kept separately, which indicates that this information could have and might have been exploited.[55]

The census on professions and the special census on Jews in Vichy France were conducted concurrently in June 1941, but they were distinct. All inhabitants between thirteen and sixty-five years of age were requested to report in the general census on professions, the completed forms processed by SNS on a national basis. In contrast, only Jews were asked to respond in the separate census. No form existed, and each individual reported by writing a letter that was kept at the police station. In several areas, because the number of Jews who reported was significantly below the expected number, it was assumed that many Jews refused to cooperate. As France had been a secular state, Jews were, indeed, able to avoid declaring themselves Jewish without the fear that they would be traced through official sources.[56] The police were requested to compile lists of the Jews in their area from the returns of the census. Even so, it was difficult to monitor Jews who moved to a different police district.

Vichy authorities could have improved monitoring of Jews by consolidating the information gathered in the two censuses in June 1941, but this was never accomplished. The reason seems to have been René Carmille's deliberately low prioritization of this contribution to a more efficient segregation of Jews, which was probably due to his resistance to the anti-Semitic policies of the Vichy regime. In late 1939, he had dissociated himself vehemently from German anti-Semitism in a publication.[57] It is true that in June 1941 he offered to commence consolidating the information gathered in the two censuses, but nothing appears to have happened before this was requested by the Commissioner for Jewish Issues in late 1941.[58] The records show that, in 1942, the SNS was working to produce a central register of Jews in Vichy France, and that the work was proceeding very slowly.[59]

One reason for this concerned the problems of identifying the same individual in both censuses. Another problem was that the national and mobilization registers were the main task of the SNS, and these diverted manpower from the creation of a punched-card-based register of Jews in France. Establishing the national register was an enormous and protracted undertaking that had not yet been completed by early 1944.[60] Moreover, as the national register had not been completed, the compilation of the register of Jews had to be based mainly on returned but as yet unprocessed forms, further impeding the process. However, these tasks should have been within the reach of the SNS, as the material had been collected and as the total number of Jews in France was only around 300,000.[61]

When the deportations started in 1942, the victims were located on the basis of lists compiled by the Gestapo and the French police. The member lists kept by Jewish communities comprised one possible source for locating the victims, but, the question is whether the Vichy national register organization also provided confidential information. So far, no case has been found of any national register disclosure for this purpose, but the problem remains as to whether any such disclosure would be recorded in the incomplete archives that have been handed down.[62] Further, if information from a punched-card register had been used, this probably would have had the form of tabulator prints; and it remains doubtful as to whether the historians who studied the various deportation records actually checked if tabulator prints were applied.

In addition, René Carmille himself was ambivalent about the confidentiality of the register information. In 1941, as already mentioned, he accepted that the information gathered in the 1941 general census could be used to supplement the information collected in the separate census on Jews in Vichy France the same year. In 1943, on at least two occasions, records were extracted from the punched-card register—50,000 men in one case—to supply people for forced labor in Germany. Further, in 1944, Carmille offered to make confidential information available for any "police purpose" without, for example, requiring a court order or warrant.[63]

The French national register after 1940 shows how a register of punched cards could be established in an autocratic state. The register was built to enhance the weakly founded Vichy state, its primary mission to improve credibility with the army. However, while the national register originally had been a shield against German detection, it gradually became a tool to monitor the individual, strengthening the Vichy state. It was the key register of all national identity numbers and was used to monitor issuing of ration cards and national identification cards.

There is no indication, however, that the endeavor to improve the conscription and mobilization administration aimed at anything other than rationalization within the French army before the summer of 1940. Only the French defeat and the German occupation of the country opened the possibility that a register of punched cards could improve the state of the nation's military position. The project to extend René Carmille's vision of a mobilization register to encompass all of France was certainly impressive, but it hardly reached beyond the technical potential of the register.

The war and the German occupation may well have provided the reasons for this project, but the Germans do not appear to have shown much interest in the national register project or to have considered its potential, limiting the danger of the register. The reason for the absence of interference from the Germans might stem from the lack of a German national register project until 1943.

Managing Resources during the War in Germany

German punched-card applications in the Second World War were distinguished by their continued reliance on nearly exclusive numeric operations and their late introduction of registers of punched cards. Only after

1939 did German authorities become interested in punched-card alphanumeric registers to control the war efforts, bringing them into conflict with the IBM subsidiary in Germany, Deutsche Hollerith Maschinen Gesellschaft mit beschränker Haftung, or Dehomag, which was not able to supply the machines needed for this purpose. Dehomag won this dispute due to the quality of their machines and the capacity of their production facilities, but this does not answer why the authorities accepted their failure.

This can be answered by the absence of an alternative punched-card producer and the chaotic power structure of the Third Reich. Dehomag succeeded in clipping the wings of its competitors in Germany through cunning use of the German patent law and its own talents in technology and sales. The chaotic power structure explains why the Nazi regime only started to develop a national register of punched cards in 1943, in spite of the technology's capacity to facilitate control. While this late date limited the options of the authorities, it did not imply a lack of control over the population in Germany. A system of local registers using simpler methods had been established in the 1930s, and their cruel efficiency was proven when Jews had to be located for deportation.

The German army staff had started their preparations for a new large-scale war in 1924. A significant component was to establish and maintain control of the German industrial capabilities crucial for modern mass warfare. For this purpose, the army staff established a Statistics Office in 1926. Around 1930, they set up index card registers of industries important to the military, one organized alphabetically according to company name, another organized according to location, and, finally, a register of machines requisite for armament production, like capstan lathes, drills and planing machines.[64] In 1937, the people in charge of this work proposed the transfer of these registers to punched cards. An obvious advantage was that the two industry registers could be reduced to one set of punched cards, to be sorted as requested. Other considerations included improved possibilities to control raw materials and semimanufactured articles for which the raw materials had previously been monitored manually. The register processing by means of punched cards was assigned to a new army punched-card service (later called Maschinelle Berichtswesen). From the outset, the punched-card registers were conceptualized as exclusively numerical, requiring all information to be coded.

The industry register demanded a new company number suitable for punched cards, the introduction of codes for geographical districts, and a number for each raw material and semimanufactured article.[65] Before the outbreak of the Second World War, the army punched-card service transferred the registers for monitoring industry, raw materials, and semimanufactured articles to punched cards. They also took charge of compiling army health statistics, statistics on the psychotechnical tests of new conscripts, and of doing inventory control on military equipment at the various units, as well as of the use of raw materials for the arms buildup. From 1939, statistics relating to armament workers were also punched-card processed.[66]

Most of this work involved processing statistics, but punched-card registers were introduced for controlling industry and raw materials. These registers differed from the Social Security register in the United States and the planned French Army service register in that they were only numerical. This enabled standard Dehomag machines to be used, but it limited the tabulator printout to numbers.

In the United States during the 1930s, public utilities were a prime field for the introduction of alphanumerical systems. From 1935, Dehomag mounted an offensive in Germany to expand its energy supply installations to cover the calculation of consumption, the amounts to be paid, and printing of invoices. This plan was predicated on the improved calculating capacity of its machines and their upgraded ability to control the movement of the forms on the tabulator during printing, although consumer bills continued to be addressed using address plates.[67]

Dehomag continued to improve the numerical capability of their machines in the late 1930s.[68] The company, which considered expanding into alphanumerical machines, had three options in this regard: (1) Dehomag could import alphanumerical tabulators produced in the United States; (2) it could start producing IBM-designed alphanumerical tabulators, as the French IBM subsidiary did or; (3) it could develop its own alphabetic tabulator (allowing numerical and alphabetic printing in separate printing positions).

The company chose to develop an alphabetic version of the numeric D_{11} tabulator, but the decision does not seem to have had much impetus. Development ran into problems, and a prototype was not finished until 1944.[69] It is not clear whether these difficulties were caused by a lack of company commitment or by an insufficient amount of engineering expertise to address the technical complexities; low demand, however, was certainly a major reason. The limited demand from government bodies has been discussed above. Only high government demand might have enabled imports, and the absence of a Dehomag production of an IBM-designed alphanumerical tabulator can only be explained by a lack of company interest. New directions during the war reversed this situation.

By the late 1930s, Dehomag's technical *Sonderweg* enhanced Heidinger's position in relation to IBM and was simultaneously strengthened by the German autarchy. Dehomag was highly profitable and had its own production, which made the German company less sensitive to the parent company's views. Ironically, however, IBM's position might have been strengthened decisively if a demand emerged in Germany for punched-card technology that went beyond Dehomag's capability—for example, for alphanumeric tabulators. It appears that IBM had only brought an alphanumeric tabulator to Germany by 1939, where the German punched-card users were starting to appreciate its improved capability.[70] Interest in this American technology grew with war needs.

The outbreak of the Second World War in September 1939 caused increased demand for punched cards to control production and to manage the warfare. Dehomag's turnover increased by 26 percent from 1939 to 1940.

In 1939, the army's economic office, headed by General Georg Thomas, stepped up its planned economic mobilization for a general war.[71] This created new tasks for the army punched-card service. Monthly statistics were established of the stock and turnover of 350 types of weapons, ammunition, and military equipment, as well as their production figures and estimates of completion within the next six months. Further, the already existing monitoring of raw materials was developed into full-scale bookkeeping of raw materials encompassing all providers and users, as well as the three armed services.[72]

During the war, the application of punched cards for control was intensified to enhance war planning. In the summer of 1941, the transition from traditional means to punched-card processing began of monthly employment statistics in important areas of the armament industry, encompassing 90,000 companies and their 10 million employees. This represented an early attempt to improve the information available to key authorities about actual industry production. The company was still the unit for keeping these statistics.[73] This application involved the numerical punched-card statistics processing of data, similar to the previous army punched-card service applications, but it also introduced the first step toward punched-card registers. Punched cards holding information on employment, raw materials, coal, and semimanufactured articles for a month were kept to compile tables projecting development over the coming months.

In the autumn of 1941, the failed Russian campaign made the low armaments production evident.[74] In December 1941, the national Armaments and Munitions Minister, Fritz Todt, established a system of entrusting eminent technicians from leading industrial firms with management of separate areas of armament production.[75] Shortly after succeeding Fritz Todt in February 1942, Albert Speer managed to improve on Todt's earlier position in relation to the other German planning authorities.[76] He extended Todt's system into a network of "industrial self responsibility." He formed thirteen "vertical" committees for managing the various kinds of weapons production, like the armor committee headed by Professor Ferdinand Porsche, and the army weapons committee headed by Krupp manager, Erich Müller.

The allocation of raw materials and intermediate goods was organized through a similar number of "horizontal" committees, supported through the formation of a special section in the Speer ministry. In addition to these vertical and horizontal committees, Speer established development commissions where army officers met the best designers in industry. These commissions were to supervise new products, suggest improvements in manufacturing, and call a halt to any unnecessary projects. A key element of the industrial self-responsibility network was to ensure that a given plant only concentrated on producing one item at a time, and on setting maximum quantities.[77]

Punched-card technology became essential to Germany's planning efforts. The transparency of available production capacity, as well as the allocation of resources and production, was crucial both for Speer to manage this network and for the various committees to function. To this end, in April 1942, the army punched-card service was transferred to the Speer ministry as a staff function.[78] By 1944 the office comprised nine branches and twenty-one offices throughout the Reich, what was then

Czechoslovakia, and Poland, employing 833 trained people and using 1,055 punched-card machines.[79]

Finding qualified available manpower was, in particular, a growing concern. By the end of May 1942, of the German workforce 9.4 million members had been conscripted, and of these 800,000 had fallen. There was strong pressure for more troops, but also for workers in industries of importance to the military. By that time, 4.2 million foreigners worked in Germany, either forced or of their own accord.[80] To improve Speer's control, an advanced system of monthly employment reports was started in the autumn of 1942. Companies of military importance were required to report how many employees they had and to divide them according to various categories, for example, German workers, workers from the "Eastern Countries," Jews, and Russian and other prisoners of war. The reports were punched-card processed, partly on alphanumerical tabulators, probably using improved table printing control. No indication was found of letter printing. Like the 1941 armaments industry report system, the punched cards from each month were kept in a numeric register, enabling control of subsequent reports and table production.[81] During 1942, the outcome of Speer's improved armaments production organization was an aggregate monthly production growth of 76 percent. Between February and December, the production of armaments grew each month.[82]

In early 1943 efforts were introduced to simplify collection and distribution of information brought on by the need to reduce administrative staff and the desire to improve control of the workforce.[83] In this process, two large alphanumeric punched-card registers were devised: an army payroll system and a national register, which for the time being was restricted to the domestic population.[84]

The start of the large alphanumeric punched-card systems in Germany can be traced via the development of the army payroll system. This was the first time that alphanumeric registers and punched cards were linked, except for in a few private companies. In the summer of 1943, a payroll system was established for the 12,000 civilian workers in the army. Each worker had a typed index card and two punched cards, enabling printing of various sorted lists including their full names, date of birth, place of work, and monthly and yearly salaries.[85]

By the end of 1943, the government started extending the system to cover all military personnel—a project that was planned for completion by 1945. For this project, thirty alphabetic tabulators were ordered from Dehomag, but they were never delivered. Perhaps this was the reason that the leader of the Speer punched-card service, Kurt Passow, made enquiries in the summer of 1943 into the possibility of building a full address printing Dehomag tabulator, resembling the alphanumeric IBM tabulator, type 405.[86] This showed that the Speer ministry punched-card people were starting to consider directly addressing individual correspondence using punched cards. They considered the full potential of alphanumeric punched-card registers.

The national register was the second major German punched-card register project. The idea was conceived in early 1943, the payroll system for the army's civilian workers acting as a test case to bring it into being. From late 1943 until late 1944, two large-scale trials were carried out.

The first trial took place in late 1943 and included the workers in selected industries in Breslau in Schlesien (today Wrocław in Poland). The aim was to replace the frequent and detailed statistical reports on workers to both the Speer ministry punched-card service and the conscription registers. The Breslau register seems to have been organized according to national company identification number. Each employee had his or her individual card, which contained this number, name, date of birth, and address. The register produced sorted lists, which were compared successfully to the many local index card registers. This kind of identification was cumbersome, as easy identification would require a unique individual identification number, along the lines of the existing national company identification number. In addition, the Breslau register only included people who were employed. The Speer ministry punched-card people realized that industry reports were not an appropriate tool on which to base a register of the entire workforce in Germany.[87]

The second large-scale experiment included all the inhabitants of the city of Ansbach in Bavaria. During this test in the autumn of 1944, the central personal register was defined, consisting of a report form for each individual and one punched card to hold a condensed version of this information—green cards for women and yellow for men. People were ordered to report, and procedures for changes were established. Each individual was identified via a national registration number containing twelve digits that was devised during this experiment.

The register held key information from the twenty or so civil registers in the city, as well as the various military registers, enabling easy comparison of information. Several errors in the existing registers became obvious, especially the information used for food rationing and people's ages. The Ansbach register covered every individual in the city and was able to provide more comprehensive information for the reports more quickly than ever before. Due to its success, it was decided to establish a national register near Berlin, but this was never implemented.[88]

The Speer punched-card people were the driving force behind the development of punched-card applications to monitor the production of military importance and this vital sector's employment of manpower. This development could have originated in several places, for example, it could be based on experiences in German industry, or it might have been inspired by information on the punched-card based French national register, established in 1940. Regardless, it originated outside Dehomag, creating a new level of demand that Dehomag had great difficulties meeting.

During the 1930s, IBM's business in Germany had involved three sets of actors: IBM, Dehomag, and the German authorities. IBM controlled Dehomag, while Dehomag maintained all relations in Germany. Only in a few instances did IBM have direct contact with the German authorities or customers. Among the rare examples were Thomas J. Watson's meetings in 1937 with Hitler and the German Minister of Economics, Hjamar Schacht, during a visit to Germany as Chairman of the International Chamber of Commerce.[89]

Between 1939 and 1941 these basic relations remained intact, as the United States was neutral. However, the expansion of punched-card applications in industry and within the armed forces during the first two years of the war, brought about by extensive warfare, the occupation of several countries, and growing industrial production in Germany, was partly facilitated by Dehomag's growing production and was reflected in its rising turnover. In addition, the German invaders compelled the various national IBM agencies or subsidiaries in the occupied countries to surrender leasing contracts on punched machines, which were and remained the property of IBM in New York.[90] The extension of punched-card applications in 1942 caused additional requisitions of IBM leases in Belgium, France, and the Netherlands.[91] These leases contributed to the Dehomag turnover as a result of punched-card purchases, as rents continued to be paid to the national IBM subsidiaries from which the machines had been confiscated.

However, the German campaign of May–June 1940 to conquer Benelux and France had provoked pressure on IBM's relations with its German subsidiary. The conquest of these countries caused Thomas J. Watson of IBM to return a German decoration that he had received in Berlin in 1937 while Chairman of the International Chamber of Commerce working for appeasement with Nazi Germany.[92] This act triggered a putsch by Dehomag's managing director and founder, Willy Heidinger, who tried to regain majority control of Dehomag and was apparently supported by German authorities. However, the IBM majority ownership was rescued by the introduction of enemy company custodianship when the United States entered the war.[93] The custodianship took over IBM's role as majority shareholder and gave Heidinger's management free reins, implying that he had regained control of his company, but not the ownership.

While Dehomag's relations with IBM disappeared for the duration of the war, new conflicts emerged with Kurt Passow about his project concerning a German-controlled punched-card industry and Dehomag's inability to supply letter printing tabulators. Passow was the leader of the army's punched-card service that had moved to the Speer ministry in April 1942.

Within the German government's autarchy policy, Passow worked to establish a German-controlled punched-card industry independent of IBM and the American Powers company. As early as 1938, he had approached Wanderer-Werke Aktiengesellschaft (Wanderer Works Limited) in Chemnitz in a drive to get them to develop a complete set of punched-card machines with the potential to replace the Dehomag and Powers machines. Passow made this approach only the year after the army punched-card service had been established.

Wanderer-Werke, which had been producing typewriters since 1902 and during the First World War, started to produce mechanical adding machines.[94] They had designed their first punched-card device in 1929 that consisted of a link between a key bookkeeping machine and a punch that facilitated punching onto cards, without extra action, of information keyed-in and processed in an ordinary bookkeeping process. This feature enhanced the modest calculation capability of a standard tabulator and resembled a similar Dehomag design also from 1929.

Since 1933, Wanderer-Werke had committed resources to developing punched-card machines, but none of them had been produced as of 1938. Moreover, Passow's approach in 1938 led to nothing. He asked Wanderer-Werke to concentrate all their resources on office machine development to generate a complete set of punched-card machines for statistics processing, while the company's directors preferred to develop bookkeeping machines.[95]

Then, in May 1940, early on in the campaign to conquer Benelux and France, Kurt Passow took a new initiative in his endeavor to establish independent, German punched-card production. As he did not have any able technician in his employ, he persuaded engineer Hermann Voigt, who had headed punched-card construction work in the 1930s at Siemens, to head his development work.[96] In 1940, tests were conducted on Powers and Bull machines.[97]

After the fall of France, the Bull company in Paris came under German control, making it the first industrial punched-card technology under complete German control. Wanderer-Werke began negotiations with Bull to gain access to its technology. A preliminary agreement that was only reached in early 1941, stayed in effect until the two companies entered a final contract nearly two years later. This contract was notably entered into between two equal partners, in spite of the German occupation of France and opened up the exchange of information about research and development, which was a one-way clause serving Wanderer-Werke, as it did not have much to share.

Wanderer-Werke was also allowed to produce Bull punched-card machines on the condition that Bull received a royalty of 2 percent of the price Wanderer received when the machines were sold.[98] Aside from that, some twenty Bull tabulators and some forty punches and verifiers had been taken as spoils of war to Germany in 1940, causing several problems, serious even during a war. Dehomag claimed that these tabulators infringed upon the German IBM and Dehomag patents, and Dehomag refused to perform maintenance on them.

In the spring of 1941, Dehomag contended that Bull tabulators produced by Wanderer-Werke would also infringe on Dehomag's patents.[99] Wanderer-Werke was up against numerous German patents that were the result of Dehomag and IBM's development work and intended to guard their prime-mover position. This was exactly the same experience that all contenders to IBM's prime-mover position in punched cards had. In reality, Dehomag used their strong patent position to avoid the German national punched-card system becoming a threat to IBM's business in German-controlled Europe, as they had successfully, via litigation, fended-off Powers in Germany from 1914 to 1923 and from 1924 to 1929—in spite of the fact that they eventually lost the cases.

Wanderer-Werke's counteraction was to try to obtain a compulsory patent license for the patents held by Dehomag. They asked the National Ministry for Trade and Industry to order a license based on the importance of punched cards for the war effort. This failed, probably because Wanderer-Werke had not yet supplied any punched-card equipment. Nevertheless, they could have obtained a compulsory license through the law courts, though this might have taken several years, as indicated by the earlier experiences of the German Powers company. Wanderer-Werke never obtained a patent license from Dehomag.[100]

Encouraged most likely by the Wanderer-Werke agreement with Bull, the German army signed, in April 1941, a contract with Wanderer-Werke to develop a series of punched-card machines, which Wanderer-Werke subsequently worked on. Two months later, this was supplemented by a contract with the army to repair the booty Bull machines brought to Germany.[101] However, development work at Wanderer-Werke did not provide results that brought into sight a substantial German production of punched-card machines outside Dehomag proper. This reduced Kurt Passow's options for obtaining, in particular, letter printing tabulators, causing two different reactions.

First, Passow reduced his stakes in Wanderer-Werke's development of punched-card machines, though they managed to complete a sorter in November 1942. He only awarded the company additional minor contracts for punched-card development work that never provided substantial results. Instead, in early 1944, Passow contracted Wanderer-Werke to make Bull machines produced at Bull's factory in Paris, which never materialized because of the liberation of Paris in August 1944.[102] However, the contract triggered claims from Dehomag for a license on imported Bull machines, which was never settled.[103]

Second, the lack of results at Wanderer-Werke caused Kurt Passow to turn his attention back to Dehomag, which erupted into a heated conflict in the summer of 1942. Passow and the army punched-card service had been transferred to Speer's successful management in April 1942, giving Passow a higher profile that he was most willing to exploit. The factor provoking the conflict was the improvements of war production control since February 1942, which caused an emerging demand for alphanumeric punched-card machines. For two years, Passow's punched-card people had gained experience by using the IBM alphanumeric tabulators requisitioned from the occupied countries and a few alphanumeric tabulators imported from the United States in 1939.[104] Passow probably applied these machines to improve the existing statistics processing and numeric register-based applications by adding letter specifications to the prints, as shown in several published examples.[105]

The heated nature of the dispute is clear from Kurt Passow's high-profile accusation in September 1942 that Dehomag withheld the alphanumeric IBM machines from the German market. Passow accused Dehomag of accepting instructions to this end from an enemy company (IBM). He raised slanderous doubts about the Dehomag management's loyalty to Germany, an extremely serious accusation in the midst of the war which, rightly, infuriated Heidinger, who had displayed his enthusiasm for the regime in the 1930s.[106] The only basis for Passow's attack was the assembly of alphanumeric tabulators by the French IBM company (Compagnie electro-comptable) for the French market.

Seen from the Dehomag perspective, because the German government had encouraged an independence from imports and liabilities in foreign currency, Dehomag had followed the principle when it developed the numerical D11 tabulator and when it established extensive production in Germany. This policy had prevented the kind of conflict with the national Ministry for Trade and Industry experienced by the German Powers company in the late 1930s due to its objection to moving the production of tabulators from the United States to Germany. Dehomag, whose customers demanded an increasing number of the numerical D11 tabulators, did not experience sufficient demand to complete building the alphabetic version of the D11 tabulator or to negotiate with IBM in New York to get the blueprints to produce the IBM alphanumeric tabulators, as the French IBM company did in the late 1930s.

As time passed, problems worsened because of air raids. Punched-card machines were destroyed and damaged, and, in August 1943, the Dehomag factory in Berlin was severely hit. Most of the production was then moved to Hechingen in Württemberg, near Dehomag's factory in Sindelfingen.[107]

By the spring of 1943, Kurt Passow's reduced number of options, however, had forced him to be content with Dehomag. Consequently, he then tried to enhance Dehomag's development of the letter printing tabulator by transferring the manager of Wanderer-Werke's punched-card development to Dehomag.[108] In 1944, Dehomag nevertheless finished the design of the letter printing version of the D11 tabulator and a prototype was built, but it never went into production. Dehomag had by far the best organizational capabilities in the German punched-card field and emerged as the major winner. They avoided the influence of the Speer ministry in their management, retained control of the vitally important large punched-card machines, and took control of the French IBM company.[109]

It seems surprising that a national register of punched cards in Germany had not been established by the end of the Second World War, particularly as France had started to build such a register just after Germany had occupied three-thirds of the country in 1940. After the war, one of Speer's punched-card managers explained the absence of a national register in Germany as being due to the lack of alphanumeric tabulators and trained staff.[110] However, by the time the national register concept was finalized in Germany in October 1944, implementation was not feasible. By then, the toll of the war had made its mark on Germany through the destruction of property and the dislocation of people. But why, only in 1943, did the Nazi authorities commence working on a national register of punched cards that could have been an effective tool applied to control the population in this totalitarian state? This can be explained by the chaotic organizational structure of the Third Reich and the many local registers that had been established using simpler methods.

After Albert Speer became minister of munitions in 1942, he accomplished the improvement of war production over the next couple of years. However, this was accomplished through the establishment of a nationwide organization and extensive punched-card processed statistics to facilitate an overview of the country. This happened at the expense of the regional party bosses (*Gauleiteren*) who resisted such a centralization of control and revolted against Speer in late 1943. In spite of his success at improving war production, Speer subsequently lost this battle.[111] A similar reaction to a national register of punched cards that would entrust Speer's punched-card people with the capability of additional nationwide control was to be expected.

In a study of the role of IBM in the Third Reich, two historians analyzed the role of IBM punched cards in the location of Jews for deportation. They searched for a substantial Dehomag complicity in the Holocaust, but their studies reduced the issue to the detailed processing of census data, which already in the 1880s had been the basis for the invention and building of Herman Hollerith's first punched-card system.[112] Information about religion was already perforated into the punched-card used to process the German census in 1933 in Prussia. The card used to process the census returns in Prussia did not have a number that could identify the individual with his or her card once punching had been completed. This made the punched card an irreversible tool to compile tables from the census data in the forms. However, the returned census forms could be used for this purpose, for example to locate Jews, but that would require manually handling every returned schedule.[113] The following census in 1939 had more detailed statistics on Jews, but like just as in 1933, the punched cards did not hold individual identification numbers that could have been used for segregating the cards of Jews by using a sorter.[114]

More important, the two historians drew attention to the importance of locating Jews in the various non-punched-card registers of people in Germany, which a second couple of historians subsequently studied.[115] The second couple noted that until 1938, the Nazi authorities used various registers of fractions of the population to control people and segregate socially undesirable individuals, like Jews and Romas (Gypsies). The Register of Jews (Judenkartei) was established in 1935 by the Gestapo (Geheime Staatspolizei, or security police) derived from Jewish communal membership lists. The Jewish communal lists were manifestations of the members' faith and cultural affiliation, not of "race," which made the Register of Jews incomplete for the Nazi segregation of Jews; as it did not, for example, encompass Jews who had converted to Christianity.[116]

Also important are the three countrywide, systematic sets of registers that were established in 1938 and 1939 to improve control of the population and segregate socially undesirables: the National Reporting System (1938), the National Register (1939), and the Ethnic Register (1939). The National Reporting System (Reichsmeldeordnung) was introduced in 1938, and its explicit purpose was social control. It replaced the vary-

SS officials monitoring people with written index card files in Germany, 1938. (Statistisches Jahrbuch der Schutzstaffel der NSDAP 1938. Berlin, 1939, 95)

ing registration systems in the different states, many of which had been established before 1933. The National Reporting System required all inhabitants to report any changes in residence to the local police, who established registration offices across the country. Each of these offices established an alphabetically organized resident register of all inhabitants in the district on index cards, subsequently updated by the use of the reports on changes.

Data on domicile was exchanged among local police forces, and all changes in residence were reported to the National Statistics Department (Statistische Reichsamt), which used the information to compile migration statistics, which contained summary data for each community.[117] The local resident registers were the basis for the regime's introduction in 1938 of obligatory identification cards. Jews were required to carry a special identification card, marked with a large, black letter *J* for *Jude* (Jew).[118]

The second couple of historians found that resident registration enabled the government to keep tabs on the physical location of all Germans, including those they desired to persecute.[119] As issuing identification cards was a local assignment, resident registration could be used at this level to locate all Jews, while the national government in Berlin had the aggregate number of Jews in every community from national statistics.

The military saw another limitation to the alphabetical organized resident register. They needed precise information about the men from each particular year, including the total number on the national level, as well as the local registers facilitating the location of each individual. The census figures provided the national level requirements, but the resident register was not well-suited to locate, for example, every male born in 1909.[120] Therefore, in early 1939, the National Register (Volkskartei) was established as locally managed registers of written cards copied from the existing resident registration register, but was—in contrast—organized by gender and year of birth to facilitate military conscription. This example displays the problems of using a register of written cards for tasks requiring it to be organized on varying criteria. A register of punched cards could either be sorted differently, or could be machine copied. However, the National Register was also linked to the project of racial enrollment from the outset, as the cards of Jewish residents were marked with a black flag.[121]

In contrast to these local-level registers, a national register of racially non-Aryan residents, the Ethnic Register (Volksturmkartei), was established between 1939 and 1942 based on data from the census in 1939. It was commissioned by the chief of the Gestapo, Reinhard Heydrich, and the Minister of the Interior, Wilhelm Frick. Each individual was registered on a card that contained surname, personal name, residence, sex, date of birth, and information on religion, mother tongue, race, profession, and for household heads, the number of children under the age of fourteen living in the household.[122]

The Register of Jews, which was used for deporting Jews to the Dachau concentration camp in late 1938, along with the Ethnic Register, was used to locate Jews for subsequent deportations.[123] These simple means proved to be capable of facilitating this inhuman task. However, machine-readable registers might have proven even more efficient, and the operators would have been spared being confronted with each victim, as represented on the register cards.

The Punched-Card Industry's Choice of Development Strategy

The punched-card producers in the United States, Great Britain, Germany, and France developed punched cards to attract bookkeeping applications between the First World War and 1935. This development success starting in the late 1920s opened up an ample expansion of the producers' business. Back in the 1910s, IBM had experienced a long and substantial expansion of its business based on its ability to process operational statistics, which had accomplished by 1907. Similarly, IBM might have chosen to reap for the next decade, or longer, the business generated from their development of punched cards for bookkeeping tasks, accomplished in 1935. However, the punched-card trade had changed, and development departments had been established at several companies, which provided a momentum calling for additional development. Further, the trade had grown competitive, in contrast to the situation in the 1910s. It is true that Remington Rand was only a minor challenger to IBM's punched-card business in the United States, but they produced reliable machines and they pursued business in a large number of countries together with their British sister company.

IBM's basic options for technical improvements on equipment within their bookkeeping model in the mid-1930s were either to improve the calculation capability of their machines by facilitating longer programs or to develop a punched-card with a larger storing capacity. Both paths were pursued in the 1930s and 1940s. However, an extended punched card never reached the market and the punched-card record remained confined to the eighty characters on a card, to be expanded only by the transition to computers with magnetic tape stations in the 1950s. In contrast, IBM produced a succession of punched-card calculators with growing capabilities to perform complex computations, including multiplication and division, IBM-600 (1931), IBM-601 (1938), IBM-602 (1945), IBM-603 (1947), and IBM-604 (1948).[124]

A major inspiration improving the calculation capacity of their equipment was scientific punched-card applications, which emerged in the late 1920s. Scientists demanded punched-card machines to execute programs of growing length and with a greater flexibility than that provided by standard machines. In addition, the extensive calculations of these tasks made scientists demand faster machines. All scientific computations were performed on electromechanical punched-card machines, as the lack of flexibility in connection-box programming made the Powers machines less qualified for this field of application.

In 1929, IBM presented a punched-card installation to Columbia

University in New York to mechanize grading the university's large number of multiple-choice tests. The following year, IBM modified a tabulator to facilitate mechanized grading, which provided IBM with the expertise to build the IBM-600 punched-card calculator, marketed in 1931.[125]

Columbia University also applied their punched-card installation for tasks from other universities in the United States. Astronomer Wallace John Eckert was one of its users. In 1933, he persuaded IBM to present several punched-card machines to an astronomy calculation center at Columbia University. During the following years, Eckert designed an electronic switch, which IBM built. It stepwise traversed about ten different adjustments of a punched-card calculator, a tabulator, and a sum punch, facilitating easier numeric computations, for example, numeric integration. Building this devise provided IBM with programming expertise. Further, there were many additional scientific punched-card installations in the United States in the 1930s, and punched cards were used for scientific computations in several fields.[126]

To improve capability for scientific computations, several advanced calculators were built in universities and research institutions in the United States and Great Britain, starting in 1935. These calculator building projects focused on improved programmability and higher speed of executing programs. To this end, electronics were introduced to replace the slower electromechanical technology that was used in punched-card machines. This development was boosted by the extreme demand for calculations during the Second World War for ballistics and cryptanalysis and for designing the atom bombs in the Manhattan Project, among other projects. Finally after the war, subsequent projects in Great Britain and the United States built the first electronic computers.

This line of development was the almost exclusive focus of most studies on the development of computing from the 1930s to the 1950s, as their authors approached this line of development with the knowledge of the importance of computers for science and industry in the 1960s and later.[127] The exceptions were the studies on individual punched-card and computer producers as well as studies based on the development of punched-card and computer applications.[128]

However, scientific calculations were only a tiny application field for punched-card producers in the 1930s. Several hundred thousand punched cards were consumed for such tasks in a few cases, but this was in contrast to the several tens of millions of cards consumed four times a year by the Social Security Administration for the old age pension program in 1938.[129] IBM was awarded the contract for this huge project after the law had been enacted. This application was not the outcome of IBM's development of technology for handling large alphanumeric registers, and it could have been based on standard equipment. However, the scope of this application was extended through the subsequent shaping of equipment for the closure of bookkeeping that ended up including building a ledger-posting attachment to an alphanumeric tabulator and a collator that merged piles of punched cards.

IBM developed several other specialized machines in the 1930s to land various contracts, and they felt that the Social Security contract fell within the scope of their current punched-card systems, which were designed for bookkeeping. They chose to consolidate their technological capabilities based on the general potential of bookkeeping with punched cards, and, particularly, the potential of punched-card technology for handling big machine-readable registers. Additional administrative programs using large registers of punched cards were established during and after the Second World War and provided extensive business.

The growth of bookkeeping and register applications explains why the sale of punched-card systems reached their zenith only in the 1950s. IBM's sale of punched-card equipment remained bigger than their computer sales until 1962, as was the situation until 1965 with the merger in Great Britain of the British Tabulating Machine Company and Powers-Samas.[130]

Conclusion

The first punched-card system grew out of problems with census processing in the United States in the late nineteenth century. This first version was extensively reshaped during the following half century in diverse ways and in the four major industrial societies of that age, the United States, Great Britain, Germany, and France. This development was not continuous but it went through four successive closures:

1. A technology for compiling counting-based census statistics, which was developed in the 1880s, stabilized in 1889, and used until about 1910
2. A technology for processing general statistics, which was developed starting in 1894, stabilized in 1907, and used until after the Second World War
3. A technology for doing bookkeeping, which was developed starting in 1906, was stabilized in 1933, and remained in use until the 1960s
4. A technology for operating large registers of people, which was developed from 1935 to 1937 and remained in use until the 1960s

This rich and complex history allows analyses of three essential aspects of how a technology is shaped in Western society—first, the shaping and reshaping of punched-card technology and business in the major industrial societies; second, concepts for studying the shaping of technology in business organizations; third, how the shaping and reshaping of punched-card technology reflected the development of Western society.

A fundamental characteristic in this story was the intimate relation between the shaping of technology and business. The invention of the first punched-card system in the United States in the 1880s grew out of a public demand for more detailed census statistics combined with the unwillingness of Congress to establish a permanent Census Office. The latter constrained the census operation due to inadequate and varying funding, which caused the frequent turnover of the administrative management. This first punched-card system was devised and built by two individuals employed in the office processing the 1880 census.

John Shaw Billings came up with the original idea, while Herman Hollerith developed the idea into a tailored system for a simple task that only required counting. In this process, Hollerith tried several designs until he found one suited for processing census returns. By adopting the first punched-card system instead of using the alternative exclusively manual systems for processing the census in 1890, the federal government came to fund the implementation of the first punched-card system. Further, this choice locked the design, because Hollerith had to produce a substantial number of punches and tabulators to fulfill the contract. Particularly, he became committed to the design of the first closure, because the many tabulators through the leasing arrangement with the census office remained his property.

When processing the United States 1890 census ended, Hollerith's punched-card business had acquired so much momentum that he chose to continue with it. In addition, his attempts had failed to solicit additional census office contracts in the United States and abroad. Therefore, to survive he had to find an additional application field for his punched-card equipment. As his new prime application field, he selected railroad auditing that reshaped his punched-card system through the introduction of a new, larger punched card and a tabulator with the ability to do addition. However, the new system did not constitute a separate closure in punched card development. Hollerith never saw the railroad system as stable, and he soon improved it through by adding a sorter and built an additional, incompatible punched-card system to facilitate processing the 1900 census.

If Hollerith had taken the alternative choice of insurance statistics as his prime application field for improving his first punched-card system in the mid 1890s, insurance users would probably have urged him first to develop a sorter, as indicated by John K. Gore's punched-card system at the Prudential Insurance Company in the 1890s. Second, they might well have encouraged him also to provide the capacity to add numbers, like the initial request from his first railway customer. Therefore, in this case, the development of a punched-card system for processing insurance statistics might also have ended in the same set of capabilities as those actually established around 1900.

The reason that the set of capabilities for statistics processing established around 1900 became the basis for Hollerith's second closure can be found in the structure of his business. Hollerith had incorporated his business as the Tabulating Machine Company in 1896 based on his first railroad customer and the Russian and French census contracts. In the years around 1900, he came to support three incompatible punched-card systems as he acquired additional customers.

Hollerith's situation became acute after losing his largest customer, the Bureau of the Census, in 1905, which had been established as the permanent federal census organization in 1902. This caused him exclusively to focus on processing general statistics—mainly operational statistics in private companies—that caused him to serve a fast rising number of customers. Hollerith's answer was his standardized punched-card system from 1907 that came to constitute the second punched-card phase.

In addition to his already established processing facilities, he added the 45-column punched card, which became an industry standard. During the second closure, Hollerith's business grew quickly. He received a generous offer from a conglomerate in 1911 to acquire his company, which he accepted and retired as a rich man. The conglomerate established an efficient business organization that provided substantial momentum to the punched-card business—including sales and a development department.

The growth of the punched-card trade between 1905 and 1920 attracted challengers to the Tabulating Machine Company's combination of monopoly and prime mover position. First, the United States government established its own production of punched-card equipment, which only ended up influencing the development of the punched-card trade by providing the expertise for James Powers to establish a company to compete against the Tabulating Machine Company. Powers became the second challenger by incorporating his business in 1913.

James Powers based his company on the competitive advantage of a tabulator with the ability to print numbers in a calculation. This was a substantial advantage over the Tabulating Machine Company's machines from their second closure, in which the operator copied the outcome from a display. Powers intended his equipment to be used for bookkeeping operations. However a number-printing feature was not sufficient to garner extensive customers in this field, and the Powers company did not add new capabilities to their products until the 1920s. Therefore, though Powers supplied a number printing tabulator six years earlier than the Tabulating Machine Company, he proved unable to capitalize on this position. The reason was his problems in producing reliable machines combined with low demand for number printing equipment in offices in the United States.

The third challenger to the Tabulating Machine Company's prime mover position was John Royden Peirce, who designed machines intended to perform bookkeeping operations. For this end, additional features were needed to print numbers and do subtraction, but several more features were introduced by various producers to attract bookkeeping customers, including extended cards, letter printing, the multiplication of numbers on punched cards, and mechanisms to control printing on preprinted forms. During the reshaping to perform bookkeeping operations between 1906 and 1933, the Tabulating Machine Company (which became IBM in 1924) gradually developed a third punched-card closure by selecting and implementing a combination of features, which, in 1933, encompassed all of the above-mentioned features, including the extension of the number of characters on a card (columns) from forty-five to eighty. This was accomplished through interaction with several users in various private companies, including policy administration in insurance companies and invoicing in public utilities and chain store warehouses.

At that time, subsidiaries of IBM and the Powers company had been established in Britain, Germany, and France, which to some degree established their own versions of the third closure as a result of interacting with local business: cheap punched-card systems with small, nonstandard cards in Britain, machines with high calculation capacity and without any alphabetical facility in Germany. In France, a local challenger, Compagnie des machines Bull was established in the early 1930s. In contrast to Germany, it had early alphanumeric printing and used a different representation of numbers and letters on punched cards than did IBM.

Large-scale record management became the prime application field for shaping the fourth punched-card closure. The alphanumeric punched-card equipment of the 1930s, for the first time, enabled national governments to have direct and individualized access to all their citizens, and to this end the ability to print people's names proved essential. However, from a machine producer's perspective, large-scale record management was only one among several bookkeeping applications. When the producer could supply alphanumeric printing tabulators, it did not require developing new basic features.

IBM developed two additional punched-card devices to improve the handling of records, but they were only minor improvements compared with the development of each of the three previous closures of punched cards. For the Bull company, the large record system for the French army's conscript administration, which failed at a test in 1935, caused the company to extend their punched card from forty-five to eighty columns, mirroring what had already been done at IBM.

In 1933, IBM began supplying alphanumeric equipment capable of handling large registers, and several large insurance companies in the United States and Great Britain had established punched-card registers of their large holding of policies in the 1920s, most of them exclusively numeric, except for the name of the policyholder, which was hand or typewritten. However, the original initiative to expand insurance company punched-card registers on a national scale came from the United States government in 1936, when they decided to establish a punched-card register of tens of millions of people to facilitate the Social Security Administration's monitoring of accumulated wages as a basis for calculating peoples' old age pensions.

Increasingly, larger punched-card registers were used from the late 1930s through the 1950s, providing significant business for IBM and enhancing their dominance of the punched-card industry. Further, this success made IBM less interested in improving the computation capability of their machines, which was the core of the development of programmable and electronic calculators that had begun in the late 1930s at universities and research institutes. Eventually, this development became the basis for the building of the first computer, which became operative in the late 1940s. While IBM applied electronic technology in the late 1940s to improve their punched-card machines, it only started to build electronic computers in 1951.

In contrast, the Powers company (acquired by Remington Rand in

1927), which was the only substantial competitor in the United States, only announced an alphanumeric line of punched-card machines in 1938, probably as a response to IBM's success the previous year supplying equipment to set up the large-scale register needed to administer Social Security. Furthermore, this line of machines was only designed and produced later. One outcome was that Remington Rand never attracted a customer for a large-scale register.

In addition to the federal government's contributions to the shaping of the first punched-card system for census processing and the fourth system for operating large registers of people, governments influenced the shape of the punched-card industry through patent and trade regulations. In the United States, the strong protection of the rights granted to inventors and patentees was aptly utilized by IBM to enhance its monopolistic position as the prime mover, reducing the impact of antitrust legislation. IBM's German subsidiary, Dehomag, also skillfully utilized patent legislation to hamstring challengers several times. The patent laws in Austria and France protected national industry against foreign companies, enabling Otto Schäffler to establish the production of punched-card equipment in Vienna in the 1890s. Moreover, patent laws facilitated production of Bull's machines in France.

The punched-card trade was also influenced by the protectionist regulations that grew out of the worldwide economic crisis that began in 1929. In France, this supported the emergence of the Bull company, though IBM still seems to have had considerable business in that country. There was no independent producer in Germany, and the German economic autarchy ended up enhancing the position of IBM's German subsidiary in its power struggle with its parent company.

Wiebe E. Bijker and Trevor Pinch's analysis of the social construction of technology—or the social shaping of technology—is the background for this book's analysis of shaping of technology in companies, but additional analysis is required to fully understand the history of punched-card systems. Bijker and Pinch's analytical strategy facilitates discussing how the development of technology is shaped. They classify the development of a closure of a technology into three phases.

The first phase is genesis, which is characterized by the interpretative flexibility of the emerging shape of the technology, contesting interpretations, and several possible directions of development. The second phase is characterized by more than one interpretation of the technology and applications built by relevant social groups, which narrows the spectrum of possible solutions in the subsequent phase. Closure and stabilization is the third and final phase, where a "lock-in" takes place on a specific technological solution and the alternatives are scrapped. The terms *closure* and *stabilization* imply that the chosen interpretation or solution of the technological subsequently remains stable for a period.

Bijker and Pinch's analysis is useful in examining how the first punched-card system was shaped in the 1880s for use in counting-based population census statistics, and it helped to locate possible methods for processing these statistics, including counting machines, various forms to be manually sorted and added up, punched strips, punched cards, and so on. However, the original version of their analysis failed to have a concept for analyzing the power relations in the shaping process. They assumed that all the parties involved in the process and the possible technological solutions had an equal say.

This limitation was addressed by Wiebe Bijker in his subsequent research into the closure and the stabilization process. He introduced the micropolitical and symbolic power located in what are known as *technological frames* as a means of understanding the choices made in determining the final closure. A technological frame comprises all elements that influence the interactions within relevant social groups and leads to the attribution of meanings to technical artifacts. These frames include artifacts stabilized in previous shaping processes.[1]

The development of punched-card technology through four distinct closure and stabilization phases provides a basis for improving the understanding of that part of the process. The first punched-card closure was shaped in the office of the United States census in 1880. Bijker's relevant social groups included several statisticians, including Herman Hollerith, who acted as prime users. As the idea of punched-card processing was first suggested by statistician John Shaw Billings, the statisticians sympathized with his sophisticated technical answer to the problems involved in organizing a new census operation every ten years and paved the way for the selection of Hollerith's first punched-card system for processing the 1890 census in a situation in which cost was not the exclusive concern.

The selection of Hollerith's first punched-card system to process the census in the United States in 1890 definitively commercialized the technology, providing weight to the "click-in" act and explaining why the technology subsequently tended to remain stable for some time. Historian Thomas P. Hughes has suggested *momentum* as a concept for understanding this process. While click-in alludes to a momentary act, momentum has duration and facilitates more nuances in an analysis through its association with the concepts of mass, velocity, and direction. The momentum of punched cards leading to the selection in 1889 of Hollerith's first punched-card system was established through his personal relations to the statisticians and their sympathy for his sophisticated method. Once the first punched-card system had been chosen, it gained additional momentum as production of the technology was established and users invested in equipment and gained expertise in managing processing census returns using this technology. For Hollerith and the successor companies, the momentum of each closure of punched-card technology was enhanced by the leasing of the big machines, as they remained the property of the producer. However, this simultaneously weakened the technology's momentum with its users. The 1890 census operation was of a limited duration and the steam fueling the momentum of the first closure soon ran out.

The punched-card system for processing general statistics differentiated fundamentally from its predecessor in that it used a standardized column punched card designed to represent numbers instead of the original field punched card, which was designed to represent varying data about people. The shaping of the technology for this closure was restricted by the choices in the previous closure, which excluded manual means and punched strips. Moreover, this shaping was generated through the selection of prime users of the technology who significantly influenced the choices in this process, thus reducing options and risks. For the Tabulating Machine Company, the substantial momentum of this closure can be observed in the company's extensive growth in the 1910s and in that the company, for six years, successfully avoided supplying a tabulator with the ability to print numbers in spite of a competitor offering this capability.

Systems for handling bookkeeping were shaped as the competing punched-card producers in the United States, Great Britain, Germany, and France tried to extend their business beyond processing statistics in public organizations and private companies. While each of the two first closures had been shaped in a single organization, now several companies shaped punched-card-based bookkeeping systems, and they never reached a common closure.

For example, each producer had its proprietary system for alphanumeric representation on punched cards, which contrasted the standard representation of digits determined by Herman Hollerith in the 1890s before any competitors emerged. Also, this shaping process was much more complex than the one that shaped general statistics, as bookkeeping operations were more divergent than statistics operations, and the critical part of this shaping process was deciding which capabilities punched-card-based bookkeeping should have. The selection of these capabilities was decided through shaping processes at many punched-card companies, with each company gradually improving technology directed at selected prime users, actual or imagined.

The fourth closure of punched cards for handling large registers had two dynamic components, the momentum of the various punched-card companies' established bookkeeping closures and their prime users. For IBM in the United States and the Bull company in France only minor technical changes were needed to achieve this closure—two new devices for IBM and the transition from 45- to 80-column cards for Bull—while basic technical changes were needed for the Powers companies. The small momentum needed explains the feasibility of establishing the new closure—while it does not explain why it actually took place. This required power that was supplied by prime users, the Social Security administration in the United States and the army's conscript administration in France.

This closure exemplifies the distinctions between the technological facilitation and the financial dynamic supplied by prime users. In determining how the four punched-card closures would be shaped, prime users provided foci for the design of technology and the selection of technical alternatives was based on assumptions of sales. This made efficient sales essential to locate and assess potential prime users, which was a key reason for IBM's success. At the same time established applications and customers lead to a status quo that curtailed development.

The primary focus in this book is how technology was shaped and produced. Just as easily the primary focus could have been sales. However, to become comprehensive using sales as an analytic strategy would also have needed to encompass the available technology. Therefore, providing a comprehensive analysis of shaping and producing technology requires

The IBM tabulator Type 405 in the punched-card installation of the Association of Danish Dairies (Mejerikontoret) in 1959. It is the machine to the left operated by the gentleman with the checked shirt and the machine in the center. It remained a core workhorse in IBM punched-card installations until the machines were replaced by computers in the 1960s. (Landbrugets EDB-Center, Risskov, Denmark)

reasonable concepts for the technical development and production, for business contributions, and for linking these two fields of analysis.

Observations of variations between punched-card systems in the United States, Great Britain, Germany, and France raise the question of what these different systems represented, how they were motivated, and their implications for the development of Western society. Several historians of technology have studied variations from one country to another, and they have used the concept of "national style" for their comparisons.[2] In an analysis of German and French diesel engines for cars in the interwar years, two historians raised the question of what constitutes a "style."[3] They stipulated reasonable stability to describe a national "distinction" in a technology as a "style," which is implied by technological closures. Further, they found that a national style, if it existed, should be the outcome of conscious choices made by members of one group in opposition to other groups. However, their study of German and French diesel engineering led them to prefer the more modest term of distinction.

The second question was what motivated the distinctions. While most technology historians have studied technical engineering practices or traditions, Thomas P. Hughes in his study of the emergence of electric power systems in Berlin, Chicago, and London found most of his distinctions in public regulations.[4]

The national statistics offices were the customers for the first shaping of punched cards for counting-based population statistics before 1900. The choice in the United States to introduce mechanized processing of census returns was contrasted by the negative reception in most European census offices to the suggestion of compiling population statistics using punched-card equipment from the first closure. Among the countries with regular census taking, except for Norway, which had a small population, only Austria and France introduced Hollerith's first punched-card system to accomplish their transition to individual records and the central processing of returns to compile more detailed and reliable statistics.[5]

Once they had completed this transition, all European census offices returned to manual means until the first punched-card system improved. First, this exemplified the importance of reliable organizations. Second, the negative reception to punched cards in the various European census offices set off the choice of a sophisticated technical solution in the office processing the United States census in 1890, instead of the reliable manual means applied in European offices, a choice that was repeated in 1936 when technically sophisticated punched cards were selected for use by for the Social Security administration. Both cases substantiate historian David Nye's observation of a preference in the United States for the technologically sublime.[6]

The second punched-card closure was developed for general statistics processing and its diffusion from about 1900s was primarily based on operational statistics tasks. This made punched-card applications a measurement of the establishment of the various industrial forms of governance in big industrial companies, as portrayed by Alfred Chandler and other business historians. There were notable distinctions between the diffusion of punched cards in the United States, Germany, Great Britain, and France in the years up to the First World War. The rapid and substantial diffusion of punched cards for operational statistics in the United States, Germany, and Great Britain confirms Chandler's conclusions, as did the slower and less extensive diffusion that took place in Great Britain.[7]

Economist Maurice Lévy-Leboyer extended this conclusion to encompass France. He found that French companies lacked industrial integration, so firm size, measured by total assets, was smaller than in the United States, Germany, and Great Britain. Also, large French firms were slow to adapt new organizational forms such as multidivisional structures.[8] These observations are corroborated by the delayed spread of punched cards in France.

Apparently, this observation provides new insight into a question concerning the substance of the claim that there was a "decline of Britain" from the end of the nineteenth century compared with Germany and the United States.[9] However, histories of the early spread of punched cards in census organizations exemplify the importance of well-established organizations. The bureaucracy at several large British office organizations in the late nineteenth century substantiated this, as did the manually operated registers established in Germany in the 1930s to control the population.

An additional explanation for national distinctions could be the varying efficiency of the sales activities in the four countries. In the United States and Germany, the Tabulating Machine Company and its German subsidiary, Dehomag, established efficient sales organizations in the 1910s, in contrast to Great Britain and France, where punched-card companies only established capable sales organizations in the 1920s.

The shaping of punched-card systems for bookkeeping tasks followed discernibly different paths in the four countries and the systems, to some extent, remained different up to the Second World War. In the United States, IBM was the first one to develop punched-card machines with a considerable ability to calculate, which by 1931, included subtraction and multiplication. Then, only two years later, it began supplying an alphanumeric punched-card system. Through this shaping process, IBM determined the essential features of punched-card closure for bookkeeping in the United States. In contrast, the Powers company tried and failed in the late 1920s to gain a competitive advantage using a system exclusively for letter printing with a reduced alphabet. Subsequently, it only followed IBM's lead, introducing the same features, though with a delay of several years.

British punched-card applications distinguished themselves via the emergence of the earliest letter printing tabulator, which was marketed by the British Powers company in 1921. However, this feature did not become a major success. Instead, the British Powers company focused on developing and marketing of cheap punched-card machines using small nonstandard cards, which found substantial success.

In Germany, the Powers agency marketed the alphabet printing tabulator that had been developed by the British Powers company, but success in Germany was also limited. German demand focused on numeric calculation capabilities, a need for alphanumeric punched-card systems only emerging during the Second World War. In contrast, French punched-card applications developed an emphasis on alphanumeric printing systems in the 1930s.

These systems for bookkeeping were shaped in the four countries by the producers' selection of actual and imagined prime users. The choices varied between countries and from one company to another. These varying paths provide insight into the process of technological development, though it is quite complicated to discern clear national distinctions because of the low number of companies in each country. In addition, the distinctions are more a question of which path of development was followed rather than of which features the punched card-based bookkeeping systems had by the Second World War. The patterns of features are most distinct in Great Britain and Germany.

In Great Britain, cheap punched-card machines using small, nonstandard cards had substantial success, which possibly indicates a demand among medium-sized companies for punched-card processing of their office tasks. Nevertheless, this strategy did not provide the British Powers company with an increase in the number of customers, nor was it pursued with much vigor by the competing British Tabulating Machine Company. The limited success of the British Powers company in using its cheap punched-card systems to attract additional customers should be contrasted to the large number of mediums-sized companies in Britain. At the same time, the British Tabulating Machine Company achieved substantially growth of its assets due to its main sales strategy of equipment using standard punched cards.

In Germany, Dehomag designed and produced exclusively numeric punched-card machines for statistics and bookkeeping tasks. This choice was based on demand from local customers, nourished by the shield against foreign products provided by Germany's policy of autarchy in the 1930s. Such a choice most likely indicates that calculation capacity was the fundamental feature preferred by German users of punched cards for bookkeeping applications, which might also have been the case for users of punched cards for bookkeeping tasks in the United States, Great Britain, and France. The alphanumeric punched feature for card systems only became crucial when large registers of people were established and as letters to customers widely began to be addressed using punched cards from the late 1930s.

However, this type of development hampered the improvement of bookkeeping jobs through letter specifications and the use of punched cards for large registers of people. The absence of a large alphanumeric register in Germany during the Second World War distinguished it from the United States, France, and Great Britain, though the major reason for this absence was the lack of demand from the Nazi regime for such registers, rooted in the chaotic organizational structure of the Third Reich and the establishment in the late 1930s of nationwide local registers of the population created by manual means.

Notwithstanding the national distinctions, a general picture arises from the histories of punched cards in the United States, Great Britain, Germany, and France. Tasks and ambitions in national governments and private companies changed as Western societies were being reshaped at the end of the nineteenth century up to the Second World War. Originally, punched-card systems were designed and applied in the late nineteenth century to produce more detailed census statistics, which national governments saw as an important contribution to facilitate industrial society.

During the following fifty years, the ambitions of the national governments grew, contributing to the shaping of society. This is manifest when comparing the census operations around 1900 with the governments' endeavor to control industrial production and transport during the two World Wars, where mass warfare required optimal performance by industrial societies.

The First World War consisted of extensive warfare that required supreme industrial efforts, making it crucial to control the use of resources and industrial production. This was facilitated by the frequent collection of statistics in narrow fields compiled by the use of punched cards in Germany, the United States, and Great Britain. For example in Germany, the supply of bread grain was closely monitored through statistics using copies of invoices from all transactions in the sector, allowing efforts to be optimized. However, the German system for controlling these efforts lacked an efficient tool for controlling sharing of resources across activities.

The absence of an ability to focus resources on diverse activities was addressed in the Second World War, where mobile tactics and open formations were applied, posing additional demand on the ability to control the use of resources. In 1942, the German minister of munitions and armament, Albert Speer, established a matrix structure of committees for this purpose. He formed thirteen vertical committees for managing the various kinds of weapons production, while the allocation of raw materials and intermediate goods were organized through a similar number of horizontal committees. To facilitate the operation of this system—and Speer's control—he had a punched-card service that produced precise and detailed statistics, which gradually came to be created by a monthly comparison of the consumption of resources and production of diverse war-related commodities in all relevant companies in the Third Reich. Close monitoring of war-related production gradually led to the introduction of register techniques, as this could improve the precision of the outcomes from traditional statistical inquiries.

In France, the United States, and Great Britain, large punched-card registers were introduced to monitor the soldiers. After the war, huge alphanumeric registers became essential for the organization of modern industrial society. Many registers were manually operated card index files, but several organizations found that punched-card registers, and later computer databases, were more efficient and provided more options. Similarly, growing punched-card-based bookkeeping applications in the 1930s showed an expansion of the scope of administrative ambitions in private companies and public utilities that differentiated from the earlier more simple applications of various key office machines of scale, like typewriters and adding machines.

By the Second World War the punched-card industry had been shaped to provide services crucial for the functioning of ambitious bookkeeping systems in many companies and public institutions and to compile and process information on large numbers of entities in large industrial societies. Furthermore, these developments took place in democracies, authoritarian regimes, and dictatorships. Punched cards were applied to monitor industrial producers of articles of military importance in Germany, soldiers for mobilization in France, and the United States soldiers in action. These jobs could have been accomplished by simpler means, but they were made easier by the rigorous standardization of the number of characters that could be stored on a punched card and by the mechanized handling of the cards. Punched cards did not shape society, but increasingly sophisticated punched-card closures came to facilitate steadily increasing complex operations in private companies and public organizations that became essential for societal changes between the late nineteenth century and the Second World War.

Acknowledgments

The idea of doing a comparative history of the punched-card industry grew out of a discussion with Arthur L. Norberg of the Charles Babbage Institute in Minneapolis, Minnesota, during a conference back in 1989. Shifting affiliations over the years would allow me to implement the idea only very slowly, but each time we met Arthur encouraged me to proceed. He invited me to the Charles Babbage Institute, provided funding for travel, and negotiated on my behalf for access to the IBM Archives in New York. This access has led to many insights into the interplay between the producers and users that serve as the basis for much of the argument of this book.

The Danish Research Council provided funding for three years of research and the extensive travel needed to collect the material for this book. I have since been appointed associate professor at the Centre for Business History at the Copenhagen Business School. Here, I have found helpful colleagues, stimulating discussion, funding for additional travel, and the support I needed to complete the English draft of this book. This proved a bigger challenge than I anticipated, and I am most grateful to the Research Council and the Business School for making this project possible.

It is both enjoyable and rewarding to study abroad, but it comes at the cost of time away from the family. I was therefore grateful to receive invitations to stay with friends while working on the book. In New York, I had the pleasure of staying in the home of Emerson W. Pugh and his wife in Cold Springs; in Paris, I enjoyed the hospitality of Pierre Mounier-Kuhn and Nathalie Kelenski; and in Berlin, Ute and Rainer Hansel graciously hosted my visits.

I collected material at thirty archives in six countries and came to rely on the amazing support of a large number of people who granted me access and helped me to locate and retrieve records. All gave superb advice and offered helpful suggestions, and I would like to mention just a few whose help proved essential in shaping my project. At IBM in New York, in 1993, I found both a gracious host and wise mentor in Emerson W. Pugh. During my subsequent visit, after IBM's Corporate Archive was relocated to Somers, New York, Paul C. Lasewicz deftly assisted my collection of material. At the Hagley Museum and Library, in Wilmington, Delaware, Michael H. Nash helped me to locate exiting material on the American Powers Accounting Machine Company, Remington Rand, and numerous punched-card users. At the National Archives in Washington, D.C., and in New York, I received extensive help from Greg Bradsher, Majorie H. Chiarlante, Tab Lewis, and Fred Romanski. In Paris, Pierre-Eric Mounier-Kuhn used his extensive knowledge of the development of computing and its application in France to help me locate material and obtain access to the Bull archives, the archives of the Contrôleur de l'Administration de l'Armée in the French army's archives, and archives of diverse French national statistical services. La Fédération des Équipes Bull granted me access to the company's archives in Paris-Bobigny, and several of its members helped me to find and recover a rich body of material on this company. I also received much-needed advice in locating material from Martin Campbell-Kelly (Warwick, United Kingdom), Hartmut Petzold (Munich), and Alain Desrosières (Paris). My analysis owes much to all of these people and their colleagues, too numerous to mention. I thank them for their help and support.

David E. Nye, Phil Scranton, Bob Brugger, Kurt Jacobsen, Håkon With-Andersen, Per Boje, Michele Callaghan, and an anonymous reader helped me to improve the book's focus and argument. In my struggle to produce a readable English prose manuscript, I was most fortunate to find Jane Horsewell and Joyce Kling of the Language Center of the Copenhagen Business School to assist me. My work with these and many other people has been both pleasurable and productive. While they provided many excellent suggestions for improvement, the sole responsibility for errors of course remains with me.

Appendix

Financial Information: Tables and Figures

Exchange Rates for Selected Currencies, 1890–1945, in $1,000 Equivalents

Date	Currency			Date	Currency		
	£	marks	francs		£	marks	francs
1890	0.2051	4.182	5.177	1918	0.2099	6.010	5.624
1891	0.2056	4.185	5.190	1919	0.2252	24.478	6.904
1892	0.2050	4.179	5.159	1920	0.2742	57.735	14.046
1893	0.2056	4.193	5.177	1921	0.2614	81.189	13.470
1894	0.2050	4.181	5.158	1922	0.2265	397.333	12.154
1895	0.2043	4.175	5.151	1923	0.2183	—	16.389
1896	0.2048	4.180	5.158	1924	0.2286	4.297	19.401
1897	0.2054	4.185	5.166	1925	0.2072	4.200	20.781
1898	0.2062	4.212	5.209	1926	0.2957	6.034	43.503
1899	0.2054	4.196	5.179	1927	0.2058	4.207	25.484
1900	0.2055	4.203	5.167	1928	0.2055	4.190	25.497
1901	0.2051	4.187	5.160	1929	0.2059	4.200	25.535
1902	0.2047	4.193	5.157	1930	0.2056	4.192	25.476
1903	0.2055	4.196	5.165	1931	0.2225	4.218	25.506
1904	0.2055	4.193	5.167	1932	0.2865	4.209	25.447
1905	0.2055	4.198	5.161	1933	0.2475	3.441	20.848
1906	0.2060	4.211	5.171	1934	0.1991	2.552	15.323
1907	0.2060	4.217	5.185	1935	0.2042	2.483	15.145
1908	0.2058	4.195	5.162	1936	0.2013	2.482	15.994
1909	0.2051	4.193	5.161	1937	0.2025	2.487	24.414
1910	0.2056	4.199	5.177	1938	0.2041	2.489	34.291
1911	0.2057	4.204	5.188	1939	0.2243	2.589	39.612
1912	0.2055	4.203	5.180	1940	0.2527	2.489	44.990
1913	0.2055	4.200	5.180	1941	0.2479	2.489	44.898
1914	0.2041	4.201	5.131	1942	0.2479	2.489	44.000
1915	0.2084	4.860	5.483	1943	0.2479	2.489	44.000
1916	0.2092	5.520	5.880	1944	0.2479	2.489	44.000
1917	0.2099	6.580	5.771	1945	0.2477	—	47.648

Sources: For the years between 1890 and 1939, this table is based on the exchange rate tables in Jürgen Schneider, Oskar Schwarzer, and Markus A. Denze, eds., *Währungen der Welt*, vol. 3.1: *Europäische und Nordamerikanische Divisenkurse 1777–1914* (Wiesbaden, Germany: Franz Steiner Verlag, 1991), 342–343, 350–351, 363, 367; and Jürgen Schneider, Oskar Schwarzer, and Friedrich Zellfelder, eds., *Währungen der Welt*, vol. 2: *Europäische und Nordamerikanische Divisenkurse 1914–1951* (Wiesbaden, Germany: Franz Steiner Verlag, 1997), 30–32, 51–52, 69–70. For the period from 1890 to July 1914, exchange rates are via Berlin. For the period from August 1914 to 1939, exchange rates come from London. No exchange rate exists in 1916 between sterling and dollars in these source tables, but a rate was calculated by linear interpolation of rates for 1915 and 1917. No exchange exists from 1915 to 1918 between marks and any of the other major currencies in this source, and the rates for these years are from *Wirtschaft und Statistik. Sonderheft zur Geldwertung in Deutschland 1914 bis 1923* (Berlin: Statistisches Reichsamt, 1925), 10. The exchange rates from 1940 to 1945 are from *Historisk statistikk 1994* (Oslo: Statistisk sentralbyrå, 1995), 658, 660. A dash indicates that information is not available.

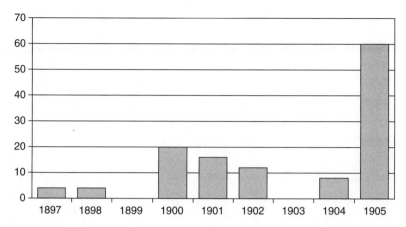

Tabulating Machine Company dividends, 1897–1905 (as percentage of nominal share value). (Undated balance sheets, folder 3, box 12, Hollerith Papers, Library of Congress)

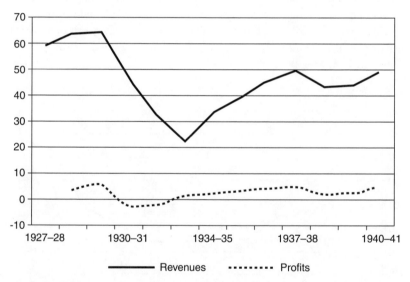

Remington Rand revenue and profit 1927–1928 to 1940–1941, prices of the year (in millions of dollars). Net profit not provided for 1927–1928. (Data for 1927–1928 to 1933–1934: Remington Rand Directors' Meeting, 23 March 1937, in vol. 5, box 15, acc. 1910, Hagley Museum and Archive, Wilmington, DE. Data for 1934–1935 to 1936: *Remington Rand Annual Report for the Year Ending March . . .* , 1936–1941, vols. 5, 6, and 17, boxes 15, 16, acc. 1910, Hagley)

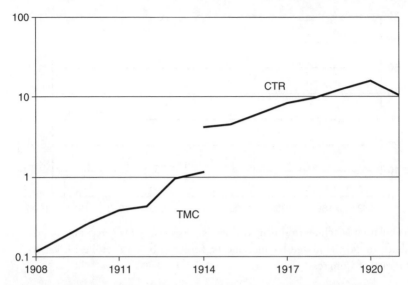

Tabulating Machine Company (TMC) and Computing Tabulating Recording (CTR) Company revenues 1908–1921 (in millions of dollars and prices of the year). (Figures in folder 7, box 10, Hollerith Papers, Library of Congress; Emerson W. Pugh, *Building IBM: Shaping an Industry and Its Technology*. Cambridge, MA: MIT Press, 1995, 23)

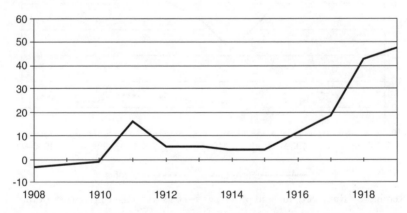

British Tabulating Machine Company pretax profits, 1908–1919 (in thousands of dollars). (Martin Campbell-Kelly, *ICL: A Business and Technical History*. Oxford: Oxford University Press, 1989, 25)

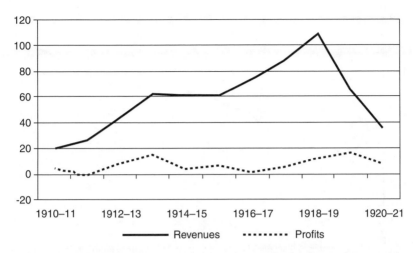

Dehomag revenues and profits, 1910–1921 (in thousands of dollars and prices of the year). (Dehomag annual accounts, B95/89, Wirtschaftsarchiv Baden-Württemberg)

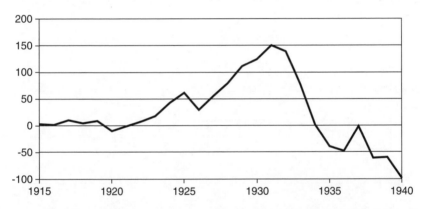

Net earnings of the French IBM companies 1915–1940 (in thousands of dollars and prices of the year). The earnings of the various French IBM companies are totaled for each year. There were four companies: International Time Recording Company (1915–1934), Société internationale de machines commerciales (1920–1934), Société française Hollerith (1935), and Compagnie électro-comptable (1936–1947). (Jaques Vernay, *Chroniques de la Compagnie IBM France*. Paris: IBM France, 1988, 266)

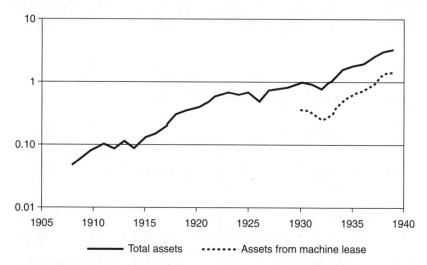

British Tabulating Machine total assets and assets from machine lease, 1908–1939 (in millions of dollars, logarithmic scale). Martin Campbell-Kelly, *ICL: A Business and Technical History*. Oxford: Oxford University Press, 1989, 25, 51, 78)

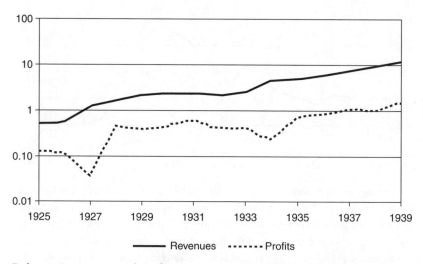

Dehomag's revenues and profits, 1924–1939 (in millions of dollars, logarithmic scale). (Dehomag annual accounts, B95/89, Wirtschaftsarchiv Baden-Württemberg)

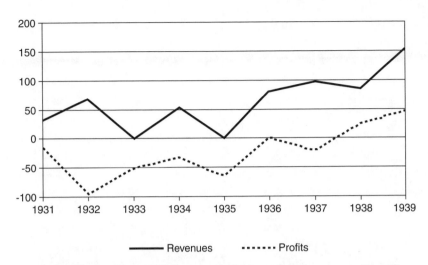

Compagnie des Machines Bull, revenues and net profits, 1931–1939 (in millions of dollars). Bull started operations on 9 March 1931. ("Rapport sur la situation financière de la Société Egli-Bull au 31 decembre 1931," 92 HIST-DGE 01/2, Archives Bull)

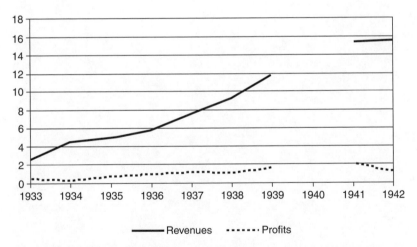

Dehomag revenues and profits, 1933–1942 (in millions of dollars). No figure for 1940 was located. (Dehomag annual accounts, B95/89, B95/110, Wirtschaftsarchiv Baden-Württemberg)

Notes

Introduction

1. Gilbert C. Fite and Jim E. Reese, *An Economic History of the United States* (Boston, MA: Houghton Mifflin, 1973), 510.
2. Not named the Social Security Administration until 1946.
3. *New York Sunday News*, 10 January 1937, 60.1.
4. Wiebe E. Bijker, "Social Construction of Technology," in *International Encyclopedia of the Social and Behavioral Sciences*, ed. Neil J. Smelser and Paul B. Baltes, vol. 23, 15,522–15,527 (Oxford, England: Pergamon Press, 2001); Michel Callon, "Actor Network Theory," in *International Encyclopedia of the Social and Behavioral Sciences*, ed. Neil J. Smelser and Paul B. Baltes, vol. 1, 62–66 (Oxford, England: Pergamon Press, 2001). For a discussion of the two approaches, see Jan Nolin, *Att kasta sten i en glashus: En översigt över den vetenskapssociologiska konstruktivismen* (Göteborg, Sweden: Institutionen för Vetenskapsteori, Göteborg University, 1990); Sergio Sismondo, "Some Social Constructions," *Social Studies of Science* 23 (1993): 515–553; Langdon Winner, "Upon Opening the Black Box and Finding It Empty: Social Constructivism and the Philosophy of Technology," *Science, Technology, and Human Values* 18 (1993): 362–378; Trevor Pinch, "Turn, Turn, and Turn Again: The Woolgar Formula," *Science, Technology, and Human Values* 18 (1993): 511–522; Hans K. Klein and Daniel Lee Kleinmann, "The Social Construction of Technology: Structural Considerations," *Science, Technology, and Human Values* 27 (2002): 28–52.
5. Michel Callon and Bruno Latour, "Unscrewing the Big Leviathan—How Do Actors Macrostructure Reality?" in *Advances in Social Theory and Methodology: Toward an Integration of Micro and Macro Sociologies*, ed. Karin Knorr Cetina and Aron Cicourel (London: Routledge and Kegan Paul, 1981), 277–303; Bruno Latour, "Mixing Humans and Nonhumans Together—The Sociology of a Door-Closer," *Social Problems* 35, no. 3 (1988): 298–310; Michel Callon, "Society

in the Making: The Study of Technology as a Tool for Social Analysis," in *Social Construction of Technological Systems: New Directions in the Sociology and History of Technology*, ed. Wiebe E. Bijker, Thomas P. Hughes, and Trevor Pinch (Cambridge, MA: MIT Press, 1989), 83–103; Bruno Latour, "Where Are the Missing Masses? The Sociology of a Few Mundane Artifacts," in *Shaping Technology/Building Society*, ed. Wiebe E. Bijker and John Law (Cambridge, MA: MIT Press, 1992), 225–258; Madeline Akrich and Bruno Latour, "A Summary of a Convenient Vocabulary for Semiotics of Human and Nonhuman Assemblies," in *Shaping Technology/Building Society*, 259–264.

6. Trevor J. Pinch and Wiebe E. Bijker, "The Social Construction of Facts and Artifacts," in *Social Construction of Technological Systems*, 17–50; Wiebe E. Bijker, "The Social Construction of Bakelite: Toward a Theory of Invention," in *Social Construction of Technological Systems*, 159–194; Wiebe E. Bijker, "The Social Construction of Fluorescent Lighting," in *Shaping Technology/Building Society*, 75–102; Wiebe E. Bijker, *Of Bicycles, Bakelites, and Bulbs* (Cambridge, MA: MIT Press, 1995); Wiebe E. Bijker, "Social Construction of Technology," in *International Encyclopedia of the Social and Behavioral Sciences*, ed. Neil J. Smelser and Paul B. Baltes (Oxford, England: Pergamon Press, 2001), vol. 23, 15,522–15,227

7. This term has been used since the mid-1980s by historians of technology. See, for instance, Donald MacKenzie and Judy Wajcman, eds., *Social Shaping of Technology. How the Refrigerator Got Its Hum* (Milton Keynes, England: Open University Press, 1985). For a reply, see Trevor Pinch, "The Social Construction of Technology: A Review," in *Technological Change: Methods and Themes in the History of Technology*, ed. Robert Fox (Amsterdam: Harwood Academic Publishers, 1996), 17–35.

8. Pinch and Bijker, "Social Construction of Facts," 28–46; Bijker, *Of Bicycles*, 269–272.

9. Bijker, *Of Bicycles*, 122–126, 143, 190–196, 260–267; Bijker, "Social Construction of Technology," 15,526a.

10. Thomas P. Hughes, *Networks of Power: Electrification in Western Society* (Baltimore: Johns Hopkins University Press, 1983).

11. Hughes, *Networks of Power*, 15–17. He has discussed the concept in "Technological Momentum," in *Does Technology Drive History? The Dilemma of Technological Determinism*, ed. Merritt Roe Smith and Leo Marx (Cambridge, MA: MIT Press, 1994), 101–113.

12. Thomas P. Hughes, *American Genesis: A Century of Invention and Technological Enthusiasm* (New York: Penguin Books, 1989); Hughes, "Technological Momentum."

13. Hughes, "Technological Momentum," 108–109.

14. Alfred D. Chandler, *Strategy and Structure: Chapters in the History of the American Industrial Enterprise* (Cambridge, MA: MIT Press, 1962); Alfred D. Chandler, *The Visible Hand: The Managerial Revolution in American Business* (Cambridge, MA: Belknap Press/Harvard University Press, 1977).

For this section I am indebted to Philip Scranton's review of Alfred D. Chandler, "Scale and Scope," *Technology and Culture* 32 (1991): 1102–1104.

15. Alfred D. Chandler, *Scale and Scope: The Dynamics of Industrial Capitalism* (Cambridge, MA: Harvard University Press, 1990).

16. Ronald H. Coase, "The Nature of the Firm," *Economica* 4 (1937): 386–405; Ronald H. Coase, "The Problem of Social Cost," *Journal of Law and Economics* 3, no. 1 (1960): 1–44.

17. Douglass C. North, *Institutions, Institutional Change and Economic Performance* (New York: Cambridge University Press, 1990), 27–35; Douglass C. North, "A Theory of Institutional Change," in *The Microfoundations of Macrosociology*, ed. Michael Hechter (Philadelphia: Temple University Press, 1983), 190–215; Douglass C. North and John J. Wallis, "Integrating Institutional Change in Economic History: A Transaction Cost Approach," *Journal of Institutional and Theoretical Economics* 155 (1994): 609–624.

18. The exceptions are the board minutes of the British producers, which are preserved but not accessible. See the Essay on Sources for more information.

19. The male pronoun is used here as no female inventor was found.

Chapter One: Punched Cards and the 1890 United States Census

1. "The Census of the United States," *Scientific American* 63:9 (30 August 1890): cover, 132; Robert P. Porter, "The Eleventh Census," *American Statistical Association*, New Series 15 (1891): 321–379; T. C. Martin, "Counting a Nation by Electricity," *The Electrical Engineer* 12 (11 November 1891): 521–530.

2. Leon E. Truesdell, *The Development of Punched Card Tabulation in the Bureau of the Census 1890–1940* (Washington, DC: Government Printing Office, 1965); Charles Eames and Ray Eames, *A Computer Perspective* (Cambridge, MA: Harvard University Press, 1973), 18–27; Geoffrey Austrian, *Herman Hollerith: Forgotten Giant of Information Processing* (New York: Columbia University Press, 1982); Martin Campbell-Kelly, *ICL: A Business and Technical History* (Oxford: Oxford University Press, 1989), 4–16; Emerson W. Pugh, *Building IBM: Shaping an Industry and Its Technology* (Cambridge, MA: MIT Press, 1995), 1–14.

3. Margo J. Anderson, *The American Census: A Social History* (New Haven, CT: Yale University Press, 1988), 99–109.

4. Carroll D. Wright and William C. Hunt, *The History and Growth of the United States Census* (Washington, DC: Government Printing Office, 1900), 8–12;

W. Stull Holt, *The Bureau of the Census: Its History, Activities and Organization* (Washington, DC: Brookings Institution, 1929), 1–2.

5. U.S. Constitution, art. I, sec. 2, cl. 3.

6. U.S. Constitution, amend.14, sec. 2 (1868).

7. Anderson, *American Census,* 242.

8. Truesdell, *Development,* 221.

9. Carroll D. Wright, "Statistics in Colleges," *Publications of the American Economic Association* 3 (1888): 5–28; Karl Theodor von Inama-Sternegg, "Der statistische Unterricht," *Allgemeines statistisches Archiv* 1 (1890): 12–13.

10. Joseph W. Duncan and William C. Shelton, *Revolution in the United States Government Statistics, 1926–1976* (Washington, DC: U.S. Department of Commerce, 1978), 1.

11. John Cummings, "Statistical Work of the Federal Government of the United States," *The History of Statistics: Their Development and Progress in Many Countries*, ed. John Koren (New York: Macmillan, 1918), 573–689; Duncan and Shelton, *Revolution,* 4–8.

12. Charles F. Gettemy, "The Work of the Several States of the United States in the Field of Statistics," in *History of Statistics,* 690–739.

13. John Koren, "The American Statistical Association, 1839–1914," in *History of Statistics,* 3–14.

14. Fernand Faure, "The Development and Progress of Statistics in France," *History of Statistics,* 217–329, on 291; Eugene Würzburger, "The History and Development of Official Statistics in the German Empire," in *History of Statistics,* 333–362, on 339; Athelstane Baines, "The History and Development of Statistics in Great Britain and Ireland," in *History of Statistics,* 365–389, on 370; Robert Meyer, "The History and Development of Government Statistics in Austria," in *History of Statistics,* 85–122, on 89.

15. Wright and Hunt, *History and Growth,* 36–40.

16. The number of families in 1840 was not published. Therefore, this number was reconstructed based on the assumption of the average number of members per family being the same in 1840 and 1850. *The Sixth Census* or *Enumeration of the Inhabitants of the United States, as Corrected at the Department of State in 1840* (Washington, DC: Blair and Rivers, 1841); *The Seventh Census of the United States: 1850* (Washington, DC: Robert Armstrong, 1853), xl.

17. *Seventh Census,* xlii–xliii; *Report on the Population of the United States of the Eleventh Census: 1890* (Washington, DC: Government Printing Office, 1895), part I, 451, and (Washington, DC: Government Printing Office: 1897), part II, 2–5.

18. Martin, "Counting a Nation," 521.

19. Hunt and Wright, *History and Growth,* 12–39; Truesdell, *Development,* 1–4.

20. Herman Hollerith, "An Electric Tabulating System," *School of Mines Quarterly* 10 (1889): 238–255, on 244; Charles F. Pidgin, *Practical Statistics. A Handbook* (Boston: William E. Smythe Company, 1888), 147–154.

21. Charles W. Seaton, Improvement in tabulating devices, [U.S.] Patent 127,435, filed and issued 1872; W. R. Merriam, "The Evolution of American Census-Taking," *Century Magazine*, LXV(1903): 831–842, on 836.

22. "Seaton, Charles William," entry in *The National Cyclopedia of American Biography*, 9 (1891), 217–128.

23. Letter from the Secretary of the Interior in relation to an appropriation to C. W. Seaton, 42nd Cong., 2d sess., *House Executive Documents*, No. 164 (1872); U.S. Congress, *Statutes at Large*, vol. 17 (1871–1873), 351.

24. Wright and Hunt, *History and Growth*, 68.

25. John Shaw Billings, "Methods of Tabulating and Publishing Records of Death," *American Public Health Association, Public Health Papers and Reports* 11 (1886), 51–66, on 55; John Shaw Billings, "Forms of Tables of Vital Statistics, with Special Reference to the Needs of the Health Department of a City," *American Public Health Association, Public Health Papers and Reports* 13 (1887), 203–221, on 205; John Shaw Billings, "Mechanical Methods Used in Compiling Data of the 11th U.S. Census," *Proceedings of the American Association for the Advancement of Science* 40 (1891), 407–409, on 407; Herman Hollerith, "The Electrical Tabulating Machine," *Journal of the Royal Statistical Society*, 57 (1894), 678–682, on 678.

26. John Shaw Billings, "Medical Libraries in the United States," 171–182, in *Special Report on Public Libraries in the United States* (Washington, DC: Department of the Interior, 1876), 176; John Shaw Billings, "The Mortality Statistics of the Tenth Census," *Transactions of the American Medical Association* 32 (1881), 297–303; John Shaw Billings, *Report on the Mortality and Vital Statistics of the United States as Returned at the Tenth Census, 1880* (Washington, DC: Government Printing Office, 1886), part I, xi–xii.

27. Pidgin, *Practical Statistics*, 28–61.

28. Hughes, "The Evolution of Large Technological Systems," in *Social Construction in Technological Systems*, 51–82, on 60.

29. *Report of a Commission Appointed by the Honourable Superintendent of Census* (Washington, DC: Government Printing Office, 1889). Reprint by H. Hollerith in box A-832-1, folder: Census, International Business Machines (IBM) Archives, New York.

30. The Pidgin system is described in Pidgin, *Practical Statistics*, 28–61. A census schedule card is preserved in box A-832-2, folder: Pidgin Company, in IBM Archives. The Hunt system is described in Truesdell, *Development*, 24–25.

31. *Report of a Commission*, 11.

32. S. N. D. North, Bureau of the Census, to W. R. Merriam, Tabulating Machine Company (TMC), 28 July 1904, RG-40, NC-54, entry 1, file 67865, National Archives (NA).

33. Anderson, *American Census,* 242. The figures are deflated through a wholesale price index on all commodities, *Historical Statistics of the United States: Colonial Times to 1957* (Washington, DC: Bureau of the Census, 1960), 115.

34. Herman Hollerith, Art of compiling statistics, [U.S.] Patent 395,781, filed 1887 and issued 1889; Hollerith, "Electric Tabulating System."

35. In this tabulator version, each counter was reset by turning the hands to zero on each counter, using a knob on the face.

36. Martin, "Counting a Nation," 528. This figure is an average based on an eight-hour workday.

37. Truesdell, *Development,* 57–83.

38. Martin, "Counting a Nation," 522.

39. Porter, "Eleventh Census," 336.

40. Truesdell, *Development,* 91, 130–131.

41. Hollerith, "Electric Tabulating System," 252. Truesdell, *Development,* 51, claims the wires were soldered to the contact points but supplies no reference.

42. On the whole, the sections on Hollerith are based on John H. Blodgett, *Herman Hollerith: Data Processing Pioneer* (unpublished thesis, Drexel Institute of Technology, Philadelphia, 1968), 4–63, and Austrian, *Herman Hollerith,* 1–73.

43. David M. Ellis, James A. Frost, Harold C. Syrett, and Harry J. Carman, *A History of New York State* (Ithaca, NY: Cornell University Press, 1967), 461.

44. Terry S. Reynolds, "The Engineer in 19th-Century America," in *The Engineer in America,* ed. Terry S. Reynolds (Chicago: University of Chicago Press, 1991), 7–26, on 20.

45. James Kip Finch, *A History of the School of Engineering, Columbia University* (New York: Columbia University Press, 1954), 27–41.

46. Blodgett, *Herman Hollerith,* appendix J; Finch, *History of the School,* 47–48; Austrian, *Herman Hollerith,* 349–350; Monte Alan Calvert, *The Mechanical Engineer in America, 1930–1910* (Baltimore: Johns Hopkins, 1967), 62.

47. Finch, *History of the School,* 7.

48. Herman Hollerith, "Report on the Steam and Water Power Used in the Manufacture of Iron and Steel," *Tenth Census 1880,* vol. 22 (1882).

49. Wright and Hunt, *History and Growth,* 67–68.

50. H. Hollerith to J. T. Wilson, 7 August 1919, in box A-23-3, IBM Archives.

51. Truesdell, *Development,* 27.

52. *Historical Statistics of the United States: Colonial Times to the Present* (New York: Basic Books, 1976), 958–959. Design applications were included with inventions until 1876.

53. H. Hollerith to J. T. Wilson, 7 August 1919, in box A-23-3, IBM Archives; K. B. Wilson to W. F. Willcox, 1926, quoted in Walter F. Willcox, "John Shaw Billings and Federal Vital Statistics," *Journal of the American Statistical Association* 21 (1926): 256–266, on 262; Merriam, "Evolution of American," 836.

54. Herman Hollerith, Apparatus for compiling statistics, [U.S.] Patent 395,783, issued 1889. No mechanical paper strip propulsion is known.

Hollerith filed a patent application for this system of compiling statistics in 1884, which he divided in 1885. Herman Hollerith, Art of compiling statistics, [U.S.] Patent 395,782, issued 1889 and Herman Hollerith, Apparatus for compiling statistics, [U.S.] Patent 395,783, issued 1889. At the 1888 renewal the applications were broadened by references to his application (serial no. 140,629), which was issued as Herman Hollerith, Art of compiling statistics, [U.S.] Patent 395,781, issued 1889.

The 1884 patent application is not preserved in the Patent Case Files, in RG-241, Patent Case File No. 395,782 (1889), NA. The division of the patent in 1885 was probably the result of Hollerith completing his patent application within two years after the filing of his initial patent application, as laid down by the U.S. Patent Act of 8 July 1870, U.S. Congress, *Statutes at Large*, vol. 16 (1869–1871), 198, sec. 32. In this way it was possible to obtain patent protection before completing an invention. U.S. patent protection lasted from the filing of the initial patent application until seventeen years after the date on which the patent was issued by the Patent Office.

55. Paul Israel, *From Machine Shop to Industrial Laboratory: Telegraphy and the Changing Context of American Invention, 1830–1920* (Baltimore: Johns Hopkins University Press, 1992), 75–78.

56. For arguments that Hollerith knew the Jacquard loom, see Austrian, *Herman Hollerith*, 17.

57. Holt, *Bureau of the Census*, 26–27.

58. H. Hollerith to A. Meyer, 19 September 1884, box 21, folder 4, Hollerith Papers, Library of Congress (LC).

59. H. Hollerith to A. Meyer, 14 July 1885, box 21, folder 4, Hollerith Papers, LC; Herman Hollerith, Art of compiling statistics, [U.S.] Patent 395,782, filed 1884 and issued 1889.

60. Wright and Hunt, *History and Growth*, 68.

61. James H. Collins, "The Story of the Printed Word," *A Popular History of American Invention*, in *A Popular History of American Invention*, ed. Waldemar

Kaempffert (New York: Charles Scribner's Sons, 1924), vol. 1, 211–261, on 233–234, 237; "Lanston, Tolbert," 611, in *Dictionary of American Biography*, 5 (1932).

62. H. Sebert, "Rapport fait par M. Sebert, au nom du Comité des Art Économiques, sur la machine à calculer, dit Arithmomètre inventée par M. Thomas de Bojano," *Bulletin de la Société d' Encouragement pour l'Industrie Nationale* series 3, 6 (1879), 393–425, on 406.

63. James W. Cortada, *Before the Computer: IBM, NCR, Burroughs, and Remington Rand and the Industry They Created, 1865–1956* (Princeton, NJ: Princeton University Press, 1993), 27–40.

64. Tolbert Lanston, Adding-machine, [U.S.] Patent 622,157, filed 1894 and issued 1889; Herman Hollerith, Adding and recording machine, [U.S.] Patent 622,470, filed 1895 and issued 1889. It appears from the drawings and the signature of the attorney that Hollerith probably prepared both patent applications. Lanston assigned his patent to Hollerith in 1893 and 1894, RG-241, entry 1009, NA.

65. Herman Hollerith, Electric calculating system, [U.S.] Patent 430,804, filed 1887 and issued 1890. Hollerith, [U.S.] Patent 395,781 (see n 54) was filed in New York in June 1887.

66. Hollerith, "Electric Tabulating System," 244; H. Hollerith to J. T. Wilson, 7 August 1919, box A-23-3, IBM Archives; *Compendium of the Eleventh Census, 1890*, 1, Washington, DC: Government Printing Office, 1892, 747.

67. Hollerith, [U.S.] Patent 430,804 (see n 65).

68. Billings, "Mechanical Methods," 409; Martin, "Counting a Nation," 529.

69. H. Hollerith to A. Meyer, 14 July 1885, in box 21, folder 4, Hollerith Papers, LC.

70. Correspondence between Clifford J. Maloney and the City Archives, Baltimore, 1965, and Clifford J. Maloney's notes from 1965, box A-832-4, folder: Census/Baltimore, IBM Archives, New York.

71. H. Hollerith, Keyboard punch, [U.S.] Patent 487,737, filed 1891 and issued 1892. This patent contains two punch constructions; Austrian, *Herman Hollerith*, 40–41.

72. The principle of a lever-based typewriter is mentioned in George Tilghman Richards, *The History and Development of Typewriters* (London: Science Museum, 1964), 9–10.

73. Albert G. Love, Eugene L. Hamilton, and Ida L. Hellman, *Tabulating Equipment and Army Medical Statistics* (Washington, DC: Office of the Surgeon General, Department of the Army, 1958), 40–42; Hollerith, "Electric Tabulating System," 247–249.

74. A. N. Kiær's report 30 October 1893, on his voyage to the United States,

Kopibok til Statistisk Sentralbyrå for 1893 (Riksarkivet, Oslo) 1426–1436, on 1433; Albert G. Love, "Medical and Casualty Statistics," in *The Medical Department of the United States Army in the World War* (Washington, DC: Government Printing Office, 1925), vol. 14, 2, 9; *Historical Statistics 1960,* 736.

75. A 1890 census clerk's late anecdotal memoirs may substantiate this. Charles A. Springer, "Data Processing 1890 Style," *Datamation,* July 1966, 44.

76. Alfred J. Chandler, *The Visible Hand: The Managerial Revolution in American Business* (Cambridge, MA: Belknap Press/Harvard University Press, 1977), 296–297.

77. *Report of a Commission, passim.*

78. Louis C. Hunter, *A History of Industrial Power in the United States, 1870–1930* (Charlottesville: University Press of Virginia, 1985), vol. 2, 265–266; Steven Lubar, *Infoculture* (Boston: Houghton Mifflin, 1993), 124.

79. Gilbert C. Fite and Jim E. Reese, *An Economic History of the United States* (Boston: Houghton Mifflin, 1973), 433.

80. H. Hollerith to G. F. Swain, 6 August 1907, folder: Tabulating Machines, George F. Swain Collection, Baker Library, Harvard Business School.

81. Blodgett, *Herman Hollerith,* 52.

82. Martin Campbell-Kelly, "Information Technology and Organizational Change in the British Census, 1801–1911," *Information Systems Research,* 7 (1996), 22–36, 35.

83. Thomas P. Hughes, *Networks of Power: Electrification in Western Society* (Baltimore: John Hopkins University Press, 1983), 14–15, 22–23.

Chapter Two: New Users, New Machines

1. Alfred D. Chandler, *The Visible Hand: The Managerial Revolution in American Business* (Cambridge, MA: Belknap Press/Harvard University Press, 1977), 81–121, 277–279, 485; JoAnne Yates, *Control through Communication: The Rise of System in American Management* (Baltimore: John Hopkins University Press, 1989), 4–9.

2. James W. Cortada, *Before the Computer: IBM, NCR, Burroughs, and Remington Rand and the Industry They Created, 1865–1956* (Princeton, NJ: Princeton University Press, 1993), 32–33, 40; Yates, *Control through Communication,* 80–85.

3. Robert P. Porter, *A Report of Examination and Review of the Census Office,* in 52nd Cong., 1st sess., 1892, ex. doc. No. 69, 11.

4. Herman Hollerith, Machine for compiling or tabulating statistics, [U.S.] Patent 26,129, filed 1892 and issued 1894; Herman Hollerith, Machine for tabulating statistics, [U.S.] Patent 526,130, filed 1892 and issued 1894.

5. For a late anecdotal story on mercury cup problems, see Charles A. Springer, "Data Processing 1890 Style," *Datamation*, July 1966, 44.

6. *Liber Patent Transfer Volumes, 1836–1954*, vol. D-46 (1893), 87–88, vol. H-50 (1894), 234–235, vol. J-56, 3–5, entry 1009, RG-241, National Archives (NA).

7. Patent Case File 622,157 (1894–1899), entry 9A, RG-241, NA.

8. Herman Hollerith, Electric calculating system, [U.S.] Patent 430,804, filed 1887 and issued 1890.

9. C. Wright to H. Hollerith, 12 February 1894, box 10, folder 1, Hollerith Papers, Library of Congress. On 7 March 1894 Hollerith filed a patent application on an adding tabulator, *Index to Assignments of Patents 1837–1923*, vol. H-24 (1897–98), 71, RG-241, National Archives. This patent was never awarded, and the application did not survive. See also annual lists of awarded patents and RG-241, NA.

The adder is described in Herman Hollerith, Electrical calculating system, [U.S.] Patent 518,604, filed 1892 and issued 1894; Herman Hollerith, Tabulating system, [U.S.] Patent 518,885, filed 1893; Herman Hollerith, Tabulating system, [U.S.] Patent 518,886, filed 1893 and issued 1894.

10. Geoffrey Austrian, *Herman Hollerith: Forgotten Giant of Information Processing* (New York: Columbia University Press, 1982), 102, 104.

11. *Bulletin de l'Institut International de Statistique* 9:2 (1895), lxii–lxiv, 249–257.

12. E. J. Moorhead, *Our Yesterdays: The History of the Actuary Profession in North America, 1809–1979* (Shaumburg, IL: Actuarial Society of America, 1989), 63; Daniel H. Wells, "Report of a Special Committee of the Council on Suggestions Regarding the Investigation of the Mortality Experience of Life Insurance Companies," *Transactions of the Actuarial Society of America* 2 (1892), 395–398.

13. Arthur Hunter, "Note on an Approximative Method of Making Mortality Investigations," *Transactions of the Actuary Society of America* 10 (1907), 361–367, on 361.

14. H. Hollerith to J. M. Craig, Metropolitan Life Insurance, 1 April 1889, box A-832-2, folder: Metropolitan Life, International Business Machines (IBM) Archives.

15. Moorhead, *Our Yesterdays*, 337. Austrian, *Herman Hollerith*, 82–83, Hollerith took this initiative of a demonstration to actuaries, which Austrian substantiates in an article in *New York Tribune*, 25 April 1890. However, this substantiation is inferior to the preceding note.

16. David Parks Fackler, "Regarding the Mortality Investigation, Instituted by the Actuarial Society of America and Now in Progress," *Journal of the Institute of Actuaries*, 37 (1902), 1–15, on 14–15; Austrian, *Herman Hollerith*, 83.

17. The author was not granted access to the archives of the Prudential Insurance Company or to the Metropolitan Life Insurance Company.

18. The card was 4½ by 2¾ inches (11.4 × 7.0 cm).

19. John K. Gore, Perforating machine, [U.S.] Patent 516,199, filed 1893 and issued 1894; Fackler, "Regarding the Mortality," 11, figure 4. Because of the differences in the cards, the Hollerith pantograph punch could not be applied.

20. Moorhead, *Our Yesterdays*, 338; Fackler, "Regarding the Mortality," 11–15; Yates, *Control through Communication*, 14–15.

21. Fackler, "Regarding the Mortality," 13; Moorhead, *Our Yesterdays*, 338, quotes the excessive figure of 250 cards per minute, which is not substantiated.

22. J. K. Gore, Prudential, to H. Hollerith, 13 May 1901, box 10, folder 1, Hollerith Papers, Library of Congress (LC).

23. J. K. Gore to H. Hollerith, 23 May 1901, in LC, Hollerith Papers, box 10, folder 1, Hollerith Papers, Library of Congress; Fackler, "Regarding the Mortality"; Moorhead, *Our Yesterdays*, 64.

24. Herman Hollerith in "The Electrical Tabulating Machine," *Journal of the Royal Statistical Society*, 57 (1894), 678–682, on 689; "Hollerith's Electrical Tabulating Machine," *The Railroad Gazette*, 19 April 1895, 246–248.

25. M. Riebenack, Assistant Comptroller, Office of Comptroller, General Office, Pennsylvania Railroad Company, reports dated 4 January 1904, 14 January 1904, Hagley Museum and Archive, Wilmington, DE (Hagley), acc. 1807: Pennsylvania Railroad Company, roll M-88, BF-172: Accounting Department, acc. 1807: Pennsylvania Railroad Company, Hagley; J. S. Donaldson, "Extract from a Paper Read before the Association of American Railway Accounting Officers" (New York: The Tabulating Machine Company, c. 1913), box A-832-2, folder: TMC/Railroads, IBM Archives.

26. A punched card is reproduced in "Hollerith's Electric Tabulating Machine," 246.

27. "Recording Waybill Statistics by Machinery," *The Railroad Gazette*, 34 (1903), 526–527.

28. Austrian, *Herman Hollerith*, 124–127, 128–129.

29. H. Hollerith to T. Talcott, 20 August 1896, box 21, folder 4, Hollerith Papers, Library of Congress; *The Railroad Gazette*, 31 January 1896, 80; 9 October 1896, 709; T. Talcott letters according to Austrian, *Herman Hollerith*, 129–130, 136–139.

30. Herman Hollerith, Tabulating apparatus, [U.S.] Patent 677,214, filed 1900 and issued 1901; A. E. Gray, *IBM Development Manual. Book I: Numerical Tabulators* (Endicott, NY: IBM, 1956), box A-25-2, folder: History/num.tab., IBM Archives; John Hayward, *IBM History* (Endicott, NY: IBM, 1957), 13, IBM

Archives. It should be mentioned that there is a contradiction between the introduction of this tabulator at New York Central and Hudson River Railway in 1896 and a patent being first filed for it in 1900. This contradiction cannot be resolved through the available sources.

31. No patent was issued in the United States.

32. Herman Hollerith, Apparatus for perforating record cards, [U.S.] Patent 682,197, filed and issued 1901.

33. Frederick S. Blackall Jr., *A History of the Taft-Peirce Manufacturing Company*, manuscript 1946, 8–9, Rhode Island Historical Society.

34. *Index to Assignments of Patents, 1837–1923*, vol. H-24 (1897–1898), 71, entry 1010, RG-241, National Archives. The patent was never granted, and the application did not survive.

35. Austrian, *Herman Hollerith*, 131–132.

36. *Liber Patent Transfer Volumes, 1836–1954*, vol. E-53, 382–385, entry 1009, NG-241, National Archives; Austrian, *Herman Hollerith*, 34.

37. Gerri Lynn Flanzraich, "The Role of the Library Bureau and Gaylord Brothers in the Development of Library Technology, 1876–1930" (PhD diss., Columbia University, 1990), 6–7, 89–93, 96–98.

38. Austrian, *Herman Hollerith*, 152–159.

39. *Liber Patent Transfer Volumes, 1836–1954*, vol. P-59, 178–181, entry 1009, RG-241, NA; Austrian, *Herman Hollerith*, 165–166; Flanzraich, *Library Bureau*, 98–99.

40. Blackall, *History of the Taft-Peirce*, 9–14.

41. A sorting box is shown in Charles F. Pidgin, Method of compiling statistics, [U.S.] Patent 719,365, filed 1899 and issued 1903; "Charles Felton Pidgin," in *National Cyclopædia of American Biography*, 13 (1906), 479.

42. Charles F. Pidgin, Keyboard for tabulating-machines, [U.S.] Patent 740,042, filed 1899 and issued 1903; Charles F. Pidgin, Apparatus for recording statistical data, [U.S.] Patent 755,168, filed 1899 and issued 1904; Charles F. Pidgin, Apparatus for compiling statistics, [U.S.] Patent 755,695, filed 1899 and issued 1904.

43. Pidgin, Patent 740,042 (see n 42 above); Charles F. Pidgin, Calculating machine, [U.S.] Patent 735,291, filed 1899 and issued 1903; *Liber Patent Transfer Volumes, 1836–1954*, vol. P-82, 491–493, entry 1009, RG-241, National Archives; Pinkerton report, box A-832-2, folder: TMC/Pinkerton investigation, 1899–1902, IBM Archives; H. Hollerith to S. G. Metcalfe, 23 May 1899, box A-832-1, folder: Correspondence/1899, IBM Archives.

44. *Report of the Commission Appointed by the Director of Census on the Competitive Test of Methods of Tabulation* (Washington, DC: Government Printing Office, 1899), file 67865, entry 1, RG-54, NA.

45. Prices in letter from S. N. D. North, Bureau of the Census, to Tabulating Machine Company (TMC), 28 July, 1904, file 67865, entry 1, RG-40, National Archives. Wholesale price index in *Historical Statistics 1960*, 115, 117.

46. Gore had no influence on the choice of his system by the Actuary Society of America for a mortality investigation (see n 23). J. K. Gore to H. Hollerith, 23 May 1901, box 10, folder 1, Hollerith Papers, LC.

47. Cortada, *Before the Computer*, 32, 40.

48. Correspondence, box 10, folder 1, Hollerith Papers, Library of Congress.

49. F. H. Wines, Census Office, to W. R. Merriam, Census Office, 11 June 1900; C. F. Pidgin to F. H. Wines, Census Office, 16 July 1900; S. N. D. North, Census Office, to W. R. Merriam, Census Office, 26 July 1900; S. N. D. North, Census Office, to W. R. Merriam, Census Office, 27 July 1900; S. N. D. North, Census Office, to F. H. Wines, Census Office, 27 July 1900; I. Hall, Census Office, to S. N. D. North, Census Office, 27 July 1900; W. F. Willer, Census Office, to W. R. Merriam, Census Office, 8 August 1900; L. G. Powers, Census Office, to W. R. Merriam, Census Office, 8 August 1900; F. H. Wines, Census Office, to W. R. Merriam, Census Office, 9 August 1900; all in box 1, folder 1-17, files No. 1-77, RG-29, NA; Harry Turner Newcomb, *Mechanical Tabulation of the Statistics of Agriculture on the Twelfth Census of the United States* (Philadelphia: Patterson & White Co., 1901).

50. Affidavits of Herman Hollerith, 22 January 1910, LeGrand Powers, 24 February 1910, in *Tabulating Machine Co. v. E. D. Durand*, box 741A, Equity Case No. 29,065, Supreme Court of the District of Columbia, Washington National Records Center, Suitland, MD.

51. Herman Hollerith, Tabulating apparatus, [U.S.] Patent 685,608, filed and issued 1901; Gray, *IBM Development Manual*, 13–16.

52. Edward W. Byrn, "The Mechanical Work of the Twelfth Census," *Scientific American*, 19 April 1902, cover, 275; affidavit of H. Momsen, 3 March 1910, box 741A, Equity Case No. 29,065, Washington National Records Center.

53. *Tabulating Machine Co. v. E. D. Durand*.

54. John H. Blodgett, *Herman Hollerith: Data Processing Pioneer* (unpublished thesis, Drexel Institute of Technology, Philadelphia, 1968), 85–107.

55. Austrian, *Herman Hollerith*, 221–237.

56. Margo J. Anderson, *The American Census. A Social History* (New Haven, CT: Yale University Press, 1988), 116–123.

57. S. N. D. North to W. R. Merriam, TMC, 28 July 1904; V. H. Metcalf to President T. Roosevelt, 13 February 1905, file 67865, entry 1, NC-54, RG-54, NA; *Annual Report of the Secretary of Commerce and Labor, 1904* (Washington, DC: Government Printing Office, 1904), 23–25; H. Hollerith to S. G. Metcalfe,

3 August 1904, 11 August 1904, two letters dated 20 September 1904, 22 September 1904, 26 September 1904, box 1, folder 2, Hollerith Papers, LC.

58. H. Hollerith to S. N. D. North, 29 September 1905, printed copy, box 34, folder 7, Hollerith Papers, LC; letter, S. N. D. North to H. Hollerith, 3 April 1905, letter, H. Hollerith to S. N. D. North, 11 April 1905, letters, S. N. D. North to V. H. Metcalf, 18 April 1905, letter, S. N. D. North to H. Hollerith, 22 April 1905, file 67865, entry 1, NC-54, RG-40, NA; *Annual Report of the Secretary of Commerce and Labor, 1904*, 24–25; *Annual Report of the Secretary of Commerce and Labor, 1905* (Washington, DC: Government Printing Office, 1905), 25.

59. S. N. D. North to C. F. Pidgin, 22 April 1905, S. N. D. North to C. F. Pidgin, 25 April 1905, S. N. D. North to Secretary of Commerce and Labor, 13 June 1905, S. N. D. North to Secretary of Commerce and Labor, 14 June 1905; S. N. D. North to Secretary of Commerce and Labor, 5 October 1905, Acting Chief Clerk, Census Bureau, to Chief Clerk, Department of Commerce and Labor, 16 October 1905, all in file 67865, entry 1, NC-54, RG-40, NA.

This was the only possible contract, as Pidgin was the offered the sole tender for processing equipment in a competition in 1905. H. Hollerith to S. N. D. North, 29 September 1905, printed copy, box A-832-1, folder: Hollerith/Corresp., IBM Archives.

60. Undated balance sheets in box 12, folder 3, Hollerith Papers, LC.

61. H. Hollerith to J. B. Lunger, New York Life Insurance Co., 20 April 1900, box 10, folder 1, Hollerith Papers, LC; D. H. Wells, Connecticut Mutual Life Insurance Co., to H. Hollerith, 12 December 1900, box 21, folder 4, Hollerith Papers, LC.

62. Correspondence between Hollerith and Library Bureau, Prudential Insurance Company, and Travelers Insurance Company, March–May 1901, box 10, folder 1, Hollerith Papers, LC.

63. *Accounting by Electricity* (Washington, DC: Tabulating Machine Company, 1903).

64. 3¼ by 7⅝ inches (8.3 × 18.7 cm).

65. Herman Hollerith, Registering apparatus, [U.S.] Patent 777,209, filed 1903 and issued 1904; Herman Hollerith, Apparatus for use in tabulating systems, [U.S.] Patent 12,523, reissued 1906.

66. Gray, *IBM Development Manual*, 16–25; Austrian, *Herman Hollerith*, 242–244.

67. Thomas P. Hughes, *American Genesis. A Century of Invention and Technological Enthusiasm* (New York: Penguin, 1989), 150–180.

68. Austrian, *Herman Hollerith*, 242.

69. H. Hollerith to S. G. Metcalfe, 27 April 1903, 18 June 1903, 29 June 1903, 30 June 1903, in box A-832-1, folder: Hollerith/Corresp., IBM Archives. In a 1930s antitrust suit, the U.S. Supreme Court ruled tie-in sales illegal, and they were abandoned by IBM. Calculation based on an undated printed balance sheet in box 10, folder 9, Hollerith Papers, LC.

70. Austrian, *Herman Hollerith*, 251–257.

71. Tabulating Machine Company, "Rentals and Card Revenues," box 10, folder 7, Hollerith Papers, Library of Congress. The 1908 figure is based on records from June to December, as only they are preserved.

72. Austrian, *Herman Hollerith*, 255. Simultaneously, the company's nominal stocks were increased from $100,000 to $500,000 to be more in accordance with the assets.

73. *Historical Statistics of the United States: Colonial Times to 1957* (Washington, DC: Bureau of the Census, 1960), 736; Oscar Theodore Barck and Nelson Manfred Blake, *Since 1900: A History of the United States in Our Times* (New York: MacMillan Publishing, 1974), 156–159.

74. Albert G. Love, *Medical and Casualty Statistics*, vol. 15: *Statistics*, part 2 of *The Medical Department of the United States Army in the World War* (Washington, DC: Government Printing Office, 1925), 327–328, 495–496; Albert G. Love, Eugene L. Hamilton, and Ida Levin Hellman, *Tabulating Equipment and Army Medical Statistics* (Washington, DC: Office of the Surgeon General, Department of the Army, 1958), 40, 52–76.

75. Barck and Blake, *Since 1900*, 156–159; Jonathan R. T. Hughes, *American Economic History* (Glenview, IL: Scott, Foresman, 1987), 413–416.

76. W. S. Gifford, "Realizing Industrial Preparedness," *Scientific American*, 114 (3 June 1916), 576, 598, 600.

77. Bernard M. Baruch, *American Industry in the War: A Report of the War Industries Board* (New York: Prentice-Hall, 1941), 38–46.

78. "Compilation of Operating Statistics Reports," *Railway Age*, 66 (1919), 1353–1360; Walter D. Hines, *War History of American Railroads* (New Haven, CT: Yale University Press, 1928), 68.

79. H. Hollerith to J. T. Wilson, 7 August 1919, in box A-23-3, IBM Archives.

80. William E. Freeman, *Automatic, Mechanical, Punching, Counting, Sorting, Tabulating and Printing Machines, Adaptable to Various Lines of Accounting and Statistical Work Essential for Public Service Corporations, with Particular Reference to Improvements in the Art of Mechanical Accounting* (New York: National Electric Light Association, 1915), 13.

81. Austrian, *Herman Hollerith*, 238–241.

82. H. Hollerith to G. F. Swain, 6 August 1907, folder: Tabulating Machines 1906-1907, George F. Swain Collection, Baker Library, Harvard Business School.

Chapter Three: U.S. Challengers to Hollerith

1. Michael Chatfield, *A History of Accounting Thoughts* (Huntington, NY: Robert F. Krieger Publishing Company, 1977), 64–76; Margery W. Davies, *Woman's Place Is at the Typewriter: Office Work and Office Workers 1870–1930* (Philadelphia: Temple University Press, 1982), 9–27; JoAnne Yates, *Control through Communication: The Rise of System in American Management* (Baltimore: Johns Hopkins University Press, 1989), 21–39, 56–63.

2. Alba M. Edwards, *Sixteenth Census of the United States, 1940: Population: Comparative Occupational Statistics for the United States, 1870 to 1940* (Washington, DC: Government Printing Office, 1943), 113, 121.

3. C. O. Price, "Calculating Machine in Railroad Accounting: A Contrast of Present Accounting Methods with Those of Ten Years Ago," *Railway Age*, 64 (1919), 972–976, on 972.

4. George Nichols Engler, "The Typewriter Industry: The Impact of a Significant Technological Innovation" (PhD diss., University of California, Los Angeles: 1969), 10–21, 24–28, 156.

5. Dorr E. Felt, Adding machine, [U.S.] Patent 371,496, filed and issued 1887; "Felt, Dorr Eugene," *The National Cyclopedia of American Biography*, New York: James T. White, 1955), 23–25; Peggy Aldrich Kidwell, "'Yours for Improvement'— The Adding Machines of Chicago, 1884–1930," *IEEE Annals of the History of Computing* 23, no. 3 (2001): 7; James W. Cortada, *Before the Computer: IBM, NCR, Burroughs, and Remington Rand and the Industry They Created, 1865–1956* (Princeton, NJ: Princeton University Press, 1993), 39–41.

6. William S. Burroughs, Calculating machine, [U.S.] Patent 388,116, filed 1885 and issued 1888; William S. Burroughs, Calculating machine, [U.S.] Patent 388,119, filed 1887 and issued 1888; Ray Abele, *The Burroughs Story*, manuscript 1975, 1.1–3.23, box 2, folder 2, Collection 90, Charles Babbage Institute, Minneapolis, ME; Peggy Aldrich Kidwell, "The Adding Machine Fraternity at St. Louis: Creating a Center of Invention, 1880–1920," *IEEE Annals of the History of Computing* 22:2 (2000): 4–21, on 8–11; Cortada, *Before the Computer*, 31–32, 35.

7. D. E. Felt, Tabulating-machine, [U.S.] Patent, 628,176, filed 1898 and issued 1899; D. E. Felt, Tabulating-machine, [U.S.] Patent 644,287, filed 1899 and issued 1890; D. E. Felt, Tabulating-machine, [U.S.] Patent 694,955, filed 1900 and issued 1902; Cortada, *Before the Computer*, 36–37.

8. Clairborne W. Gooch, Adding machine, [U.S.] Patent, 825,205, filed 1905 and issued 1906; Abele, *Burroughs Story*, 4.6, 4.23.

9. Charles J. Bashe, Lyle R. Johnson, John H. Palmer, and Emerson W. Pugh, *IBM's Early Computers* (Cambridge, MA: MIT Press, 1986), 12, for more information on the use of complements in subtraction.

10. H. G. Schnackel and Henry C. Lang, *Accounting by Machine Methods: The Design and Operation of Modern Systems* (New York: Ronald Press Company, 1929), 24.

11. Abele, *Burrough's Story,* 5.9–5.11.

12. Werner Lange, *Buchungsmaschinen: Meisterwerke feinmechanischer Datenverarbeitung 1910 bis 1960* (Munich: Oldenbourg Verlag, 1986), 63; Kidwell, "Adding Machine Fraternity," 15–16.

13. Louis Couffignal, *Les Machines à Calculer, leur Principes, leur Évolution* (Paris: Gauthier-Villars, 1933), 63.

14. H. S. McCormack, "Keeping Books by Machine: The Punched Card as a Saver of Brain Energy," *Scientific American,* 108 (1913): 194–195, on 195.

15. Thomas P. Hughes, *Networks of Power: Electrification in Western Society* (Baltimore: John Hopkins University Press, 1983), 335–341; Giuliani Amato, *Antitrust and the Bounds of Power: The Dilemma of Liberal Democracy in the History of the Market* (Oxford: Hart Publishing, 1997), 7–19.

16. Consolidated Patent Act of 1870, U.S. Congress, *Statutes at Large,* vol. 16 (1869–1871), 198, sec. 22; Ulf Anderfelt, *International Patent-Legislation and Developing Countries* (The Hague, the Netherlands: Martinus Nijhoff, 1971), 10–25.

17. For example, Secretary of Commerce and Labor to the U.S. President, 13 February 1905, and Census Director to Secretary of Commerce and Labor, 15 April 1905, both in file 67865, NC-54, entry 1, RG-40, National Archives (NA).

18. During the first two years, the Census Bureau machine constructions took place in an existing machine shop at the Bureau of Standards, but in 1907 a new machine shop was equipped in the Census Bureau.

19. *Annual Report of Secretary of Commerce and Labor, 1905* (Washington, DC: Government Printing Office, 1905), 25.

20. Regarding the invention of the sorter, see H. Hollerith, Tabulating apparatus, [U.S.] Patent 685,608, filed and issued 1901; regarding the invention of the punch, see H. Hollerith, Apparatus for perforating record cards, [U.S.] Patent 682,197, filed and issued 1901. These key patents eventually became the basis for patent litigation in Germany.

21. Agreement by James Powers, 19 July 1907, file: James Powers, National Personnel Records Center, St. Louis, MO (NPRC).

22. "Memorandum Re. Persons Employed under the Appropriation 'Tabulating Statistics,' Census Office," 3 May 1909, file 67865, NC-54, entry 1, RG-40, NA.

23. Charles W. Speicer, Record element for tabulating systems, [U.S.] Patent 1,212,727, filed 1911 and issued 1917; Charles W. Speicer, Cleaning-tank, [U.S.] Patent 1,929,407, filed 1914 and issued 1919; correspondence between Spicer Tabulating Machine Company, and His Majesty's Stationery Office (HMSO), London, 1912, "Hire of Sorting, Punching and Tabulating Machines, 1912–1915, STAT 12/14/1, Public Record Office, London.

24. Powers became a United States citizen in 1907. James Powers to Census Bureau, 9 March 1909, file: James Powers, NPRC; James Powers' passport application 7 October 1913, in NA, RG-59, Passport Records, Certificate No. 16,686, roll 196, M-1490, RG-59, NA.

25. James Powers, Coin controlled photographic apparatus, [U.S.] Patent 565,297, filed 1895 and issued 1896; James Powers, Machine for cutting toothpicks, [U.S.] Patent 517,625, filed and issued 1894; James Powers, Bread or cake box, [U.S.] Patent 690,874, filed 1901 and issued 1902; James Powers, Pad-support, [U.S.] Patent 748,005, filed and issued 1903; James Powers, Pad-holder, [U.S.] Patent 748,006, filed and issued 1903; James Powers, Automatic photographic apparatus, [U.S.] Patent 754,090, filed 1903 and issued 1904.

26. According to the printed patents, Powers assigned two out of his six patents issued before 1907 to different companies (1896 and 1904). Before 1911, there is no entry for Powers in the Patent and Trademark Office assignment records. In his application for work in the Census Bureau, Powers did not mention an ability to get inventions developed for production, which is notable because of his subsequent weakness in this field; James Powers to Census Bureau, 9 March 1909, in his NPRC file.

27. NPRC file: James Powers; James Powers, Machine for compiling and sorting statistics, [U.S.] Patent 1,061,118, filed 1909 and issued 1913; M. D. Davies, General Services Administration, to R. H. Schellenberg, IBM, 11 August 1969; Richard H. Schellenberg, "Final Report on James Powers" (1969), copies of both in box 1, B/5/B, National Archive for the History of Computing, Manchester, England.

28. Census Director to the Secretary of Commerce, 18 April 1905, file 67865, NC-54, entry 1, RG-40, NA.

The next section is mainly based on *Report of Director of the Census Bureau, 1906* (Washington, DC: Government Printing Office [GPO], 1906), 24; *Report of Director of the Census Bureau, 1907* (Washington, DC: GPO, 1907), 20; *Report of Director of the Census Bureau, 1908* (Washington, DC: GPO, 1909), 19–25; *Report of Secretary of Commerce and Labor for 1909*, House Doc. No. 109, 61st Cong., 2nd sess., December 1909, 29–30; *Report of Director of the Census Bureau for 1909* (Washington, DC: GPO, 1909), 18–20; *Report of Secretary of Commerce and Labor for 1910*, House Doc. No. 1008, 61st Cong., 3rd sess., November

1910, 41–42; *Report of Director of the Census Bureau for 1910*, House Doc. No. 1008, 61st Cong., 3rd sess., December 1910, 139–141; *Report of Secretary of Commerce and Labor for 1911*, House Doc. No. 122, 62nd Cong., 2nd sess., January 1912, 51–52; *Report of Director of the Census Bureau, 1911*, House Doc. No. 122, 62nd Cong. 3rd sess., December 1911, 134–142; *10th Annual Report of Secretary of Commerce and Labor, Fiscal Year 1912* (Washington, DC: GPO, 1912), 54–55; *Annual Report of Director of the Census Bureau, Fiscal Year 1912* (Washington, DC: GPO, 1913), 10–15; Edward Dana Durand, "Tabulation by Mechanical Means: Their Advantages and Limitations," *15th International Congress on Hygiene and Demography. Transactions* (Washington, DC: Author, 1912), vol. 6, 83–90; H. Hamilton Talbot, "Counting Our People by Machine," *Scientific American* 11 September 1909, 176; "Handling the Census Records of the Whole United States," *American Machinist* 33 (1910), 809–811.

29. Herman Hollerith, Art of compiling statistics, [U.S.] Patent 395,781, filed 1887 and issued 1889; Herman Hollerith, Art of compiling statistics, [U.S.] Patent 395,782, filed 1884 and issued 1889; Herman Hollerith, Apparatus for compiling statistics, [U.S.] Patent 395,783, filed 1884 and issued 1889.

30. *Memorandum for the Secretary [of Commerce and Labor]*, 6 July 1907, copy in *Report of Progress on Tab. Mach. and Integrating Counter*, undated, RG-29, box 1, folders 1–6, PI-161, entry 149, RG-29, NA.

31. S. N. D. North to Secretary of Commerce and Labor, 3 May 1907, J. Powers to S. N. D. North, 28 January 1908, both in file: James Powers, NPRC; "Memorandum for the Director [of the Census]," 25 April 1909, file 67865, NC-54, entry 1, RG-40, NA.

32. The four remaining columns indicated the enumeration district and were punched separately.

33. "An Interesting Bending Fixture Job," *American Machinist* 33 (1910), 959–960; James Powers, Combined punching and counting mechanism, [U.S.] Patent 992,245, filed 1908 and issued 1911; James Powers, Perforating-machine, [U.S.] Patent 992,246, filed 1908 and issued 1911.

34. Reports in box 2, folders 2–9, PI-161, entry 149, RG-29, NA.

35. James Powers, Machine for compiling and sorting statistics, [U.S.] Patent 1,061,118, filed 1909 and issued 1913.

36. "Memorandum for the Solicitor," Bureau of the Census Disbursing Clerk, 21 May 1909, box A-832, folder: Census, International Business Machine (IBM) Archives.

37. Opinion of Court, 14 March 1910, and Judgment, 23 May 1912, in *Tabulating Machine Co. v. E. D. Durand*, Supreme Court of the District of Columbia, Equity Case No. 29,065, box 741A, Washington National Records Center, Suitland, MD. The patent is Herman Hollerith, Registering apparatus, [U.S.] Patent

777,209, filed 1903, issued 1904, and reissued as Herman Hollerith, Apparatus for use in tabulating systems, [U.S.] Patent, 12,523, filed and issued 1906.

38. "Memorandum for the Secretary [of Commerce and Labor]," 6 July 1907, copy in *Report of Progress on Tab. Mach. and Integrating Counter,* undated, box 1, folder 6, PI-161 Entry 149, RG-29, NA.

39. "Memorandum Relative to Purchase and Rental of Machines for Tabulating Agricultural Data," by W. F. Willoughby, 9 May 1910, file 67001/10, NC-54, entry 1, RG-40, NA; "Mechanical Equipment of the Census Bureau," 1 November 1910, box 15, folder 7, PI-161, entry 145, RG-29, NA; "Report on Calculating Machines for Use in the Bureau of the Census," 5 December 1910, box 14, folder 3, PI-161, entry 145, RG-29, NA.

40. J. A. Stewart, "Electricity and the Census: A Glance at the Machines that Make Our Decennial Inventory a Matter of Weeks Instead of Years," *Scientific American,* 31 January 1920, 109.

41. Herman Hollerith, Tabulating apparatus, [U.S.] Patent 685,608, filed and issued 1901, basic sorter patent; Herman Hollerith, Registering apparatus, [U.S.] Patent 777,209, filed 1903 and issued 1904, horizontal sorter; for court case annotation, see note 37. The last patent was reissued as Herman Hollerith, Apparatus for use in tabulating systems, [U.S.] Patent 12,523, filed and reissued 1906.

42. O. P. Braitmayer to S. C. Metcalf, 27 March 1914, box 10, folder "Business Corresp. June–August 1914," Hollerith Papers, Library of Congress (LC). "Re whether the work of developing card tabulating machines in the Bureau of the census machine shop should be continued," 24 June to 3 July 1924, file 67001/24, entry 1, RG-40, National Archives; *Report of Progress on Tab. Mach. and Integrating Counter* (1913–1918), box 1, folder 6, PI-161 entry 149, RG-29, NA.

43. E. M. LaBoiteaux, "Memorandum for the Director [of the Census]," 10 August 1917, box 1, folder 6, PI-161, entry 149, RG-29, NA; Leon E. Truesdell, *The Development of Punched Card Tabulation in the Bureau of the Census 1890–1940* (Washington, DC: Government Printing Office, 1965), 139–142, 161–169, 195–196.

44. A. Ross Eckler, *The Bureau of the Census* (New York: Praeger Publishers, 1972), 112–115.

45. Friedrich W. Kistermann, "Locating the Victims: The Nonrole of Punched Card Technology and Census Work," *IEEE Annals of the History of Computing* 19 (1997): 31–45, on 33–34.

46. J. Powers to E. D. Durand, 16 March 1911, file: James Powers, National Personnel Records Center; Powers assigned his patents to Powers Accounting Machine Company, *Liber Patent Transfer Volumes, 1836–1954,* vol. X-87 (1911), 48–50, vol. C-91, 107–113, entry 1009, RG-241, NA; J. T. Ferry, ed., "A History

of the Sperry Rand and Remington Rand Corporations, and Their Predecessors, With Emphasis on Tabulating Machine Equipment," typescript 1964, acc. 1825, Hagley Museum and Archive, Wilmington DE.

47. James Powers, Keyboard for perforating-machines and the like, [U.S.] Patent 1,086,397, filed 1912 and issued 1914; James Powers, Perforating machine, [U.S.] Patent 1,138,314, filed 1913 and issued 1915.

48. William E. Freeman, "Automatic Mechanical Punching, Counting, Sorting, Tabulating and Printing Machines Adaptable to Various Lines of Accounting and Statistical Work Essential for Public Service Corporations with Particular Reference to Improvements in the Art of Mechanical Accounting." Paper presented at the annual convention of the National Electric Light Association, San Francisco, CA, 7–11 June 1915.

49. Freeman, "Automatic Mechanical Punching," 24–28.

50. James Powers, Tabulating machine, [U.S.] Patent 1,224,411, filed 1912 and issued 1917; "Remington Rand Inc.—Development and Growth," in Ferry, "A History of the Sperry Rand," 8.

51. "Powers Systemet," *System*, 1930: 5, 6–7; Lars Heide interviews with Keld Pedersen, formerly the Danish Powers agency, 1985–1986.

52. *The American-Canadian Mortality Investigation, Based on the Experience of Life Insurance Companies of the United States and Canada during the Years 1900 to 1915, Inclusive on Policies Issued from 1843 to 1914, Inclusive* (New York: Actuarial Society of America, 1918), 1–10.

53. Freeman, "Automatic Mechanical Punching," 30; H. Nielsen, "Hulkort," *Købstadforeningens Tidsskrift*, 1950/51, 416.

54. From 1908 to 1911, W. W. Lasker filed at least fourteen applications for patents on typewriter inventions, *Index to Assignments of Patents 1837–1923*, vol. L-18 (1908–10), 149, 211, 268, vol. L-19 (1910–1912), 273, RG-241, National Archives.

55. See, for example, Hans Goerlitz (Berlin, Germany), Adding and sorting machine, [U.S.] Patent 1,223,690, filed 1915 and issued 1917; Robert Neil Williams (London, England), Tabulating machine and cards therefor, [U.S.] Patent 1,274,484, filed 1915 and issued 1918; Robert Neil Williams (London, England), Perforating-machine, [U.S.] Patent 1,311,544, filed 1915 and issued 1919; Charles Foster (Croyden, England), Tabulating-machine, [U.S.] Patent 1,274,528, filed 1917 and issued 1918; Harold D. Penney, Total-printing offset mechanism, [U.S.] Patent 1,317,454, filed 1916 and issued 1919; Harold D. Penney and Joshua E. Davidson, Electrical non-stop card-tabulator printer, [U.S.] Patent 1,376,572, filed 1917 and issued 1921.

56. James Powers, Printing tabulator, [U.S.] Patent 1,245,503, filed 1914 and issued 1917. Lasker's role in building the improved Powers tabulator from

1913 is not clear and no definitive source is known. Powers filed the patent, which does not prove that Lasker did not play an important role in the process. The company history attributes a big role to Lasker but this could have been influenced by Lasker's subsequent role in the company. "Remington Rand Inc.—Development and Growth," in Ferry, *A History of the Sperry Rand*, 8.

57. James Powers, Statistical-card-verifying machine, [U.S.] Patent 1,203,263, filed 1914 and issued 1916; Freeman, *Automatic Mechanical Punching*, 24.

58. James Powers, Tabulator printer for statistical purposes (stop card), [U.S.] Patent 1,245,502, filed 1913 and issued 1917; James Powers, Automatic total-taking mechanism of accounting-machines (stop card and total card), [U.S.] Patent 1,245,506, filed 1915 and issued 1917; Freeman, "Automatic Mechanical Punching," 31; "Remington Rand Inc.—Development and Growth," in Ferry, "History of the Sperry Rand," 9–10.

59. James Powers, Combined type-writer and perforating machine, [U.S.] Patent 1,271,614, filed 1914 and issued 1918; James Powers, Twelve-key hand punch, [U.S.] Patent 1,272,089, filed 1915 and issued 1918; James Powers, Perforating-machine, [U.S.] Patent 1,299,022, filed 1916 and issued 1919; "Remington Rand Inc.—Development and Growth," in Ferry, "History of the Sperry Rand," 8.

60. William W. Lasker, Sorting-machine, [U.S.] Patent 1,315,370, filed 1918 and issued 1919; Thomas O'Keefe, "Chronological History of the Tabulating Machines Division," in Ferry, "A History of the Sperry Rand." The year 1919 is provided from "A Short History of Powers Samas 1919–1950," box 1, B/5/5, National Archive for the History of Computing, Manchester, England.

61. William W. Lasker, Rotary card sorter, [U.S.] Patent 1,198,971, filed 1917 and issued 1919.

62. Patent assignments between Powers Accounting Machine Company and Edward P. Meany, Convent, NJ, 31 July 1913, *Liber Patent Transfer Volumes, 1836–1954*, vol. H-93 (1913), 58–61, entry 1009, RG-241, NA; W. Heiddinger to F. Kraus, 13 June 1918, 6–7, file: I.HA/Rep. 120 C VIII 1, Nr.84, adhib.18, Beiheft 2104, Geheimes Staatsarchiv, Preußischer Kulturbesitz, Berlin; letter, Deutsche Gesellschaft für Addier- und Sortiermaschinen to Preußische Handelsministerium (Prussian Ministry of Commerce), 13 March 1918, file: I.HA/Rep. 120 C VIII 1, Nr.84, adhib.18, Beiheft 2106, Geheimes Staatsarchiv; Powers Accounting Machine Company assigned patents to Peoples Trust Company, New York, as mortgage, *Liber Patent Transfer Volumes, 1836–1954*, vol. Z-99 (1916), 84–87, entry 1009, RG-241, NA.

63. "Chronological History of the Migrations-Management-etc.," in Ferry, "History of the Sperry Rand."

64. James Powers, Wheel-chain for vehicles, [U.S.] Patent 1,391,402, filed 1919 and issued 1922; James Powers, Toothpick cutter and dispenser, [U.S.] Patent

1,436,528, filed 1919 and issued 1922; James Powers, Trench digger, [U.S.] Patent 1,580,570, filed 1921 and issued 1926; James Powers, Self starting automatic total taking mechanism, [U.S.] Patent 1,836,039, filed 1924 and issued 1931; James Powers, Selecting machine, [U.S.] Patent 1,782,442, filed 1926 and issued 1930.

65. Assignments, James Powers to Powers Accounting Machine Company, 21 October 1924, 11 May 11 1925, *Index to Assignments of Patents, 1837–1923,* vol. P-32 (1924–1925), 145, 291, entry 27, RG-241, NA; J. Powers to W. S. James, Powers Accounting Machine Co., 7 December 1925, H. J. Pritchard to J. Powers, 5 January 1926, J. Powers to H. J. Pritchard, 26 January 1926, H. J. Pritchard to J. Powers, 10 February 1926, object no. 1992.3214.03, Smithsonian Institution, Washington, DC.

66. James Powers, Tool for applying glaziers points, [U.S.] Patent 1,028,778, filed 1911 and issued 1912; James Powers, Apparatus for making and vending sanitary cups, [U.S.] Patent 1,077,295, filed 1912 and issued 1913; James Powers, Sanitary drinking-cups, [U.S.] Patent 1,107,347, filed 1912 and issued 1914; James Powers, Manufacturing sanitary drinking-cups, [U.S.] Patent 1,107,348, filed and issued 1914; James Powers, Apparatus for conveying the speech of performers simultaneously and synchronously with the displaying of pictures, [U.S.] Patent 1,280,542, filed 1915 and issued 1918; James Powers, Transferring master drill device, [U.S.] Patent 1,310,034, filed 1917 and issued 1919; James Powers, Apparatus and method for formulating and translating codes, [U.S.] Patent 1,315,406, filed 1916 and issued 1919; James Powers, Auxiliary steering post, patent application 1916, never granted.

67. O'Keefe, "Chronological History," in Ferry, "A History of the Sperry Rand."

68. Diverse assignments in *Liber Patent Transfer Volumes, 1836–1954,* vol. A-117, 77–86 (1922); vol. H-117, 378 (1923); vol. S-131, 445–453 (1927); vol. E-136, 107–122 (1928), entry 1009, RG-241, NA; O'Keefe, "Chronological History;" "Vickers Limited. A Short Account of the Subsidiary Companies: Powers Samas" (1959), 11, Document 771, Vickers Archives, Cambrige University Library.

69. "James Henry Rand Dead at 81: A Co-founder of Sperry Rand, *New York Times,* 4 June 1968, 47:1; assignments from National Bank of New York to Powers Accounting Machine Company, *Liber Patent Transfer Volumes, 1836–1954,* vol. S-131, 445–453 (1927), vol. E-136, 107–122 (1928), entry 1009, RG-241, NA; assignments from Powers Accounting Machine Company and the National Bank of New York to Remington Rand (RR), *Liber Patent Transfer Volumes, 1836–1954,* vol. L-136, 504 (1928), vol. M-136, 261–162 (1928).

70. Executive Committee of Directors of RR, meeting 9 February 1942, vol. 3, 104, box 17, acc. 1910, Hagley; Ferry, "A History of the Sperry Rand."

71. RR, *Eleventh Annual Report* (1938), vol. 6, box 15, acc. 1910, Hagley.

72. L. E. Brougham of Powers-Samas, "Confidential Report of a Visit to the United States March–April 1935," box 1, ICL/Powers-Samas Papers, National Archives for the History of Computing, Manchester, England.

73. RR, *Ninth Annual Report* (1936), *Tenth Annual Report* (1937), vol. 6, box 15, acc. 1910, Hagley.

74. The agreement was confirmed in 1922 within the version of the agreement between IBM and RR, 4 March 1931, printed in "Petition, *USA v. IBM, RR,* etc," 14 April 1932, 17–27, equity 66–215, District Court of the Southern District of New York, NA, Northeast Region, New York.

75. Herman Hollerith, Automatic control for tabulating machines, [U.S.] Patent 1,830,699, filed and issued 1931.

76. RR Directors Meeting, 3 March 1931, RR Minutes, box 15, vol. 3, 161, acc. 1910, Hagley; Executive Committee of Directors of Remington Rand, 31 March 1931, box 17, vol. 1, 231–233, acc. 1910, Hagley; Brougham, "Confidential Report."

The RR information agrees with the 1931 agreement as reported in the in "Petition, *USA v. IBM, RR*"; see note 74 above.

77. William W. Lasker, Printing tabulator mechanism, [U.S.] Patent 1,668,916, filed 1924 and issued 1928; William W. Lasker, Type selector, [U.S.] Patent 1,705,983, filed 1925 and issued 1929.

78. O'Keefe, "Chronological History."

79. *Historical Statistics of the United States: Colonial Times to 1957* (Washington, DC: Bureau of the Census, 1960), 519–520, 527.

80. "Machine-Made Records," *Factory and Industrial Management* 82 (1931): 492–493; L. F. Woodruff, "A System of Electric Remote-Accounting," *American Institute of Electrical Engineers Transactions* 57 (1938): 78–87; "Chronological History of Punched Card Equipment," and O'Keefe, "Chronological History."

81. James Powers, Self starting automatic total taking mechanism, [U.S.] Patent 1,836,039, filed 1924 and issued 1931; Harold D. Penney and Joshua E. Davidson, Electrical non-stop card-tabulator printer, [U.S.] Patent 1,376,572, filed 1917 and issued 1921; Robert Edward Paris, Perforated card controlled machine, [U.S.] Patent 1,864,053, filed 1927 and issued 1932; O'Keefe, "Chronological History."

82. William W. Lasker, Tabulating machine, [U.S.] Patent 1,780,621, filed 1929 and issued 1930; H. L. Tholstrup, Perforated storage media, *Electrical Manufacturing*, December 1958, 58.

83. "Model 2 Tabulator Controls," *Keeping Tabs* 2:1A (1932); "Model Two Tabulator: A Few Tips on Direct Subtracting Features," *Keeping Tabs* 2:2A (1932);

"Powers Ninety Column Equipment," *Keeping Tabs* 2:3A (1932); "Powers 90 Column Sorter," *Keeping Tabs* 2:3C(1932); "Model 2 Tabulator: Y-Wire Grand Totals," *Keeping Tabs* 2:5 (1932); all in box 14, acc. 1825, Hagley; "Introduction to UNIVAC Punched-Card Data Processing," (Remmington Rand or Sperry-Rand, c. 1955), 25–27, box 347, accession 1825, Hagley; "Chronological History of Punched Card Equipment" and O'Keefe, "Chronological History of the Tabulating Machines Division,"in Ferry, "A History of the Sperry Rand."

84. "Introduction to UNIVAC," 27–28, box 347, accession 1825, Hagley.

85. "Introduction to UNIVAC," 29.

86. "Introduction to UNIVAC," 30–34.

87. "J. Royden Peirce IBM Engineer Dies Suddenly," *Business Machines*, 12 January 1933, 1; e-mails from Nydia Cruz, S. C. Williams Library, Stevens Institute of Technology, Hoboken, NJ, to Lars Heide, 21 December 1998 and 26 January 1999.

88. James E. Brittain and Robert C. McMath Jr., "Engineers and the New South Creed: the Formation and Early Development of Georgia Tech," in *The Engineer in America*, ed. Terry S. Reynolds (Chicago: University of Chicago Press, 1991), 123–149, on 124; Terry S. Reynolds, "The Engineer in 19th-Century America," in *Engineer in America*, 7–26, on 21.

89. For example, J. R. Peirce, Stone-working machine, [U.S.] Patent 819,080, filed 1904 and issued 1906; J. R. Peirce, Stoneworking-machine, [U.S.] Patent 862,933, filed 1904 and issued 1907; J. R. Peirce, Stone rubbing machine, [U.S.] Patent 798,587, filed and issued 1905; J. R. Peirce, Polishing-machine, [U.S.] Patent 817,798, filed 1905 and issued 1906.

90. John Royden Peirce, Bookkeeping machine, [U.S.] Patent 1,233,699, filed 1907 and issued 1917. This patent mentions two additional not-granted patent applications, filed 1906, which contained a detailed description of the system and were discarded by the Patent Office, as not-granted applications.

91. John Royden Peirce, Perforating machine, [U.S.] Patent 998,631, filed 1907 and issued 1911; John Royden Peirce, Bookkeeping machine, [U.S.] Patent 1,233,699, filed 1907 and issued 1917.

92. John Royden Peirce, Machine for check-making, &c, [U.S.] Patent 1,260,704, filed 1909 and issued 1918; John Royden Peirce, Recording mechanism, [U.S.] Patent 1,260,705, filed 1909 and issued 1918; John Royden Peirce, Listing-machine for pay-rolls and the like, [U.S.] Patent 1,260,706, filed 1909 and issued 1918; John Royden Peirce, Recording apparatus, [U.S.] Patent. No.1,110,643, filed 1908 and issued 1914; Freeman, "Automatic, Mechanical Punching," 14–15; John Royden Peirce, Bookkeeping Machine, [U.S.] Patent. No. 1,219,766, filed 1909 and issued 1917; *The Royden Automatic Accounting Machines*, brochure marked 1912. During the years that followed, Peirce carried on this work and

eventually made two extensive descriptions of his system, both with the title *The Royden Systems of Perforated Cards*, n.d. (probably 1915–1920). All three in box A-919-2, IBM Archives.

93. John Royden Peirce, Distributing machine, [U.S.] Patent 1,316,461, filed 1911 and issued 1919; John Royden Peirce, Sorting machine, [U.S.] Patent 1,219,767, filed 1911 and issued 1917; John Royden Peirce, Recording-machine, [U.S.] Patent 1,245,500, filed 1911 and issued 1917; John Royden Peirce, Perforating and printing machine, [U.S.] Patent 1,182,309, filed 1912 and issued 1916; John Royden Peirce, Machine for recording or listing items on cards, [U.S.] Patent 1,248,902, filed 1913 and issued 1917; Peirce called the punch a "perforating machine" and the tabulator a "distributing machine."

94. John Royden Peirce, Perforated card for accounting systems, [U.S.] Patent 1,236,475, filed 1912 and issued 1917.

95. M.O. Chance, F. H. Tonsmeire, and E. H. Halding, "Report on the Royden System of Perforated Cards" to the United States President's Commission on Economy and Efficiency, 9 March 1912; H. S. McCormack, "Report on Royden-Peirce Machine. Compiled for Mr. Gershom Smith by the Business Burse," 13 May 1912, copies in box A-919-2, IBM Archives; Prospectus for the Royden Company to be established with 24 September 1912 cover letter from the company, box 10, folder 7, Hollerith Papers, LC.

96. Prospectus for the Royden Company to be established with 24 September 1912, covering letter from the company, box 10, folder 7, Hollerith Papers, LC.

97. Mutual Benefit ranked eighth in value of insurance in force in 1910, J. Owen Stalson, *Marketing Life Insurance: Its History in America* (Cambridge, MA: Harvard University Press), 1942, 800.

98. Percy C. H. Papps, "The Installation of a Perforated Card System with a Description of the Peirce Machines," *Transactions of the Actuarial Society of America* 15 (1914): 49–61.

99. "Description of the Peirce Listing and Adding Machine as Designated for Compilation of the Reports of the Metropolitan Life Insurance Company," prepared by the Peirce Patents Company, undated but accompanied with comments by Metropolitan actuary J. M. Craig, dated 7 May 1913, copy in box A-919-2, IBM Archives; *Metropolitan Life Insurance Company: Its History, Its Present Position in the Insurance World, Its Home Office Building, and Its Work Carried on Therein* (New York: Metropolitan Life Insurance Company, 1914), 71–72.

100. Report by a Metropolitan assistant actuary, 1 May 1913, copy in box A-919-2, IBM Archives.

101. Series of letters between J. R. Peirce and J. M. Craig, Metropolitan Life actuary, 29 May 1913, through 23 June 1913; commentary by J. M. Craig on

Papps, 1914; correspondence between Peirce and Metropolitan and Metropolitan commentaries, from 1 October 1915, to 5 October 1916; all in box A-919-2, IBM Archives.

102. JoAnne Yates, "Co-evolution of Information-Processing Technology and Use: Interaction between the Life Insurance and Tabulating Industries," *Business History Review* 67 (1993), 1–51, on 36.

103. Peirce's 150-column card was organized in six punched code strips, each consisting of twenty-five columns. Specimen cards attached to letter from F. T. Hines, U S. War Veterans Bureau to IBM, 17 February 1925, box A-919-2, IBM Archives. An earlier code was based on ten rows and is found in Royden System, see note 92.

104. Correspondence between Peirce and Metropolitan Life with internal Metropolitan Life comments, contract between Metropolitan Life and Peirce, 23 April 1918; copy annexed to memorandum of agreement, 13 July 1926, between IBM and Metropolitan Life; all three references copies in box A-919-2, IBM Archives; John Royden Peirce, Method and apparatus for perforating record sheets, [U.S.] Patent 1,506,381, filed 1922 and issued 1924; Yates, "Co-evolution of Information-Processing," 38.

105. Memorandum of agreement, 17 June 1918, between the Peirce Accounting Machine Company, New York, and the Prudential Insurance Company of America, copy in box A-919-2, IBM Archives.

106. U.S. Veterans Bureau to IBM, 17 February 1925 and 15 April 1925, box: A-919-2, IBM Archives.

107. In 1921, the Veterans Bureau was established in a merger that included the Bureau of War Risk Insurance.

108. Agreement, 14 August 1920, between Thomas Jackson Felder, Archduke Frederick, and General Real Estate & Trust, Switzerland (copy); Cooper, Kerr & Dunham to IBM, 2 July 1924; both in box A-919-2, IBM Archives.

109. Assignment of Peirce Patents to Tabulating Machine Company (TMC), *Liber Patent Transfer Volumes, 1836–1954,* vol. B-116 (1921), 72, entry 1009, RG-241, National Archives; contract between Computing Tabulating Recording Company (CTR) and J. R. Peirce, 5 January 1933; unsigned agreement between J. R. Peirce and CTR, 14 March 1922; assignment of J. R. Peirce's patents to CTR, 5 May 1922; all in box A-919-2, IBM Archives.

110. IBM to War Veterans Bureau, 17 February 1925, 26 March 1925, and War Veterans Bureau to IBM, 15 April 1925, 18 May 1925; memorandum of agreement, 13 July 1926, between IBM and J. R. Peirce; all in box A-919-2, IBM Archives.

111. W. J. Banett, "A Study of the Peirce Machines," *Metropolitan Life Report,* 1 March 1929, copy in box A-919-2, IBM Archives.

112. John Royden Peirce, Method and apparatus for perforating record sheets, [U.S.] Patent 1,506,381, filed 1922 and issued 1924; John Royden Peirce, Perforation reading instrumentalities, [U.S.] Patent 1,506,383, filed 1922 and issued 1924.

113. Alfred D. Chandler, *Scale and Scope: The Dynamics of Industrial Capitalism* (Cambridge, MA: Harvard University Press, 1990), 34–36.

114. Herman Hollerith, Apparatus for compiling statistics, [U.S.] Patent 395,783, filed and issued 1889; Herman Hollerith, Art of compiling statistics, [U.S.] Patent 395,782, filed and issued 1889.

115. Herman Hollerith, Apparatus for perforating record cards, [U.S.] Patent 682,197, filed and issued 1901; Herman Hollerith, Tabulating apparatus, [U.S.] Patent 685,608, filed and issued 1901.

116. National Archives, RG 241, entry 9A, Patent Case File 1,830,699; Herman Hollerith, Automatic group control for tabulating machines, [U.S.] Patent 1,830,699, filed 1914 and issued 1931; James Powers, Automatic total-taking mechanism of accounting-machines, [U.S.] Patent 1,245,506, filed 1915 and issued 1917; "Consolidated Patent Act of 1870," U.S. Congress, *Statutes at Large*, vol. 16 (1869–1871), 1198, sections 22, 23.

Chapter Four: The Rise of International Business Machines

1. Saul Engelbourg, *International Business Machines: A Business History* (completed as his Ph.D. diss. 1954; New York: Arno Press, 1976), 60–62; Alfred D. Chandler, *Scale and Scope: The Dynamics of Industrial Capitalism* (Cambridge, MA: Harvard University Press, 1990), 76, 81.

2. Geoffrey Austrian, *Herman Hollerith: Forgotten Giant of Information Processing* (New York: Columbia University Press, 1982), 306–307.

3. Engelbourg, *International Business Machines*, 27–30, 32–52; Emerson W. Pugh, *Building IBM: Shaping and Industry and Its Technology* (Cambridge, MA: MIT Press, 1995), 25–26.

4. Charles R. Flint, *Memories of an Active Life: Men and Ships, and Sealing Wax*, ed., Irvin S. Cobb (New York: G. P. Putnam's Sons, 1923), 312.

5. Computing Tabulating Recording Company (CTR), Executive Committee Meetings, 11 September 1911, 1 February 1912, CTR Minutes, International Business Machines (IBM) Archives.

6. James W. Bruce, Fred M. Carroll, and Clair D. Lake provide examples of inventors working across the four companies that constituted CTR. *The Official Gazette of the United States Patent Office: Register, 1912–1929.*

7. CTR, Executive Committee Meeting, 13 December 1918.

8. Pugh, *Building IBM*, 28.

9. CTR, Executive Committee Meeting, 1 February 1912.

10. CTR Board Meetings, 24 October 1911, 11 January 1913.

11. Copy of agreement between H. Hollerith and CTR, 15 July 1911, box 10, folder 6, Hollerith Papers, Library of Congress (LC).

12. CTR Board Meeting, 24 October 1911; CTR, Executive Committee Meeting, 9 November 1911.

13. M. O. Chance, F. H. Tonsmeire, and E. H. Halding, "Report on the Royden System of Perforated Cards to the United States President's Commission on Economy and Efficiency," 9 March 1912; H. S. McCormack, "Report on Royden-Peirce Machine. Compiled for Mr. Gershom Smith by the Business Burse," 13 May 1912, box A-919-2, IBM Archives; Prospectus for the Royden Company to be established with 24 September 1912, covering letter from the company, box 10, folder 7, Hollerith Papers, LC.

14. F. N. Kondolf to H. Hollerith, 16 February 1914, box 10, folder 8, Hollerith Papers, LC; Herman Hollerith, Card testing device, [U.S.] Patent 1,132,345, filed 1914 and issued 1915.

15. Drawing attached to letter, H. Hollerith to S. G. Metcalf, 23 May 1899, box A-832-1, folder: Corresp./1899, IBM Archives; Herman Hollerith, Registering apparatus, [U.S.] Patent 1,030,304, filed 1904 and issued 1912; Herman Hollerith, Apparatus for use in tabulating systems," [U.S.] Patent 1,109,841, filed 1913 and issued 1914; CTR Board Meeting, 28 October 1913.

16. Herman Hollerith, Registering and recording apparatus for tabulating systems, [U.S.] Patent 1,295,167, filed 1913 and issued 1919.

17. J. M. Hiznay, "Interview with C. D. Lake," 1954, box A-25-3, folder: History, IBM Archives.

18. Chance et al., "Report on the Royden System"; McCormack, "Report on Royden-Peirce Machine"; CTR to H. Hollerith, 8 November 1912, with two enclosures, box 10, folder 7, Hollerith Papers, LC. The McCormack report reverses JoAnne Yates' assertion that Peirce caught Hollerith's attention as early as 1912, whereas the Tabulating Machine Company (TMC) management only reacted in 1914. JoAnn Yates, "Co-evolution of Information-Processing Technology and Use: Interaction between the Life Insurance and Tabulating Industries," *Business History Review*, 67 (1993), 1–51, on 24, 29.

19. Herman Hollerith, Tabulating machine, [U.S.] Patent 1,087,061, filled 1912 and issued 1914; James Powers, Tabulator printer for statistical purposes, [U.S.] Patent 1,245,502, filed 1913 and issued 1917. As early as 1906 Hollerith worked on controlling tabulating operations by punched cards (Herman Hollerith, Tabulating apparatus," [U.S.] Patent 998,095, filed 1906 and issued 1914).

20. Patent Case File 1,830,699, entry 9A, RG-241, National Archives (NA); Herman Hollerith, Automatic group control for tabulating machines, [U.S.] Patent

1,830,699, filed 1914 and issued 1931; "Consolidated Patent Act of 1870," U.S. Congress, *Statutes at Large*, vol. 16 (1869–1871), 198, sections 22, 23.

21. Herman Hollerith, Apparatus for perforating record-cards, [U.S.] Patent 1,110,261, filed and issued 1914; *Historical Record of Card and Machine Development*, IBM 1955, 4, IBM Archives.

22. CTR Executive Committee Meeting, 15 June 1914.

23. CTR Executive Committee Meetings, 23 December 1913, 30 December 1913, and 2 January 1914.

24. CTR, Executive Committee Meeting, 2 January 1914.

25. Engelbourg, *International Business Machines*, 77–78.

26. CTR, Board Meeting, 19 May 1914.

27. Engelbourg, *International Business Machines*, 67.

28. CTR, Board Meetings, 20 April 1914, 19 May 1914, and 15 March 1915.

29. Engelbourg, *International Business Machines*, 81–84; Thomas Belden and Marva Belden, *The Lengthening Shadow: The Life of Thomas J. Watson* (Boston: Little, Brown and Company, 1962), 3–58; Thomas J. Watson Jr. and Peter Petre, *Father, Son & Co.: My Life at IBM and Beyond* (New York: Bantam Books, 1990), 8–12.

30. Samuel Crowther, *John H. Patterson, Pioneer in Industrial Welfare* (New York: Doubleday, Page & Co, 1924), 73–74.

31. Belden and Belden, *The Lengthening Shadow*, 133–134; Gunnar Nerheim and Helge W. Nordvik, *'Ikke bare maskiner': Historien om IBM i Norge 1935–1985* (Oslo: Universitetsforlaget, 1986), 20–21.

32. Belden and Belden, *The Lengthening Shadow*, 135–137; Engelbourg, *International Business Machines*, 188–191; Gerald Bordman, *American Musical Theatre: A Chronicle* (New York: Oxford University Press, 1992), 441–442.

33. George F. Daly, Minute of Discussion with Connolly, 12 October 1953, box TAR-119, IBM Archives.

34. Copy of letter, G. Smith, TMC, to T. J. Watson, CTR, 26 June 1914, box 10, folder 8, Hollerith Papers, LC.

35. CTR, Board Meeting, 25 August 1914; Pugh, *Building IBM*, 39–40.

36. Letter with enclosures, F. N. Kondolf to H. Hollerith, 16 February 1914; E. A. Ford to H. Hollerith, 10 March 1914; E. A. Ford to H. Hollerith, 17 March 1914; E. A. Ford to H. Hollerith, 30 March 1914; letter with enclosure, F. N. Kondolf to H. Hollerith, 7 April 1914; all box 10, folder 8, Hollerith Papers. E. A. Ford to H. Hollerith, 15 June 1915, box 11, folder 2, Hollerith Papers, LC; O. Braimaier to E. A. Ford, 17 March 1916; T. J. Watson to H. Hollerith, 19 December 1916; copy of letter, E. A. Ford to T. J. Watson, 9 December 1916; all in box 11, folder 3, Hollerith Papers, LC. H. Hollerith to C. D. Lake, 1 May 1917;

box 11, folder 4, Hollerith Papers, LC; H. Hollerith to C. D. Lake, 27 November 1917; H. Hollerith to C. D. Lake, 30 November 1917; all in box 11, folder 5; C. D. Lake to H. Hollerith, 5 April 1916, 1 May 1916, 18 June 1916, 27 April 1917, box 21, folder 2, Hollerith Papers, LC.

37. CTR Board Meeting, 23 May 1916; Fred M. Carroll, Computing and printing machine, [U.S.] Patent 1,091,482, filed 1906 and issued 1914, assigned to the White Adding Machine Company, New Haven, CT; Fred M. Carroll, Printing mechanism for cash-registers, [U.S.] Patent 1,245,191, filed 1915 and 1917, assigned to the National Cash Register Company.

38. Pugh, *Building IBM*, 40–42, 45–47.

39. J. M. Hiznay, "Interview with C. D. Lake, 1954," A-25-3, folder history, IBM Archives; "Daly's Poughkeepsie Talk, June 1, 1955," 16, box TAR 119, IBM Archives.

40. Clair D. Lake, Improvement in automatically-controlled tabulating machine, [U.S.] Patent 1,570,264, filed 1921 and issued 1926; Alvin E. Gray, *IBM Development Manual, Book I: Numerical Tabulators*, typescript 1956, 39–41, box A-25-2, folder: History/Num.tab., IBM Archives.

41. Clair D. Lake, Printing tabulator, [U.S.] Patent 1,379,268, filed 1919 and issued 1921; Gray, *IBM Development Manual*, 41–56; John Hayward, "Historical Development of IBM Products and Patents," typescript 1957, 9; box A-25-2, folder: History/Num.tab, IBM Archives.

42. Gray, *IBM Development Manual*, 41, in IBM Archives, box A-25-2 (Endicott Engineering), folder: History/Num.tab; Fred M. Carroll, Listing machine, [U.S.] Patent 1,516,079, filed 1922 and issued 1924 (later machine version with the rotary drum printer).

43. Charles J. Bashe, Lyle R. Johnson, John H. Palmer, and Emerson W. Pugh, *IBM's Early Computers* (Cambridge, MA: MIT Press, 1986), 8–9.

44. C. N. Dutton to F. N. Kondolf, CTR, 1 April 1914; F. N. Kondolf, CTR, to C. N. Dutton, 1 April 1914; both in LC, Hollerith Papers, box 10, Hollerith Papers; CTR Executive Council Meeting, 26 May 1914; CTR Board Meetings, 6 June 1914, 25 July 1916, Executive and Finance Committee Meeting, 7 September 1920; Hayward, *Historical Development*, 101–102.

45. Patent Case File 1,830,699, entry 9A, RG-241, NA; James Powers, Automatic total taking mechanism, [U.S.] Patent 1,245,506, filed 1915 and issued 1917; Charles A. Tripp, Stop card inserting machine, [U.S.] Patent 1,208,051, filed and issued 1916; John Royden Peirce, Distributing machine, [U.S.] Patent 1,316,461, filed 1911 and issued 1919; Herman Hollerith, Automatic group control for tabulating machines, [U.S.] Patent 1,830,699, filed 1914 and issued 1931; Felix Thomas, "Interference Litigation between Hollerith, Peirce, Tripp, and Powers, 1917 to 1924," 68–77, in "Historical Development of IBM Products and

Patents; Accounting Machine Development," ed. John Hayward, IBM typescript, 1948, 1, box A-1091-1, IBM Archives.

46. "Daly's Poughkeepsie Talk," 14

47. CTR Board Meeting, 28 October 1913; Herman Hollerith, Apparatus for perforating record-cards, [U.S.] Patent 1,110,261, filed and issued 1914; *Historical Record of Card and Machine Development,* 4; G. F. Daly, "Historical Data—Electric Key Punch Development," 27 June 1962, box A-25-1, folder: History/calculators, IBM Archives.

48. CTR Executive Committee Meeting, 10 May 1914; Stephen B. Tily, John G. Rehfuss and Martin O. Rehfuss, Card punching machine, [U.S.] Patent 1,061,883, filed 1911 and issued 1914; Hayward, *Historical Development,* 101.

49. Contract between J. T. Schaaff and TMC, 17 January 1918, in Smithsonian, object no. 191.3180.04; "The Detailed Explanation of Exclusive Licenses: Schaaff" (1953), "Further Notes on Discussion with Thomas Re. Early History" (1953), and "Daly's Discussion with Connolly October 12, 1953," in IBM, box TAR-119 (Daly) IBM Archives; G. F. Daly, "Historical Data."

50. *Historical Record of Card and Machine Development,* 5, 9.

51. *Historical Record of Card and Machine Development,* 4; Bashe, Johnson, Palmer, and Pugh, *IBM's Early Computers,* 7–8.

52. *Historical Record of Card and Machine Development,* 5.

53. Belden and Belden, *The Lengthening Shadow,* 1962, 96.

54. Executive and Finance Committee Meeting, 21 July 1914; Executive Committee Meeting, 15 December 1914; in IBM, CTR minutes. The agreement was for fifteen years.

55. Belden and Belden, *The Lengthening Shadow,* 1962, 95–97.

56. Martin Campbell-Kelly, *ICL: A Business and Technical History* (Oxford: Oxford University Press, 1989), 35, parallels the German and U.S. licenses but has no reference to the German rate.

57. Belden and Belden, *The Lengthening Shadow,* 97; Watson and Petre, *Father, Son & Co.,* 141–142, 215–217.

58. CTR Executive Committee Meetings, 28 January 1919, 21 June 1922, 7 June 1922, 26 September 1922; Executive and Finance Committee Meetings, 26 October 1920, 8 June 1926, 8 September 1926.

59. "Petition," filed 26 August 1935, in *Remington Rand v. IBM* at Supreme Court of New York, Equity 66-215, District Court of the Southern District of New York, NA, Northeast Region, New York.

60. Chandler, *Scale and Scope.*

61. "Memo for Mr. Watson," 9 June 1922; copy of letter, F. G. Idler, Prudential Insurance Company, to CTR, 5 February 1924; J. R. Peirce, CTR, to F. G. Idler, Prudential Insurance Company; all in box A-163-3, IBM Archives.

62. CTR Executive Committee Meeting, 5 January 1922.

63. John Royden Peirce, Record sheet and apparatus controlled thereby, [U.S.] Patent 1,506,382, filed 1922 and issued 1924 (probably this is the Veterans Bureau machine); H. J. Ekelund, J. A. Ruth, and F. Thomas, "Memorandum dated October 29, 1952, on Alphabetical Accounting Machines," in Hayward, "Historical Development," 52–53.

64. Ekelund, Ruth, and Thomas, "Memorandum on Alphabetical Accounting Machine," 52–53; John Royden Peirce, [U.S.] Patent 1,506,382 (see n 63 above); John Royden Peirce, Perforated card controlled machine, [U.S.] Patent 1,750,191, filed 1927 and issued 1930; John Royden Peirce, Perforated record-controlled accounting machine, [U.S.] Patent 1,946,915, filed 1929 and issued 1934; "Engineering Report July-August 1927," 2.5, box A-163-3, IBM Archives.

65. Fred M. Carroll, Card-controlled printing machine, [U.S.] Patent 1,623,163, filed 1922 and issued 1927; George F. Daly and Ralph E. Page, Tabulating machine, [U.S.] Patent 1,762,145, filed 1925 and issued 1930; Pugh, *Building IBM,* 47–48.

66. C. D. Lake, Record sheet for tabulating machines, [U.S.] Patent 1,772,492, filed 1928 and issued 1930.

67. Felix Thomas, "Historical Review," in Hayward, "Historical Development," 26–27; Gray, *IBM Development Manual,* 59–62.

68. Letter printed by use of this tabulator, C. D. Lake to T. J. Watson Sr., 4 October 1928, box A-163-3, IBM Archives.

69. Sherwin K. Decker and Gordon Roberts, Accounting means, [U.S.] Patent 1,916,969, filed 1929 and issued 1933; Sherwin Decker and Gordon Roberts, Accounting means, [U.S.] Patent 1,930,253, filed 1929 and issued 1933; Einar Lawrence Kirkegaard, Tabulator, [U.S.] Patent 2,059,797, filed 1930 and issued 1936; Gray, *IBM Development Manual,* 38–39, 65–67; memorandum with enclosures by G. F. Daly, 20 December 1961, box A-25-1, folder: History/Calculators, IBM Archives.

70. Boris Emmet and John E. Jeuck, *Catalogues and Counters: A History of Sears, Roebuck and Company* (Chicago: University of Chicago Press, 1950), 319–320.

71. "Daly's Poughkeepsie Talk," 24–25.

72. Darmstätter und Nationalbank, Auswechselungseinrichtung für die Typenträger an schribenden Lochkarten-Machinen, [German] Patent 452,233, filed and issued 1927; Wilhelm August Hoffmann, Type holding appliance, [U.S.] Patent 1,853,211, filed 1927 and issued 1930.

73. Regarding the later model, IBM Type 375: George F. Daly, Tabulating machine, [U.S.] Patent 2,107,143, filed 1930 and issued 1938; George F. Daly and Jonas E. Dayger, Billing machine, [U.S.] Patent 1,954,041, filed 1931 and issued

1934; Ekelund, Ruth, and Thomas, "Memorandum on Alphabetical Accounting Machines," in *Historical Development,* 53–54.

74. Pugh, *Building IBM,* 50, dates RR's announcement to 1930. The correct year was 1929.

75. Ekelund, Ruth, and Thomas, "Memorandum on Alphabetical Accounting Machines," 54–55.

76. Ekelund, Ruth, and Thomas, "Memorandum on Alphabetical Accounting Machines," 55–62.

77. Lars Heide, *Hulkort- og edb i Danmark, 1911–1970* (Århus, Denmark: Systime 1996), 40.

78. Ekelund, Ruth, and Thomas, "Memorandum on Alphabetical Accounting Machines," 59–60.

79. J. W. Bryce, Multiplying machine, [U.S.] Patent 2,178,950, filed 1928 and issued 1939; J. W. Bryce, Calculating machine, [U.S.] Patent 2,271,248, filed 1928 and issued 1942; J. W. Bryce, Multiplying machine, [U.S.] Patent 2,178,951, filed 1931 and issued 1939; Hayward, "Historical Development," 116, 121; Bashe, Johnson, Palmer, and Pugh, *IBM's Early Computers* 1986, 14–15.

80. Gustav Tauschek, Maschine zur rechnerischen Auswertung von Lochkarten, [German] Patent 459,168, filed 1926 and issued 1928.

81. Hartmut Petzold, *Rechnende Machinen. Eine historische Untersuchung ihrer Herstellung und Anwendung vom Keiserreich bis zur Bundesrepublik* (Düsseldorf: VDI Verlag, 1985), 221–222.

82. IBM Executive and Finance Committee Meeting, 30 September 1930; Hayward, "Historical Development," 104–106, 116–122.

83. "Endicott Laboratory: Research and Invention Budget for 1935," summary, box A-163-1, IBM Archives; Gustav Tauschek, Printing mechanism, [U.S.] Patent 2,010,652, filed 1933 and issued 1935; Gustav Tauschek, Machine for simulating hand-writing, [U.S.] Patent 2,049,675, filed 1935 and issued 1936; Gustav Tauschek, Printing mechanism for tabulating machines, [U.S.] Patent 2,066,748, filed 1935 and issued 1937; Gustav Tauschek, Sound recording and reproducing machine, [U.S.] Patent 2,089,309, filed 1933 and issued 1937; Gustav Tauschek, Tabulating machine, [U.S.] Patent 2,093,529, filed 1932 and issued 1937; Gustav Tauschek, Tabulating machine, [U.S.] Patent 2,113,634, filed 1935 and issued 1938; Gustav Tauschek, Luminous sign, [U.S.] Patent 2,20,378, filed 1934 and issued 1938; Gustav Tauschek, Tabulating machine, [U.S.] Patent 2,182,006, filed and issued 1934.

84. Gustav Tauschek, Record filing and selecting apparatus, [U.S.] Patent 2,013,012, filed 1932 and issued 1935; Gustav Tauschek, Reading machine, [U.S.] Patent 2,115,563, filed 1932 and issued 1938.

85. H. S. McCormack, "Keeping Books by Machine: The Punched Card as a Saver of Brain Energy," *Scientific American*, 108 (1913): 194–195, on 195.

86. Thomas P. Hughes, *American Genesis: A Century of Invention and Technological Enthusiasm* (New York: Penguin, 1989), 53–55.

Chapter Five: Decline of Punched Cards for European Census Processing

1. H. Hollerith to Census Office, 10 March 1889, printed copy, box 34, folder 7, Hollerith Papers, Library of Congress (LC).

2. Geoffrey Austrian, *Herman Hollerith: Forgotten Giant of Information Processing* (New York: Columbia University Press, 1982), 47.

3. Herman Hollerith, Apparat zur Ermittlung statistischer Ergebnisse und zum Sortieren von Zählkarten, [Austrian] Patent 39/2707, filed and issued 1889, K. K. Privelege Archiv, Vienna; Herman Hollerith, Improvements in the method and apparatus for compiling statistics, [British] Patent, 1889:376, filed and issued 1889); Herman Hollerith, Procédé et appareil permettant d'obtenir des résultats statistiques de toute nature, et de classer les cartes de comptage d'une manière exacte, [French] Patent 195,265, filed and issued 1889, Archives d'Institut national de la propriété industrielle, Paris; Herman Hollerith, Verfahren und Apparat zur Ermittlung statistischer Ergebnisse und zur Sortierung von Zählkarten, [German] Patent 49,593, filed and issued 1889.

4. Herman Hollerith, Hilfsventiel für Luftdruck- und Vakuumbremsen, [German] Patent 45,524, filed 1887 and issued 1888; Herman Hollerith, Elektrisch betriebenes Ventil für Luftdruck- und Vakuumbremsen, [German] Patent 45,525, filed 1887 and issued 1888.

5. "Privilegiengesetz," 15 August 1852, [Austrian] *Reich Gesetz Blatt* 184 (1852), section 3; "Loi sur les brevets d'invention," *Le Moniteur Belge* 25(145; 25 May 1854), 1622, Article 25; the French "Loi du 5 juillet 1844 sur les brevets d'invention," clause 32; the French "Loi du 20 mai 1856 modifiant l'art. 32 de la loi du 5 juillet 1844 sur les brevets d'invention."

6. Austrian, *Herman Hollerith*, 48–49.

7. Adolf Adam, *Vom himmlischen Uhrwerk zur statistischen Fabrik. 600 Jahre Entdeckungsreise in das Neuland Österreicher Statistik und Datenverarbeitung* (Vienna: Verlag Herbert O. Munk, 1973); Heinz Zemanek, "Otto Schäffler. Ein vergessener Österreicher. Die Biographie eines genialen Unternehmers und Erfinders," *Österreichischer Gewerbeverein. Jahrbuch*, 92 (1974), 71–92.

8. Zemanek, "Otto Schäffler," 88, wrongly asserts that Schäffler's approach caused Hollerith to file an Austrian patent application. The application was filed on 8 January 1889, Herman Hollerith, Apparat zur Ermittlung statistischer Ergeb-

nisse und zum Sortieren von Zählkarten, [Austrian] Patent 39/2707, filed and issued 1889, K. K. Privelege Archiv.

9. "Privilegiengesetz," 15 August 1852, [Austrian] *Reich Gesetz Blatt* 184 (1852), sec. 3.

10. The Austrian census was processed starting October 1891. Heinrich Rauchberg, "Die elektrische Zählmaschine und ihre Anwendung insbesondere bei der Österreicher Volkszählung," *Allgemeines Statistische Archiv*, 2 (1892): 78–126, on 79; Heinrich Rauchberg, "Erfahrungen mit den elektrischen Zählmaschinen," *Allgemeines Statistisches Archiv*, 4 (1896): 131–163; V. Klepacki, "Die Hollerith'sche elektrische Zählmaschine für Volkszählungen," *Polytechnische Zentralblatt*, Berlin, 57:11 (1896): 121–125, on 121.

11. Otto Schäffler, Neuerungen an statistischen Zählmachinen, [Austrian] Patent 46/3182, filed and issued 1895, K. K. Privelege Archiv.

12. Schäffler was never granted a patent in the United States for plugboard programming.

13. Anders Nicolai Kiær (Norway), Nicolas Troïnitsky (Russia), Victor Turquan (France) participated according to *Bulletin de l'Institut International d'Statistique*, 6 (1892): xxx; Frédéric Probst, "Description de la machine électrique servant au dépouillement du recensement autrichien de 1890," *Bulletin*, 19–25.

14. A. N. Kiær's report dated 30 October 1893, on his voyage to the United States, "Kopiboka til Statsitisk Sentralbyrå for 1893," 1426–1436, on 1432, Statistisk Sentralbyrås arkiv, Riksarkivet, Oslo; Signy Arctander, "Den offisielle statistikks historie," *Arbeidsnotater. Statistisk Sentralbyrå*, No. IB 70/1 (1970), 61–62, 73–74, 75–77.

15. *Stortingstidende, Stortinget*, 44. *samling, 1895, Kongeriget Norges ordinære stortingsforhandlinger*, 965–966.

16. A. N. Kiær, "En reform i den norske handelsstatistik. Elektricitet anvendt i statistikkens tjeneste," *Statsøkonomisk tidsskrift*, 1910, 234–243.

17. Austrian, *Herman Hollerith*, 115–163; Henning Bauer, Andreas Kappeler, and Brigitte Roth, ed., *Die Nationalitäten des Russischen Reiches in der Volkszählung von 1897* (Stuttgart, Germany: Franz Steiner Verlag, 1991), vol. A, 9, 54–87.

18. Klepacki, "Die Hollerith'sche elektrische Zählmaschine," 221.

19. *Statistique générale de la France. Histoire et travaux de la fin du XVIIe siècle au début du XXe* (Paris: Imprimerie National, 1913), 17.

20. Robert Ligonnière, *Préhistoire et histoire des ordinateur, Des origines du calcul aux premiers calculateurs électronique* (Paris: Robert Laffont, 1987), 143–144, Béatrice Touchelay, *L'I.N.S.E.E. Des origines à 1961: Évolution et relation avec la réalité économique, politique et sociale* (Thèse de Doctorat en Histoire Économique présentée à la Faculté de Lettre de Paris XII, Paris, 1993), 20–28. No

part of the Statistique générale de la France archive from this period has survived.

21. Lucien March, "Les procédés du recensement des industries et professions en 1896," *Société des Ingénieurs civils de France. Bulletin*, 1899, 396–424; Lucien March, Système de dépouillement de renseignements collectifs, sans transcription préalable, sur des cartes perforées, dit classicompteur, [French] Patent 288,628, filed and issued 1899, 1; both in Archives d'INPI, Paris.

22. *Die Volkszählung am 1. Dezember 1890 im Deutschen Reich, Statistik des Deutschen Reichs*, new series, 68 (1894), 7*.

23. Georg von Mayr, "Die Einrichtung der Bevölkerungsaufnahme vom 1. Dezember 1890 in den größeren deutschen Staaten," *Allgemeines Statistisches Archiv*, 2 (1892): 349–368; Georg von Mayr, "Die für die deutsche Volkszählung von 1. Dezember 1890 und deren reichsstatistische Ausbeutung getroffenen Bestimmungen," *Allgemeines Statistisches Archiv*, 1 (1890): 371–398.

24. Klepacki, "Die Hollerith'sche elektrische Zählmachine for Volkszählungen," 124–125.

25. Separate censuses were conducted in Ireland and Scotland.

26. A separate census institution was established in 1920. *Guide to Census Reports. Great Britain 1901–1966* (London: His Majesty's Stationery Office [HMSO], 1977), 1–4, 11–23.

27. Martin Campbell-Kelly, "Information Technology and Organizational Change in the British Census, 1801–1911," *Information Systems Research* 7 (1996): 22–36.

28. 1890 figure. Separate censuses for Bosnia, Herzegovina, and Hungary.

Chapter Six: Punched Cards for General Statistics in Europe

1. Cecilia Pahlberg, *Subsidiary: Headquarters Relationships in International Business Networks* (doctoral thesis, Uppsala University, 1996).

2. "Robert Percival Porter (1852–1917)," *Dictionary of American Bibliography*; "Death of Mr. R.P. Porter," *The Times*, 1 March 1917, 10:4; *New York Times*, 1 March 1917, 13:5; *New York Times*, 2 March 1917, 10:4.

3. Martin Campbell-Kelly, *ICL: A Business and Technical History* (Oxford: Oxford University Press, 1989), 13–14.

4. Geoffrey Austrian, *Herman Hollerith: Forgotten Giant of Information Processing* (New York: Columbia University Press, 1982), 212–213; Campbell-Kelly, *ICL*, 16–18.

5. Campbell-Kelly, *ICL*, 19–20, 27–28. The 1908 contract is published on 387–391. On 26, Campbell-Kelly wrongly asserts a 25% royalty of total revenues. There was no royalty charged on sale of punches or punched cards.

6. B.T.M. Co.: *Rentals and Accounts 1916–1919*, National Archives for the History of Computing (NAHC), Manchester, England; figures from 1908 and 1914 in Campbell-Kelly, *ICL*, 24, 27.

7. Computing Tabulating Recording Company (CTR) Board Meeting, 25 July 1912, CTR Board Minutes, International Business Machine (IBM) Archives, New York; Campbell-Kelly, *ICL*, 33–34, 37–38.

8. C. A. E. Greene, *The Beginnings*, 1958, 1–7, ICC/A1k, NAHC; Campbell-Kelly, *ICL*, 17–18.

9. Greene, *The Beginnings*, 9, 11–14, 21–22; "The Hollerith Tabulating System," *The Engineer*, 1934, 274; *The British Tabulating Machine Company Limited, 1907–1957*, 3, typescript, NAHC.

10. Greene, *The Beginnings*, 8–11, 17; "The British Tabulating Machine Company Limited, 1907–1957," 3, typescript, NAHC; "B.T.M. Co.: Rentals and Accounts 1916–1919," NAHC.

11. "B.T.M. Co.: Rentals and Accounts"; Greene, *The Beginnings*, 11–13.

12. Registrar General, "Memorandum re. Census of 1901," no date, Census returns: Correspondence and papers, 1908, RG 19/45, Public Record Office (PRO), London.

13. "The Tabulator," *The Engineer*, 1911, 96–97, 146–148, 196–197, 279–280; G. H. Baillie, Adding &c. machines, [British] Patent 29,686, filed and issued 1910; C. A. E. Greene, Tabulating systems, [British] Patent 10,681, filed and issued 1910; H. J. Norballe, Counting-machines, [British] Patent 10,918, filed and issued 1910; British Tabulating Machine Co., Sorting statistics &c., [British] Patent 28,544, filed and issued 1913; Report, 15 November 1910, on purchase of tabulating machines on behalf of Census Office, by R. Bailey, and supplemental report from 29 November 1910, STAT 12/10/9, Public Record Office; Treasury Chambers to Controller, HMSO, 7 January 1911, STAT 12/35/20, PRO, London; *Guide to Census Reports: Great Britain 1801–1966* (London: HMSO, 1977), 24.

14. Campbell-Kelly, *ICL*, 30, 38. I applied in vain for access to the minutes of the board meetings and other board documents from the British Tabulating Machine Company and the British Powers Company, now both in custody of International Computers Limited (ICL). Martin Campbell-Kelly had access to these papers when he wrote his book, and he kindly offered his photocopies for my work.

15. "Report on the 'Powers' Accounting and Tabulating Machines," June 1915, and Controller, HMSO, to Treasury, 25 June 1915, T-1/11800, PRO.

16. "Powers-Samas Accounting Machines Com Ltd. in Vickers Limited. A Short Account of the Subsidiary Companies," Document No. 771, Vickers Archives, Cambridge University Library. British Tabulating Machine Company (BTM) figures from "B.T.M. Co. Rentals and Accounts."

17. Herman Hollerith, Improvements in tabulating machines, [British] Patent Specification 117,985, filed and issued 1918; James Powers, Improvement in automatic total taking means for recording tabulating machines, [British] Patent Specification 104,703, filed 1916 and issued 1917; Herman Hollerith, Automatic group control for tabulating machines, [U.S.] Patent 1,830,699, filed 1914 and issued 1931.

18. Campbell-Kelly, *ICL*, 89–90. Unlike the present author, Campbell-Kelly had access to the board minutes and papers of the two British punched card companies.

19. Herman Hollerith, Automatic group control for tabulating machines, [U.S.] Patent 1,830,699, filed 1914 and issued 1931.

20. Herman Hollerith, Improvements in tabulating machines, [British] Patent 117,985, filed and issued 1918.

21. The [UK] Patents and Designs Act, 1907, 7 Edward 7, c. 29, sec. 13, 17.

22. The [UK] Patents and Designs Act, 1907, 7 Edward 7, c. 29, sec. 27. The [UK] Patents and Designs (Convention) Act, 1928, 18 & 19 George 5, c. 80, sec. 2, reduced the time limit to three years.

23. Herman Hollerith, Tabulating apparatus, [U.S.] Patent 685,608, filed and issued 1901. See the [UK] Patents Act, 1902, 2 Edward 7, c. 34, sec. 3.5.

24. The [UK] Patents and Designs Act, 1907, 7 Edward 7, c. 29, sec. 7.

25. *Statistics of the Military Effort of the British Empire during the Great War, 1914–1918* (London: HMSO, 1922), 30–32.

26. Martin Campbell-Kelly, "Large Scale Data Processing in Prudential, 1850–1930" [in England], *Accounting, Business and Financial History* 2 (1992): 117–139; Martin Campbell-Kelly, "The Railway Clearing House and Victorian Data Processing," in *Information Acumen: The Understanding and Use of Knowledge in Modern Business*, ed., Lisa Bud-Frierman (London: Routledge: 1994), 51–74; Martin Campbell-Kelly, "Data Processing and Technological Change: The Post Office Savings Bank, 1861–1930," *Technology and Culture* 39 (1998): 1–32.

27. See, for example, various reports and correspondences 1912–1914, STAT 12/14/1 and STAT 12/16/3, PRO.

28. Controller, HMSO, to Treasury, 25 June 1915, T 1/11800, PRO.

29. *History of the Ministry of Munitions*, 12 vols., London, 1921–1922, vol. 1, part 2, 57–80.

30. "B.T.M. Co.: Rentals and Accounts"; *History of the Ministry of Munitions*, vol. 9, part 1, 22–25, 55.

31. "B.T.M. Co.: Rentals and Accounts."

32. Company size according to assets in 1913, Alfred D. Chandler, *Scale and Scope: The Dynamics of Industrial Capitalism* (Cambridge, MA: Harvard University Press, 1990), 698.

33. W. Heidinger to Zwangsverwalter, 13 June 1918, 1–2, *Anlage No. 3* to "Zwangsverwaltung der feindlichen Beteiligung an der Deutschen Hollerith GmbH in Berlin, 1918–1932," file I.HA/Rep. 120 C VIII 1 Nr.84 adhib. 18 Beiheft 2104, Geheimes Staatsarchiv, Berlin (GS); H. Goerlitz, *Zur Geschichte der Powers-Maschine*, 1934, History Archive of IBM Deutschland; Hartmut Petzold, *Rechnende Maschinen. Eine historische Untersuchung ihrer Herstellung und Anwendung vom Kaiserreich bis zur Bundesrepublik* (Düsseldorf: VDI Verlag, 1985), 197–198.

34. Agreement of 22 November 1910, between the Tabulating Machine Company (TMC) and Deutsche Hollerith Maschinen Gesellschaft (Dehomag), mbH, B95/95, Wirtschaftsarchiv Baden-Württemberg; "An die Gesellschafter der Deutschen Hollerith Maschinen Gesellschaft m.b.H.," 7 April 1913, B95/89, Wirtschaftsarchiv Baden-Württemberg.

35. CTR, Board Meeting, 25 June 1911; CTR, Executive Committee Meetings, 9 September 1913, 14 and 21 October 1913, 30 January 1914; agreement between TMC and Deutsche Hollerith-Maschinen Gesellschaft m.b.H. (Dehomag), 24 November 1919, B95/95, Wirtschaftsarchiv Baden-Württemberg.

36. Processing census returns was a responsibility of the states.

37. *Festschrift zur 25–Jahresfeier der Deutschen Hollerith-Maschinen Gesellschaft* (Berlin: Dehomag, 1935), 69.

38. Alfred D. Chandler, *Scale and Scope: The Dynamics of Industrial Capitalism* (Cambridge, MA: Harvard University Press, 1990), 476–477.

39. Dehomag to Der Treuhänder für das feindliche Vermögen, 18 March 1918; Dr. Lucas, "Mechanisches Abrechnungsverfahren in der Fabrikbuchhaltung," manuscript, 24 July 1918; both in I.HA/Rep. 120 C VIII 1 No. 84 adhib. 18 Beiheft 2104, GS; *Hollerith Mitteilungen* 1 (May 1912), 21–22, 2 (December 1912), 84–86, 4, 1–6, 16–31; letter, "An die Gesellschafter der Deutschen Hollerith Maschinen Gesellschaft m.b.H.," 7 April 1913, B95/89, Wirtschaftsarchiv Baden-Württemberg.

40. *Hollerith Mitteilungen*, 1 (1912); R. N. Williams, "Die Hollerithsche Sortier- und Addiermaschine und ihre Verwendung im Eisenbahndienste," *Zeitung des Vereins deutscher Eisenbahnverwaltungen* 52 (1912), 195–199, 211–214.

41. *Hollerith Mitteilungen*, 2 (1912).

42. Pavel, "Zur Feststellung des Gütersolls durch Verkehrskontrollen," *Zeitung des Vereins deutscher Eisenbahn Verwaltungen*, 1912, 470; Otto Müller, "Das Lochkartenverfahren und seine Verwendung im Eisenbahndienst," *Zeitschrift für Verkehrswissenschaft*, 3 (1925): 222–223; *Festschrift*, 68–70.

43. *Hollerith Mitteilungen* 4 (1914), 3–6.

44. R. Lorant, "Maschinelle Kontokorrentbuchhaltung bei Großbanken," *Organisation* (Berlin) 9 (1925), 403–410; Max Zander, "Die Lochkarte im Bankbetrieb," *Die Lochkarte und das Powers-System* 10 (1932), 69–76.

45. Danish examples were the Danish Statistics (Statistisk Departement) and the Danish State Railways (Statsbanerne), Lars Heide, *Hulkort- og edb i Danmark, 1911–1970* (Århus, Denmark: Systime 1996), 36.

46. Polizei-Präsident, Abteilung I, to Minister for Trade and Industry, 11 April 1918, regarding the Deutsche Gesellschaft für Addier- und Sortiermaschinen, and Geschäftsaufsicht über die Deutsche Gesellschaft für Addier- und Sortiermaschinen, June 15, 1918, both in I/HA/Rep. 120 C VIII 1, No. 84 adhib. 18 Beiheft 2106, GS; CTR, Executive Committee Meeting, 30 January 1914; CTR Executive Committee Meeting Minutes, IBM; Goerlitz, *Zur Geschichte der Powers-Maschine*.

47. Goerlitz, *Zur Geschichte der Powers-Maschine*, cites a letter from Dehomag to their customers noting the declining importance of a printing tabulator.

48. *Blatt für Patent-, Muster- und Zeichenwessen* 23 (1917): 33–41. The two contested patents were H. Hollerith, Vorrichtung zum Sortieren von Zählkarten für statistische Angabe u.dgl., [German] Patent 139,022 (1903, valid from 1901), and H. Hollerith, Statistische Maschine mit Stiftkasten, dessen bewegliche Stifte bei der Bearbeitung der Angaben durch Lochungen einer Angabekarte hindurchtreten Order von der Karte zurückhalten werden, [German] Patent 228,316 (1910, valid from 1909). Patent 139,022 expired in 1916, and Patent 228,316 expired in 1924. (Hollerith's German Patent 139,022 corresponded to his U.S. Patent 685,608 of 1901.) The German patents were valid for fifteen years from the date of filing ("Patentgesetz" from 7 April 1891, *Reichs-Gesetzblatt*, Berlin, 1891:12, article 7).

49. Goerlitz, *Zur Geschichte der Powers-Maschine*.

50. Stephan H. Lindner, *Das Rechtskommissariat für die Behandlung feindlichem Vermögens im Zweiten Weltkrieg* (Stuttgart, Germany: Franz Steiner Verlag, 1991), 16–18. During the First World War, this surveillance was a state assignment. The archives of the custody of Dehomag and the German Powers agency are in GS.

51. Report, Regarding the Geschäftsaufsicht über die Deutsche Gesellschaft für Addier- und Sortiermaschinen, 15 June 1918; F. Krause, Zwangsverwalter, Schlussbericht über die Deutsche Gesellschaft für Addier- & Sortiermaschinen, 1 August 1919; P. Roeder, Zwangsverwalter, to Minister for Trade and Industry, 15 September 1923; all three in I./HA/Rep. 120 C VIII 1, No. 84 adhib. 18 Beiheft 2106, GS.

52. Karl Hardach, *Wirtschaftsgeschichte Deutschland im 20. Jahrhundert* (Göttingen, Gemany: Vandenhoech & Ruprecht, 1976), 20–21.

53. Gennerich, "Erste vollständige Kontrollbuchhaltung mit Hollerith in Deutschland 1915–1922," (1950) [manuscript], B95/88, Wirtschaftsarchiv Baden-Württemberg.

54. Krause, *Bericht*, 24 June 1918.

55. Note, Treuhänder für das feindliche Vermögen to Minister für Handel

und Gewerbe, March 25, 1918, I./HA/Rep. 120 C VIII 1, Nr. 84 adhib. 18 Beiheft 2104, GS.

56. W. Heidinger to F. Krause, 13 June 1918, I.HA/Rep. 120 C VIII 1 Nr.84 adhib. 18 Beiheft 2104, GS; *Festschrift*, 11; Petzold, *Rechnende Maschinen*, 205.

57. H. Tolle and F. Bauer, Einstellvorrichtung für Rechenmaschinen und ihrliche Maschine, [German] Patent 316,962, filed 1914 and issued 1920; H. Tolle, Kontaktvorrichtung für elektromotorisch angetriebene Maschinen mit begrenztem Stellweg, [German] Patent 300,935, filed 1915 and issued 1917; H. Tolle, Vorrichtung für elektrisch angetriebene Maschinen mit begrenztem Arbeitsweg, [German] Patent 306,519, filed 1915 and issued 1918; H. Tolle, Maschine zum Sortieren gelochter Karten für statistische Zwecke mit einer Abtastvorrichtung für nur eine beliebig einstellbare Kartenspalte, [German] Patent 324,110, filed 1916 and issued 1921; *Festschrift*, 11, 18.

58. H. Tolle, Tastenmaschine mit Kraftantrieb, [German] Patent 335,645, filed 1917 and issued 1921; H. Tolle, Schreib-, Druck- u.dgl. Maschine mit Kraftantrieb, Zusatz zu Patent. 335,645, [German] Patent 345,366, filed 1920 and issued 1921.

59. See, for example, H. Tolle, Zehnerschuttwerk für Zählwerke, [German] Patent 346,478, filed 1920 and issued 1922; H. Tolle, Einrichtung zur Steuerung von statistischen und ähnlichen Maschinen durch zwischen Walzpaaren hindurchgeführte Lochkarten oder Lochstreifen, [German] Patent 367,982, filed 1920 and issued 1923; H. Tolle, Zählwerk für Addition und Substraktion mit Wendegetriebe, [German] Patent 375,563, filed 1920 and issued 1923; H. Tolle, Typensegmentschreibvorrichtung für Rechenmaschinen, [German] Patent 371,085, filed 1920 and issued 1923; H. Tolle, Sortiermaschine, besonders für gelochte Zählkarten, [German] Patent 375,189, filed 1921 and issued 1923); Dehomag, Tabelliermaschine mit Kraftantrieb, [German] Patent 391,467, filed 1922 and issued 1924.

60. *Festschrift*, 11.

61. F. Krause, Bericht No. 1, 24 June 1918; W. Heidinger, "Bericht," 18 June 1943, R87/6248, Akte 921/45, Bundesarchiv, Berlin. In this document, Heidinger purported that the 1922 debt was 450 trillion marks, which was based on a later exchange rate, probably from August or September 1923. Exchange rates from Jürgen Schneider, Oskar Schwarzer, Friedrich Zellfelder, ed., *Währungen der Welt*, vol. 2: *Europäische und Nordamerikanische Divisenkurse 1914–1951* (Stuttgart, Germany: In Kommission bei F. Steiner, 1997).

62. Gunnar Nerheim and Helge W. Nordvik, *'Ikke bare maskiner': Historien om IBM i Norge 1935–1985* (Oslo: Universitetsforlaget, 1986), 26–27

63. CTR Special Executive and Finance Committee Meeting minutes, 15 June 1922, and CTR Executive and Finance Committee Meeting minutes, 13 February

1923, in IBM Archives; *Festschrift,* 12; W. Heidinger, "Bericht," 18 June 1943, 313–314.

64. Edwin Black, *IBM and the Holocaust: The Strategic Alliance between Nazi Germany and America's Most Powerful Corporation* (New York: Crown, 2001), for example, the notes on 432–435.

65. Agreement between TMC and Dehomag, 24 November 1919, B95/95, Wirtschaftsarchiv Baden-Württemberg; CTR Special Executive and Finance Committee Meeting, 15 June 1922, CTR Executive and Finance Committee Meeting, 13 February 1923; *Festschrift,* 12; W. Heidinger, "Bericht," 18 June 1943.

66. Michel Huber, "Lucien March, 1859–1933," *Journal de la Société de statistique de Paris* 74 (1933): 269–280, on 269; Michel Huber, "Quarante années de la statistique générale de la France, 1896–1936," *Journal de la Société de statistique de Paris* 78 (1937): 179–213, 179–182.

67. Eda Kranakis, *Constructing a Bridge: An Exploration of Engineering Culture, Design, and Research in Nineteenth-Century France and America* (Cambridge, MA: MIT Press, 1997), 217–219, 220–223.

68. Huber, "Lucien March"; Alain Desrosières, *La politique des grand nombres: Hhistoire de la raison statistique* (Paris: Éditions de la découverte, 1993), 197–198.

69. Lucien March, Système de dépouillement de renseignement collectifs, sans transcription préalable, sur des cartes perforées, dit classi-compteur, [French] Patent 288,618, filed and issued 1899; Lucien March, Une classi-compteur-imprimeur (système March), [French] Patent 303,965, filed and issued 1900, both in Archives d'INPI; Michel Huber, *Cours de démographie et de statistique sanitaire, I: Introduction à l'étude démographiques et sanitaires* (Paris: Harmann et Cie., 1938), 50–51; Robert Ligonnière, *Préhistoire et histoire des ordinateurs* (Paris: Robert Laffont, 1987), 144–146.

70. Pietra Gaetano, "La prima classificatrice meccanica è stata ideata da un italiano," *Barometro economico italiano,* 1934, 461–465; Huber, *Cours de démographie,* 49–50.

71. *Résultats statistiques du Dénombrement de 1891* (Paris: Imprimerie Nationale, 1894), table 7; *Résultats statistiques du recensement générale de la population effectué le 5 mars 1911* (Paris: Imprimerie Nationale, 1913), 1:1, table 4.

72. Huber, "Quarante années," 206.

73. Huber, "Quarante années, " 181–182, 206; Alfred Sauvy, "Statistique Générale et Service National de Statistique de 1919 à 1944," *Journal de la Société de statistique de Paris* 116 (1975): 34–43, on 35. Only a minor part of the archives of Statistique générale de la France from the 1930s and 1940s is preserved. Information by Alain Desrosières, Institut National de la Statistique et des Études Économiques (INSEE), in May 1998.

74. Report on the classi-compteur machine in the French Statistical Bureau, 7 October 1912; Ministère de l'Intérieur, Bruxelles, to HMSO, London, 4 November 1912; both STAT 12/16/3; Jan van der Ende, *Knopen, kaarten and chips. De geschiedenis van de automatiseing bij het Centraal Bureau voor de Statistiek* (Voorburg/Heerlen, the Netherlands: Centraal Bureau voor de Statistiek, 1991), 28–29.

75. Emil Borel, "La statistique et l'organisation de la présidence du conseil ministres," *Journal de la Société de statistique Paris* 61 (1920): 9–13; Huber, "Quarante années," 180–207; Henri Brunle, "A propos de l'article de A. Sauvy . . . ," *Journal de la Société de statistique de Paris* 116 (1975): 244–246; Béatrice Touchelay, *L'INSEE: Des origines à 1961: Évolution et relation avec la réalité économique, politique et sociale* (Thèse de Doctorat en Histoire Économique présentée à la Faculté de Lettre de Paris XII, Paris, 1993), 24–37; Mireille Moutardier, ed., *Cinquante ans d'INSEE . . . ou la conquête du chiffre* (Paris: INSEE, 1996), 19–25.

76. Jean-Charles Asselain, "La stagnation économique," in *Entre l'état et le marché: L'économie française des années 1880 à nos jours*, ed. Maurice Lévy-Leboyer and Jean-Claude Casanova (Paris: Gallimard, 1991), 199–250; Maurice Lévy-Leboyer, Introduction, *Entre l'état et le marché*, 251–255; Michel Lescure, "L'intervention d'État: Myths et realités. La manque de ressources: Les années 1880–1935," *Entre l'état et le marché*, 256–274.

77. Gerd Hardach, *Der Erste Weltkrieg* (Munich: Deutsche Taschenbuch Verlag, 1973), 96–101, 143–144.

78. Lescure, "La manque de ressources," 267–273

79. Lescure, "La manque de ressources," 256.

80. Aimée Motet, *Les logiques de l'entreprise: La rationalisation dans l'industrie française de l'entre-deux-guerres* (Paris: Éditions de l'écoles des hautes études en sciences sociales, 1997), 15–110; Delfine Gardey, *Un monde en mutation: Les employés de bureau en France 1890–1930: Féminisation, mécanisation, rationalisation* (PhD diss., Université Paris 7, 1995), 824–836; Ludovic Cailluet, "Accounting and Accountants as Essential Elements in the Development of Central Administration during the Inter-War Period: Management Ideology and Technology at Alais, Froges et Camargue," *Accounting, Business and Financial History* 7 (1997): 295–314.

81. Jaques Vernay, *Chroniques de la Compagnie IBM France* (Paris: IBM France, 1988), 17, 23, 24. In 1935, the company was called Société française Hollerith, machines comptable et enregistreuse; from 1936–1947, it was called Compagnie électro-comptable de France; after 1948, Compagnie IBM France.

82. Vernay, *Chroniques,* 17–18, 20.

83. The Caisse des Dépôts is a public financial institution that was created in 1816 to manage private funds that the public authorities wanted placed under

special protection, for example, escrow accounts or funds deposited with lawyers and notaries.

84. List, "Administration publiques et privées où les machines française Bull on remplacé les machines concurrentes étrangères," August 1934, folder DARC 193/51, box 92VEN 08-1, Archives Bull; Vernay, *Chroniques,* 42–43.

85. The French "Loi du 5–8 juillet 1844, sur les brevets d'invention." This law was valid until 1959, Albert Chavanne and Jean-Jacques Burst, *Droit de la propriété industrielle* (Paris: Editions Dalloz, 1998), 25–26.

86. Tabulating Machine Company, Dispotif contrôlant automatiquement les appareils enregistreurs d'un tabulateur, [French] Patent 487,667, filed 1917 and issued 1918.

87. René Carmille, "Société anonyme des machines à statistique, Rapport particulier d'enquête No. 80," 23 January 1935, 3–4; box N77, folder 6, Archives du Service historique de l'Armée de terre, Château de Vincennes, Paris; Campbell-Kelly, *ICL,* 68.

88. "Effects on Samas/Cimac of the new arrangements with R. R. Inc.," dated 4 April 1950, Martin Campbell-Kelly's personal archive; Campbell-Kelly, *ICL,* 70, 75.

89. John Connolly, *A History of Computing in Europe* (New York: IBM), E-6.

90. Précis of a talk given by H. R. Russell on 14 January 1931 on "Worldwide Use of Powers Machines," B/5/B, Powers-Samas box 1, ICL archives, NAHC.

91. Campbell-Kelly, *ICL,* 22–23.

92. Chandler, *Scale and Scope.*

93. Maurice Lévy-Leboyer, "The Large Corporation in Modern France," in *Managerial Hierarchies: Comparative Perspectives on the Rise of the Modern Industrial Enterprise,* ed., Alfred J. Chandler and Herman Daems (Cambridge, MA: Harvard University Press, 1980), 117–160, on 118–119.

Chapter Seven: Different Roads to European Punched-Card Bookkeeping

1. Derek Fraser, *The Evolution of the British Welfare State: A History of Social Policy since the Industrial Revolution* (London: Palgrave Macmillan, 2003), 164–168; Pat Thane, *The Foundations of the Welfare State* (London: Longman, 1982), 75–77.

2. Fraser, *The Evolution of the British Welfare State,* 176–182, 188; Thane, *The Foundations of the Welfare State,* 78–88; Mary MacKinnon, "Living Standards, 1870–1914," in *The Economic History of Britain since 1700,* vol. 2: *1860–1939,* ed. Roderick Floud and Donald McCloskey (Cambridge: Cambridge University Press, 1994), 265–290, on 290.

3. Arnold Wilson and Hermann Levy, *Industrial Assurance: A Historical and Critical Study* (London: Oxford University Press, 1937), 16–39; R. W. Bernard, *A Century of Service: The Story of the Prudential, 1848–1948* (London: Prudential Assurance Company, 1948), 33–46, 91; Martin Campbell-Kelly, "Large Scale Data Processing in Prudential, 1850–1930," *Accounting, Business and Financial History* 2 (1992): 117–139, on 118–129.

4. Bernard, *A Century of Service*, 90, 139; Campbell-Kelly, "Large Scale Data Processing," 129–130.

5. Charles Foster and the Accounting and Tabulating Machine Corporation of Great Britain Ltd., Improvements in and relating to tabulating machines, [British] Patent 108,942, filed 1916 and issued 1917; "Powers-Samas Accounting Machines Com Ltd." (typescript), 9, History Document no .771, Vickers Archives, Cambridge University Archives; *Powers-Samas Gazette* 22 (1953), 2–3; Lars Heide's interview with K. Petersen, formerly the Danish Powers agency, 1986.

6. "Powers-Samas Accounting Machines," 9.

7. "Powers-Samas Accounting Machines," 2–3; Martin Campbell-Kelly, *ICL: A Business and Technical History* (Oxford: Oxford University Press, 1989), 44–45.

8. Bernard, *A Century of Service*, 139; Campbell-Kelly, *ICL*, 46, 394–395.

9. F. G. S. English, "The Measure of Progress," *Powers-Samas Gazette* 22 (1953): 2–5, on 2–3; L. E. Brougham of Powers-Samas interviewed by B. G. Bellringer (typescript, 1977), 7, International Computers, Ltd. (ICL)/Powers-Samas, box 1, National Archive for the History of Computing, Manchester, England (NAHC); "Powers-Samas Accounting Machines," 9–10; Campbell-Kelly, *ICL*, 47–48.

10. Campbell-Kelly, *ICL*, 54–55, 69; Brougham interviewed by Bellringer, 3–4.

11. Frank P. Symmons, "Mechanical Aids for Life Assurance Records," *The Insurance Records* February 1929, 86–93.

12. "How Ordinary Branch Bonus Certificates Are Prepared at the Prudential Assurance Co., Ltd.," *The Powers-Samas Magazine*, August 1937, 5–8; B. H. Cooper, "How the Powers System Handles Free Policies at the Prudential Assurance Co., Ltd.," *The Powers-Samas Magazine* March 1939, 2–6.

13. C. Ralph Curtis, "The 'Punched Card System': Mechanisation in Gas Offices," *Gas Journal*, 14 June 1933, 807–809; A. Newling, "The System of Costing and Accountancy Adopted by R. & H. Green & Silley Weir, Ltd., by Means of Powers' Machines," *Powers-Samas Magazine* October 1935, 2–5; C. Ralph Curtis, "Bankers' Payments and the Punch Card," *Powers-Samas Magazine* December 1936, 3–6; O. S. Ross, "Complete Accounting on Powers Machines," *Powers-Samas Magazine* February 1938, 4–9.

14. Campbell-Kelly, *ICL*, 53.

15. "Powers-Samas Accounting Machines," 15.

16. "Powers Accounting Machine Manufacturing System," *American Machinist*, European Edition, 26 (1927), 171–174, 182–184, on 171.

17. "A Short History of Powers-Samas 1919–1950," 5–6, B/5/5, Powers-Samas, box 1, NAHC; English, "The Measure of Progress," 3–4.

18. English, "The Measure of Progress," 5.

19. Brougham interviewed by Bellringer, 9.

20. Campbell-Kelly, *ICL*, 73–76, 77–78, 94. In 1936 the manufacturing company was renamed Powers Accounting Machines Limited, which was considered preferable to the previous name: The Accounting and Tabulating Corporation of Great Britain Limited.

21. "Powers-Samas Accounting Machines," 29.

22. Campbell-Kelly, *ICL*, 82.

23. *The Office Machine Manual* (London: International Office Machines Research Limited, 1937–1940), vol. 1, sec. 1–2; Curtis, "Punched Card System," 23. In Campbell-Kelly, *ICL*, 85, this card is said to have been 2 by 4¾ inches.

24. Campbell-Kelly, *ICL*, 84.

25. *Powers-Samas Magazine* July–August 1935, 4–5; February 1938, 5.

26. "Electricity Billing and Statistics on Powers-Four at the Westminster Electric Supply Corporation, Limited," *Powers-Samas Magazine* September 1935, 4–7; M. G. Middleton, "Stock Control at Messrs. Joseph Lucas, Ltd.," *Powers-Samas Magazine* December 1935, 3–7; B. A. Shotman, "Purchase Ledger and Purchase Analysis on Powers-Four," *Powers-Samas Magazine* January 1936, 2–6.

27. Campbell-Kelly, *ICL*, 86.

28. C. D. Lake, Improvements in or relating to record-cards for use in statistical machines, [British] Patent 315,881, filed and issued 1939, equivalent to C. D. Lake, Record sheet for tabulating machines, [U.S.] Patent 1,772,492, filed 1928 and issued 1930.

29. "Some New Powers Developments," *Powers-Samas Magazine* July 1936, 8.

30. Powers-One, *The Office Machine Manual*, vol. 1, sec. 1-2, 1–5; Powers-Samas, section 1-2, 2.

31. Campbell-Kelly, *ICL*, 40–43; Computing Tabulating Recording Company (CTR) Executive Committee Meeting 25 June 1912, International Business Machines (IBM) Archives, New York.

32. "The British Tabulating Machine Company Limited 1907–1957," 6–7, typescript, NAHC; Campbell-Kelly, *ICL*, 50–51, 62–64.

33. "A New Type of Tabulating Machine," *The Engineer*, 1924, 274–276.

34. Campbell-Kelly, *ICL*, 55–57, 392–393.

35. "The Hollerith Tabulator," *Engineering*, 26 January 1934, 102–103; "Punched-Hole Accounting. Notes and Illustrations Describing the Operation and

Application of 'Hollerith' Machines," British Tabulating Machine Company (BTM) typescript 1935/36, 107e-113, ICL/B/5/A/BTM, box 2, NAHC.

36. "A Short History of the British Tabulating Machine Company," BTM 1952 (typescript), 5, ICL/B/5/A/BTM, box 5, NAHC; Campbell-Kelly, *ICL*, 40-51.

37. "A New Type of Tabulating Machine," *The Engineer*, 1924, 274-276.

38. "A Short History of the British Tabulating Machine Company," 4-6; "The British Tabulating Machine Company Limited," 3:1, 3:2.

39. Campbell-Kelly, *ICL*, 76.

40. Harry W. Richardson, *Economic Recovery in Britain, 1932-9* (London: Weidenfeld and Nicolson, 1967), 236-265; Campbell-Kelly, *ICL*, 79-80.

41. Campbell-Kelly, *ICL*, 80-81.

42. Harold Hall Keen, Improvements in or relating to record-card-controlled statistical machines, [British] Patent Specification 314,928, filed 1928 and issued 1929; Harold Hall Keen, Improvements in or relating to record-card-controlled statistical machines, [British] Patent Specification 339,935, filed 1929 and issued 1930; Harold Hall Keen, Improvements in or relation to record-card-controlled statistical printing machines, [British] Patent Specification 359,037, filed 1930 and issued 1931; Harold Hall Keen, Improvements in or relation to subtracting mechanisms, [British] Patent Specification 392,962, filed 1930 and issued 1931; Harold Hall Keen, Improvements in or relating to record-card-controlled statistical machines, [British] Patent Specification 422,135, filed 1933 and issued 1935; Harold Hall Keen, Improvements in or relating to record-card-controlled statistical machines, [British] Patent Specification 422,179, filed 1933 and issued 1935; Harold Hall Keen, "The Hollerith Tabulator," *Engineering*, 26 (1934): 102; Campbell-Kelly, *ICL*, 90-92.

43. BTM Board Minutes, 29 January 1935, copy supplied by Martin Campbell-Kelly; Harold Hall Keen, Improvements in or relating to record-card-controlled printing mechanism, [British] Patent Specification 446,104, filed 1934 and issued 1936; Harold Hall Keen, Improvements in or relating to record-card-controlled statistical printing machines, [British] Patent Specification 446,521, filed 1934 and issued 1936; Harold Hall Keen, Improvements in or relating to multiplying machines, [British] Patent Specification 436,136, filed 1934 and issued 1935; "The British Tabulating Machine Company Limited," 3:5; Campbell-Kelly, *ICL*, 100-101.

44. Harold Hall Keen, Improvements in or relating to record-card-controlled statistical machines, [British] Patent Specification 422,135, filed 1933 and issued 1935; Harold Hall Keen, Improvements in or relating to record-card-controlled printing mechanism, [British] Patent Specification 446,104, filed 1934 and issued 1936; Harold Hall Keen, Improvements in or relating to record-card-controlled statistical

printing machines, [British] Patent Specification 446,521, filed 1934 and issued 1936; Campbell-Kelly, *ICL*, 92.

45. *The Tabulator* 1936–1940; *Accounting by Electricity* (London: BTM, c.1940).

46. Exceptions were the use of punched cards for addressing in the following cases: "Invoicing," *The Tabulator* May 1937, 2–4; "Rent-Roll and Rent Collection," *The Tabulator* July 1938, 2–4; "Service," *The Tabulator* August–September 1940, 3–5; "The Kraft Cheese Co.'s 'Hollerith' method," *The Tabulator* October 1940, 2–3.

47. *The Tabulator* November 1937, 2; "Invoices and Ledgers," *The Tabulator* October 1938, 2–4; "Alphabetical Pay-Roll," *The Tabulator* November 1938, 2–4; "'Hollerith' in representative industries: Cork Manufacturing Company Ltd.," *The Tabulator* November–December 1939, 4–7; "Shipbuilding and Engineering Costs: 'Hollerith' Procedure at J. Samuel White & Co. Ltd," *The Tabulator* May 1940, 2–4.

48. Max Sering, *Deutschland unter dem Dawes Plan* (Berlin: Walter de Gruyter & Co., 1928), 48–81; Karl Hardach, *Wirtschaftsgeschichte Deutschland im 20. Jahrhundert* (Göttingen, Germany: Vandenhoech & Ruprecht, 1976), 41.

49. *Festschrift zur 25–Jahresfeier der Deutschen Hollerith-Maschinen Gesellschaft* (Berlin: Deutsche Hollerith Maschinen Gesellschaft [Dehomag], 1935), 72–73.

50. Gary Herrigel, *Industrial Constructions: The Sources of German Industrial Power* (Cambridge: Cambridge University Press, 1996), 117–118; Wilhelm True, *Gesellschaft, Wirtschaft und Technik Deutschlands im 19. Jahrhundert, Gebhardt Handbuch der deutschen Geschichte* (Munich: Deutscher Taschenbuch Verlag, 1975), vol. 17, 218–219.

51. Ursula-Maria Ruser, *Die Reichsbahn als Reparationsobjekt* (Freiburg, Germany: Eisenbahn-Kurier Verlag, 1981), 202, 205.

52. James W. Angell, *The Recovery of Germany* (New Haven, CT: Yale University Press, 1929), 68–69.

53. Dorpmüller, "Rationalisierung bei der Reichsbahn," *Verkehrstechnische Woche* 22 (1928): 1–9; Erich Dombrowski, *Die Wirkungen der Reparationsverpflichtugen auf die Wirtschaftsführung der Deutschen Reichsbahn-Gesellschaft* (Berlin: Dochterdruch, 1930), 125–137; Gaier, "Die Lochkarte zur Abrechnung und Kontrolle des Güterverkehrs," *Zeitung des Vereins Deutscher Eisenbahnverwaltungen*, 1924, 960; Stinner, "Das Lochkartenverfahren und seine Verwendung im Eisenbahndienst," *Verkehrstechnische Woche* 18 (1924): 422–427; *Festschrift*, 79–80, 98.

54. Pankraz Görl, "Technisierung der Administration. Maschinelle Datenverarbeitung und die Rationalisierung der Verwaltung in Deutschland 1924–1945" (unpublished M.A. thesis, Ludwig-Maximilians-Universität, Munich, 1993), 72–

82; Fritz Reinhardt, "Lochkartenmässige Betriebsrechnung in den Anstaltsapotheken," *Hollerith-Nachrichten*, 17 (1932): 202–208

55. Görl, "Technisierung," 82–85.

56. *Festschrift*, 78.

57. Direktionsbericht der Zentralwerksverwaltung 1924/25, 12–13, 15/Lg/562, Siemens Archiv; WWF Jahresbericht 1926/27, 4; WWF Jahresbericht 1927/28, 5–6; the latter two in 15/Lc/816, Siemens Archiv.

58. 1928 contracts between Dehomag and Siemens-Schuchertwerke, Nuremberg, 34/Ld/ 6, Siemens Archiv; Einführung in die Arbeitsgebiete der Hollerith-Abteilung, 9 July 1937, 34/Ld/6, Siemens Archiv; Aufteilung der Belegschaft der Abteilungen AP-MAR-VG nach dem Stand vom 7. Mai 1942, 10, 34/Ld/6, Siemens Archiv; Wilfred Feldenkirchen, *Siemens 1918–1945* (Munich: R. Piper, 1995), 231–232.

59. Alfred Böhme, "Bericht über die Entstehung und Werdegang der ZWR," 1966, 21/Lg/889, Siemens Archiv.

60. Robert Feindler, *Das Hollerith-Lochkarten-Verfahren für maschinelle Buchhaltung und Statistik* (Berlin: Verlag von Reimar Hobbing, 1929), 332–341. For similar descriptions, see Albert Schad, *Das Lochkartenverfahren im industriellen Rechnungswesen* (Stuttgart: C. E. Poeschel Verlag, 1930), 66–76, 102–109; J. C. Eichenauer, *Analyse der Wirtschaftlichkeit des Hollerith-Lochkarten-Systems* (Stuttgart: C. E. Poelschel, 1933), 26–52.

61. Report, Dehomag (Rotke/Humel) to Reichskommissar für die Behandlung feindlichen Vermögens, 20 January 1942, R-87/6248, Bundesarchiv, Berlin-Lichterfelde.

62. IBM, Executive and Finance Committee meeting, 29 April 1924, IBM Archives; *Festschrift*, 11; Report, Dehomag (Rottke/Hummel) to Reichskommissar für die Behandlung feindlichen Vermögens, 20 January 1942, R-87/6248, Bundesarchiv, Berlin-Lichterfelde.

63. *Festschrift*, 19.

64. *Hollerith Mitteilungen*, 4 (1914): 3–6.

65. Max Zander, "Die Lochkarte im Bankbetrieb," *Die Lochkarte und das Powers-System*, 10 (1932): 69–76.

66. Rudolf Lorant, "Maschinelle Kontokorrentbuchhaltung bei Großbanken," *Organisation* (Berlin), 9 (1925): 403–410; Rudolf Lorant, "Die Lochkartenmaschinen im Dienste der Bank-Buchhaltung," *Der Zahlungsverkehr*, 8 (1926): 228–237; undisclosed punched-card brand at Provinzialbank Pommern, in Range, "Lochkartensystem bei der Provinzialbank," 171–179. Powers punched cards were used in Dresdner Bank: Friedrich-Wilhelm Henning, "Innovation und Wandel der Beschäftigen Struktur im Kreditgewerbe von der Mitte des 19. Jahrhunderts bis 1948," *Bankhistorisches Archiv* 12 (1986): 52–53; Meyen, *120 Jahre Dresdner Bank*, 95.

67. Heinrich Adalbert Weinlich, Debit and credit tabulator, [U.S.] Patent 1,917,002, issued 1933 (filed in Germany in 1926); Friedrich W. Kistermann, "The Way to the First Automatic Sequence-Controlled Printing Calculator: The 1935 DEHOMAG D11 Tabulator," *IEEE Annals of the History of Computing* 17:2 (1995): 33–49, on 40–42.

68. Alvin E. Gray, *IBM Development Manual, Book I: Numerical Tabulators* (New York: IBM, 1956), 59–65, box A-25-2, folder: History/num. tab., IBM Archives; *Historical Records of Card and Machine Development* (New York: IBM, 1955), 16.

69. "Das Hollerith-Lochkartenverfahren. Die Hilfsmittel zur Formularbeschriftung an der Hollerith-Tabelliermaschine Type D11," *Hollerith-Nachrichten* 93 (1939): 1403–1410.

70. "Die schreibende Tabelliermaschine Type BK, Universalausführung," *Hollerith Nachrichten* 38 (1934): 439–440.

71. "Die schreibende Tabelliermaschine," 485–490.

72. "Hollerith-Tabelliermaschinen Type D11 und Type D9," *Hollerith Nachrichten* 74 (1937): 1022–1024; Kistermann, "The Way to the First," 43–46.

73. "Powers," *Organisation* (Berlin), 25 (1923), 138–140; H. Goerlitz, "Zur Geschichte der Powers-Machine," (typescript), 1934, History archive of IBM Deutschland.

74. R. Lorant, "Maschinelle Kontokorrentbuchhaltung bei Großbanken, Organisation" [Berlin], 27 (1925): 403–410; R. Lorant, "Die Lochkartenmaschinen im Dienste der Bank-Buchhaltung," *Der Zahlungsverkehr* 8 (1926): 228–237; "Der Powers Buchstabendrucker," *Die Lochkarte und das Powers-System* 13 (1930): 105–106.

75. Ralph Lorant, "Lohnabrechnung der Großbetriebe mit Hilfe des Powers-Lochkarten-Systems," *Organisation* (Berlin) 28 (1926): 677–684.

76. Jan Büchter, Kartenlochvorrichtung zum Anbau an Transportmitteln z. B. Eisenbahnwagen, [German] Patent 431,112, filed 1924 and issued 1926; Jan Büchter, Vorrichtung zum Einstellen einer Addiermaschine durch Lochkarten, [German] Patent 438,093, filed 1924 and issued 1926; Jan Büchter, Fühlvorrichtung für mit Lochkarten arbeitende Sortier, Tabellier- und ähnliche Maschinen, [German] Patent 442,236, filed 1926 and issued 1927; Johannes Josephus [Jan] Büchter, Martin Lebeis, and Rezö Lorant, Statistische Tabelliermaschine mit selbsttätiger Summensteuerung, [German] Patent 464,197, filed 1926 and issued 1928; Rezö Lorant, Einrichtung an Lochmaschinen für statistische Zählkarten zum Einstellen von Komplementärzahlen, [German] Patent 451,940, filed 1927 and issued 1928; Jan Büchter, Martin Lebeis, and Rezö Lorant, Maschine zur Prüfung der Lochungen in Lochkarten für statistische Zwecke, [German] Patent 461,264, filed 1927 and issued 1928.

77. Frederick Reed, "The World Tour: Germany," *Keeping Tabs, Powers Accounting Machine Corporation* 18 (August 1933): 1–4; Goerlitz, "Zur Geschichte der Powers-Machine."

78. "Die 90stellige Powers-Lochkarte," *Die Lochkarte und das Powers System* 27 (1931): 263–264.

79. German Powers balance sheets 1939–1940, 1940–1941, 23, vol. 1, R87/6198; Dehomag balance sheets 1939, 1940, 19, 22, vol. 1, R87/6248; Bundesarchiv, Berlin-Lichterfelde.

80. James Powers, Statistische Tabelliermaschine, bei welcher die nach verschiedenen Gruppen gelochten Karten nur innerhalb der aufeinanderfolgenden Gruppen zur Tabellierung und Summierung gelangen, [German] Patent 333,413, issued 1921 (valid from 1915); The Tabulating Machine Company, Selbsttätige Tabelliervorrichtung für Zählkarten, [German] Patent 406,744, issued 1924 (valid from 1914); *Blatt für Patent-, Muster- und Zeichenwessens*, 1930, 6–7.

81. For 1928 prices, see Howard Sylvester Ellis, *Exchange Control in Central Europe* (Westport, CT: Greenwood Press, 1971), 381–382.

82. Norbert Frei, *Führerstaat: Nationalsozialistische Herrschaft 1933 bis 1945* (Munich: Deutscher Taschenbuch Verlag, 1987), 85–90.

83. IBM, Executive and Finance Committee minutes, 5 June 1933, IBM Archives; *Festschrift*, 16, 42.

84. W. Heidinger, in *Denkschrift zur Einweihung der neuen Arbeitsstätte der deutschen Hollerith Maschinen Gesellschaft m.b.H. in Berlin-Lichterfelde am 8. Januar 1934* (Berlin: Dehomag, 1934), 40.

85. *Festschrift*.

86. H. Rottke, in *Denkschrift*, 35–37.

87. Petra Bräutigam, *Mittelständische Unternehmer in Nationalsozialismus: Wirtschaftliche Entwicklungen und Soziale Verhältnisse in der Schuh- und Lederindustrie Baden und Württembergs* (Munich: R. Oldenbourg Verlag, 1997), 116, 118; Anne Sudrow, "Das 'deutsche Rohstoffwunder' und die Schuhindustrie. Schuhproduktion unter der Bedingung der nationalsozialistischen Autarkiepolitik," *Blätter für Technikgeschichte* 60 (1998): 63–92; Feldenkirchen, *Siemens*, 663, 680.

88. *Festschrift*, 14; Bericht, Dehomag to Reichskommissar für die Behandlung feindlichen Vermögens, 20 January 1942, R-87/6248, Bundesarchiv, Berlin-Lichterfelde.

89. Black, *IBM and the Holocaust*, 71–175, 209–217, 282–294, 346–347, 371–372, 390–391.

90. Frei, *Führerstaat*, 257; *Festschrift*, 74.

91. Organisation und Kapital der Powers GmbH, 1 July 1941, and Bilanz, R-87/6198, Bundesarchiv, Berlin-Lichterfelde; *Die Lochkarte und das Siemens-Powers-System*, 69 (1935), 709.

92. Minutes of the Meeting of the Remington Rand (RR) Board of Directors, 16 October 1934, RR Directors' Minutes, vol. 3, 157, box 15, acc. 1910, Hagley Museum and Archive, Wilmington, DE (Hagley).

93. The high-voltage Siemens-Schuckertwerke in Nuremberg simultaneously established a Dehomag installation.

94. For example, Paul Mansel, Elektrisch gesteuerte Sortiermaschine für mit Lochsymbolen versehene Karten oder Papierstreifen, [German] Patent 550,916, filed 1928 and issued 1932; Paul Mansel, Steurungsvorrichtung für die Typen- oder Zählwerksschreiben von Lochkarten-Auswertmaschinen oder Rechenmaschinen, [German] Patent 551,150, filed 1928 and issued 1932; Fritz Winkler, Vorrichtung zum Selbstätigen Rücktransport des Einstellwagens an Kartenloch- und Druckmaschinen, [German] Patent 558,186, filed 1928 and issued 1932; Fritz Winkler, Verbindung einer Tabelliermaschine mit einer Lochmaschine, [German] Patent 582,632, filed 1929 and issued 1933; Paul Mansel, Auswertungsvorrichtung für mit Lochkombinationen versehene Lochkarten, [German] Patent 550,959, filed 1930 and issued 1932.

95. "Die Gemeinschaftsarbeit Siemens-Powers," *Die Lochkarte und das Powers-System*, 64/65 (1934): 654–655, 660–663.

96. Feldenkirchen, *Siemens*, 243–447, 663.

97. RR, Board of Directors meeting, 7 June 1934, RR Directors Minute Book, no. 3, box 15, acc. 1910, Hagley; RR, Executive Committee of the Board of Directors meetings, 6 July 1935, 8 September 1936, RR Executive Committee Minute Book, no. 2, box 17, acc. 1910, Hagley; Contract between Powers and Siemens and Halske, 29 August 1934, 21/Lg/889, Siemens Archiv.

98. Note, Aufträge, 22 June 1939, 21/Lg/889, Siemens Archiv; B. Drost and F. Winckler, "Siemens-Powers-Kupplung Cordt-Universal," *Die Lochkarte und das Siemens-Powers System* 70/71 (1935): 712–717; "Siemens-Powers Sortiermaschine," *Die Lochkarte und das Siemens-Powers System* 75/76 (1935): 769; Hermann Fauth, "Die Lochkartenmässige Bearbeitung der Rechnungsstatistik in der Finanzstatistischen Abteilung der Reichshauptstadt Berlin," *Die Lochkarte und das Siemens-Powers-System* 121 (1939): 1385–1392, on 1389–1390; "Der Powers-Hamenn Rechenautomat," *Die Lochkarte und das Siemens-Powers-System* 138–140 (1940): 1481–1482.

99. Hahn and Roloff, "Aktenvermerk, Betr. Auslegung des Vertages S&H/Powers," 10 February 1938; Voigt, "Die Entwicklung unseres Verhältniss zur Powers-GmbH, Berlin," 9 April 1938; both in 21/Lg 889, Siemens Archiv.

100. "Die Gemeinschaftsarbeit Siemens-Powers," 655.

101. "Die Grundzüge des Powers-Verfahrens," *Die Lochkarte und das Powers-System*, 132–134 (1940): 1458–1469; Kurt Nonner, "Das Powers-Lochkarten-Verfahren in der Versicherung," *Die Lochkarte und das Powers-System*, 135–137

(1940): 1469–1476, on 1474–1475; Werner Funk, "Die 90 selbige Lochkarte," *Die Lochkarte*, 147 (1941): 1423–1427.

102. Aufträge, 22 June 1939 (see n 98).

103. Note, Voigt to Storch, Betr. Powers-GmbH., c. 6 April 1938; note, Voigt, Die Entwicklung unseres Verhältniss zur Powers-GmbH, Berlin, 9 April 1938; both 21/Lg889, Siemens Archiv.

104. Modern bank accounting on Powers Machines, as adopted by Banque d'Alsace et de Lorraine, France, Legal Correspondence, agreements. RR Law Department, box 1, folder: Bank of Alsace-Lorraine, France, 1930; acc. 1825, Hagley; Frederick Reed, "The World Tour: France," *Keeping Tabs*, 17 (July 1933): 2–4.

105. List, Administration publiques et privées où les machines française Bull ont remplacé les machines concurrentes étrangères, August 1934, folder: DARC 193/51, box 92VEN 08-1, Archives Bull, Paris; Précis of a talk given on 14 January 1931, by H. R. Russell on "World-Wide Use of Powers Machines," box 1, B/5/5 (Powers-Samas), NAHC; "Quelques références de la Compagnie electro-comptable de France," Vincennes, 8 N 77, folder 6: attachment to letter, Compagnie electro-comptable to Ministry of War, 1 July 1938, 8N/77, Archives du Service historique de l'Armée de terre, Château de Vincennes, Paris; Jaques Vernay, *Chroniques de la Compagnie IBM France* (Paris: IBM France, 1988), 42–43.

106. List, Administration publiques et privées où les machines française Bull ont remplacé les machines concurrentes étrangères, August 1934, folder DARC 193/51, box 92VEN 08-1, Archives Bull; Précis of a talk by H. R. Russell.

107. Circular, "District 'A'—concurrence Bull," 18 November 1937, folder "Dossier reunion. box 26-12-51," box 2.12, Archives Bull.

108. Marc Olivier Baruch, *Servir l'État français* (Paris: Fayard, 1997), 21–41.

109. Dominique Pagel, interview with Elie Doury, c. 1973, box 3, 94HIST-COM03, Archives Bull.

110. Circular 34, from P. Laval to the Minister of War, 24 December 1935, folder 6, box 8 N77, series N, Archives de l'Armée.

111. Accounting and Tabulating Machine Company board meetings, 26 September 1935, 2(?) December 1935, copies of minutes supplied by Martin Campbell-Kelly; René Carmille, "Rapport Particulier No. 80: Concernant la Société anonymes des machines à statistique," 1936, 4–7, folder 6, box N77, Archives de l'Armée; History Document no. 771, 30–31.

112. Powers-Samas, Accounting and Tabulating Machine Company board meetings, 25 January 1938, 16 March 1938, copies of minutes supplied by Martin Campbell-Kelly; RR Directors Meeting, June 27, 1939, Directors Minute Books no. 5, 58, box 15, series 1 (minute books), acc. 1910, Hagley; RR Executive Com-

mittee Meeting, 14 May 1940, Executive Committee Minute Book no. 3, 31–32, box 17, series 1, acc. 1910, Hagley; History Document No. 771, 34, 36. The development work at SAMAS did not result in its being awarded any patent in France in the 1930s.

113. Vernay, *Chroniques*, 27–28.

114. Vernay, *Chroniques*, 264.

115. IBM, Executive and Financial Committee meeting, 29 April 1924, IBM Archives.

116. Georges Ziguelde, Perfectionnements aux machines comptables ou autres machines analogues, [French] Patent 705,529, filed 1930 and issued 1931; Louis Momon, Perfectionnements aux machines à statistiques ou autres machines analogues, [French] Patent 711,031, filed and issued 1931; Société internationale des machines commerciales (SIMC), Perfectionnements aux machines trieuses de cartes enregistreuses dans le but d'obtenir une sélection automatique de groupe de cartes, [French] Patent 719,566, filed and issued 1931; SIMC, Perfectionnement aux machines d'impression et de mise en liste commandées par des cartes perforées, [French] Patent 731,213, filed and issued 1932; SIMC, Dispositif de commande d'une machine à écrire par l'intermédiaire d'une perforatrice de cartes enregistreuses, [French] Patent 731,774, filed and issued 1932; SIMC, Perfectionnements aux dispositifs de contrôle automatique dans les machines tabulatrices, [French] Patent 748,466, filed and issued 1933; Georges Ziguelde, Perfectionnements aux machines comptables, [French] Patent 749,306, filed 1932 and issued 1933; SIMC, Perfectionnements aux machines pour le triage des cartes perforées, [French] Patent 792,997, filed 1934 and issued 1935; SIMC, Perfectionnements aux machines comptables, [French] Patent 796,071, filed 1934 and issued 1936. In addition, IBM filed many separate patents in France that corresponded to their patents in the United States.

117. John Connolly, *A History of Computing in Europe* (New York: IBM, 1968), 21.

118. Vernay, *Chroniques*, 24, 25.

119. René Carmille, "Rapport Particulier, No. 78: Fourniture d'un équipement mécanographique au Service de la Liquidation des Transports," 16 December 1935, 19, folder 6, box 8 N77, series N, Archives de l'Armée.

120. Gaston Marie, "René Carmille. Son Oeuvre" [obituary], *Journal de la Société de statistique de Paris* 86 (1945): 145–148; [Robert Ligonnière], "Petite histoire des statistique, " *Ordinateurs*, 25 November 1985, 44–48; Jean-Pierre Azéma, Raymond Lévy-Bruhl, and Béatrice Touchelay, *Mission d'analyse historique sur le système de statistique français de 1940 à 1945* (Paris: INSEE, 1998), 11–12.

121. Jean Delmas, "L'organisation militaire en France—l'armée," in *Histoire militaire de la France*, ed. Guy Pedroncini (Paris: P.U.F., 1992), vol. 2, 429–430.

122. M. Conquet, "Rapport général sur l'évolution de l'application des procédés mécanographique dans les Services de la Guerre," 11 March 1937, folder 5, box 8 N77, series N, Archives de l'Armée.

123. René Carmille, "Rapport Particulier No. 71: Fabrications de machines à statistique qui peuvent être utilisées par les services du Département de la Guerre," 14 May 1935, 2–4; René Carmille, "Rapport Particulier No. 105 concernant le position de la Compagnie Bull vis-à-vis du Département de la Guerre," 11 March 1938, 18; both in folder 6, box 8 N77, series N, Archives de l'Armée.

This development was simultaneous with the failed attempt starting in and after early 1935 to mechanize the administration of conscription and mobilization using Bull punched card machines, see chapter 8.

124. René Carmille, *La mécanographie dans les administrations*, 1st ed. (Paris: Recueil Sirey, 1936), 77–96.

125. "Rapport sur l'Atelier de Construction de la Compagnie des Machines Bull," 12 July 1936, folder DAC I 193/51, box 92VEN 08-1, Archives Bull.

126. Carmille, *La mécanographie* (1936), 77, 96.

127. Lars Heide, "Facilitating and Restricting a Challenger: Patents and Standards in the Development of the Bull-Knutsen Punched Card System 1919–1938," *Business History* 51 (2009), 28–44.

128. Minutes of Assemblée Constitutive, 21 February 1931, 9 March 1931, box 1, 93DJFG-DDS02, Archives Bull; Société Egli-Bull, June 1932, box 2, 92HIST-DGE01, Archives Bull; Minutes of Réunion Conseilles d'Administration, 9 March 1931, vol. 1, box 44, 93DJFG-DDS02, Archives Bull.

129. "Rapport sur la situation financier de la Société Egli-Bull au 31 décembre 1931," box 2, 92HIST-DGE01, Archives Bull.

130. Pierre-Eric Mounier-Kuhn, "Bull: A World-Wide Company Born in Europe," *Annals of the History of Computing*, 11 (1989): 279–297, on 282.

131. Telegrams from U.S. Powers to Bull or Egli company in Zürich, 11–12 December 1931; letter from AITEC (Vieillard, Genon, Vindevoghel), Paris, to RR, New York, 15 December 1931, both in box 2.3 (Remington Rand, 1931–1960), Archives Bull.

132. Letter from AITEC (Vieillard, Genon, Vindevoghel), Paris, to RR, New York, 15 December 1931, box: 2.3 (Remington Rand, 1931–1960), Archives Bull.

133. "Rapport sur la situation financier de la Société Egli-Bull au 31 décembre 1931," box 1, 92HIST-DGE, Archives Bull; Assemblée Générale Extraordinaire: 16 April 1932, procès verbal, box 1, 93DJFG-DDS02.

134. Assemblée Générale, 31 March 1933, box 1, 93DJFG-DDS02, Archives Bull; Assemblée Générale Extraordinaire, 25 April 1933, box 2, 93DJFG-DDS02.

135. Liste des equipments vendue (list of sold equipnment), June 1935, Liste des equipments loués (list of leased equipment), June 1935, box 2, 92HIST-DGE01, Archives Bull; note, G. Vieillard to M. Bassot, 14 May 1934; note, Bosman to G. Vieillard, 28 January 1935; note, Guerner to G. Vieillard, 26 July 1935; all box 3.1 (folder Products 1931–1946), Archives Bull.

136. P. Cailles, Aussedat, to M. Bassot, Compagnie de Machines Bull (CMB), 12 September 1934, ring-folder: *Fisches . . . annuelles 1931–1934*, Archives Bull.

137. Conseils d'administration, 12 June 1935, 18 October 1935; exchange of telegrams between E. Genon in New York and the Bull board, 7–18 May 1935, box 3, 92HIST-HUR01, Archives Bull; D. Pagel's interview with G. Vieillars, box 1, 94HIST-COM03.

138. Mounier-Kuhn, "Bull," 284.

139. These companies do not appear to have aspired to be admitted to this society, as SAMAS did not have any production of punched-card machines in France and IBM only started theirs in 1934. René Carmille, "Rapport Particulier No. 80: Société anonymes des machines à statistique," 1936; René Carmille, "Rapport Particulier No. 71: Fabrication des machines à statistique qui peuvent être utilisées par les Services du Department de la Guerre," 1935, 4; both in folder 6, box N77, series N, Archives de l'Armée.

140. Conseil d'administration, 20 December 1931, box 1, 93DJFG-DDS02, Archives Bull.

141. Mounier-Kuhn, "Bull," 282.

142. Conseil d'administration, 20 January 1933, box 1, 93DJFG-DDS02, Archives Bull.

143. René Carmille, "Rapport Particulier No. 72: Fabrications de machines à statistique et situation particulière de l'industrie française en cette matière," 11 June 1935, 31–40, 60–63, folder 6, box 8 N77, series N, Archives de l'Armée; minutes, "Résumé de l'entretien de M. Vieillard avec M. le Contrôleur Carmille, et M. Essig, Inspecteur des Finances," 23 January 1935, and "Note pour M. Jourdain en vue de l'entretien avec M. Guinant, secrétaire Général du Ministère de la Guerre," 24 January 1935, folder DAC I 193/51, box 92VEN08, Archives Bull.

144. René Carmille, "Rapport Particulier No. 78: Fourniture d'un équipement mécanographique au Service de la Liquidation des Transports, 16 December 1935, 19, folder 6, box 8 N77, series N, Archives de l'Armée.

145. Liste des equipments vendue, juin 1935, Liste des equipments loués, juin 1935, box 2, 92HIST-DGE01, Archives Bull.

146. Minutes from technical conferences at Egli-Bull, 28–29 April 1931, 5 June 1931, June 13, 1931 and June 20, 1931, box 1, 93DJFG-DDS02, Archives Bull; Note de M. Vieillard à M. Bassot, Vice-Président Délégué, May 14, 1934, ring-folder, CMB. *Fisches . . . annuelles 1931–1934*; Conseil d'Administration,

25 February 1935, box, 93DJFG-DDS02; note, Bosman to G. Vieillard, 28 January 1935, and note Guerner to G. Vieillard, 26 July 1935, folder, Products 1931–1946, box 3.1, Archives Bull.

147. Dossier justicatif des budgets, 1 June 1932, box 2, 92HIST-DGE01, Archives Bull.

148. Heide, "Facilitating and Restricting."

149. Knut Andreas Knutsen, Dispositif imprimeur particulièrement applicable aux machines à tabuler pour fiches perforées, [French] Patent 41,549, filed and issued 1932; Knut Andreas Knutsen and Anders Eirikson Vethe, Printing device, [U.S.] Patent 1,971,860, filed and issued 1934; Knutsen interview 1978, 89, box 3, 92HIST-DGE07, Archives Bull.

150. Indicated in Henrik Hartzner Diaries, custody of Hartzner family, Denmark, 1 February 1931.

151. List of Bull customers, August 1935, in Archives Bull, folder DARC, 92–VEN-08, Archives Bull; F. R. Bull to Rentenanstalt, 16 March 1922, and note, Vorstellung der RA, 1973, both kindly supplied from Swiss Life; "In memoriam Prof. Dr. Emile Marchand 1890–1971," *Mitteilungen der Vereinigung schweizerischer Versicherungsmathematiker*, 71 (1971): 1–8.

152. Hartzner Diaries, 28 October 1931, 26 November 1931, 30 May 1932, author's copies; Knut Andreas Knutsen, Dispositif imprimeur particulièrement applicable aux machines à statistiques, machines à calculer ou machines à tabuler pour fiches perforées, [French] Patent 734,945, filed and issued 1932; Knut Andreas Knutsen, Printing device for statistical, calculating, and tabulating machines, [U.S.] Patent 2,046,464, filed and issued 1936; Knut Andreas Knutsen, Dispositif imprimeur particulièrement applicable aux machines à statistiques, machines à calculer ou machines à tabuler pour fiches perforées, [French] Patent 43,173E, filed 1933 and issued 1934; Knut Andreas Knutsen, Card controlled device, [U.S.] Patent 2,046,465, filed and issued 1936.

153. W. W. McDowell to F. M. Carroll, 10 July 1936, box A-22-3, IBM Archives.

154. Charles J. Bashe, Lyle R. Johnson, John H. Palmer, and Emerson W. Pugh, *IBM's Early Computers* (Cambridge, MA: MIT Press, 1986), 481–482.

155. Knut Andreas Knutsen, Dispositif imprimeur, spécialement pour machines contrôlées par cartes enregistreuses, [French] Patent 795,586, filed 1935 and issued 1936; Knut Andreas Knutsen, Printing device, particularly for tabulating machines controlled by record cards or bands, [U.S.] Patent 2,175,530, filed and issued 1939; Dominique Pagel, *Histoire de Compagnie des machines Bull*, manuscript 1979, Archives Bull.

156. Conférence technique [seminar on technology], 28–29 April 1931, minutes, box 1, 93DJFG-DDS02, Archives Bull.

157. The Tabulating Machine Company, Perfectionnements aux enregistreuses pour machines tabulatrices, [French] Patent 677,427, filed and issued 1929, the equivalent of Clair D. Lake, Record sheet for tabulating machines, [U.S.] Patent 1,772,492, filed 1928 and issued 1939.

158. Bureau de Recrutement de Versailles to CMB, 30 May 1936, in Archives Bull, box 92VEN08, box 4, folder Ministère de la Guerre (1934–1944).

159. Cour d'Appel de Paris, Greffe Civil, no. H-1440 (23 April 1947), *Compagnie Machines Bull v. Compagnies Electro Comptable*, in Archives de Paris. More explicit argumentation in previous, parallel case, Cour d'Appel de Paris, Greffe Civil, no. H-1441 (23 April 1947), *Compagnie Piles Wonder v. Compagnies Electro Comptable*, Archives de Paris.

160. Interference in Patent Case File 1,772,492, entry 9A, RG-241, National Archives.

161. Minutes from Conférence technique, 28–29 April 1931, Archives Bull, folder Conférences technique, box 1, 93DJFG-DDS02, Archives Bull; minutes from Conseil d'administration, 15 January 1934, box 25, 93DJFG-DDS02.

162. "Rapport sur l'Atelier de Construction de la Compagnie des Machines Bull," 12 July 1936, folder DAC I 193/51, box 1, 92VEN08, Archives Bull.

163. See previous note; Mr. Delaage interview with K. A. Knutsen, 5 November 1976, box 3, 92HIST-DGE07, Archives Bull; Dominique Pagel interview with R. Faucon, c. 1976, box 1, 94HIST-COM03.

Chapter Eight: Keeping Tabs on Society with Punched Cards

1. Public Law 74-271 (49 Stat. 620), 74th Cong., 1st sess. Background on the Social Security Act can be found in Clarke Chambers, *Seedtime of Reform: American Social Service and Social Action 1918–1933* (Minneapolis: University of Minnesota Press, 1963); Roy Lubove, *The Struggle for Social Security, 1900–1935* (Cambridge, MA: Harvard University Press, 1968): 45–65; Theda Skocpol, *Protecting Soldiers and Mothers: The Political Origins of Social Policy in the United States* (Cambridge, MA: Harvard University Press, 1992); Price V. Fishback and Shawn Everett Kantor, *Prelude to the Welfare State: The Origins of Workers' Compensation* (Chicago: University of Chicago Press, 2000); Gary M. Wilton and Hugh Rockoff, *History of the American Economy* (Stamford, CT: Thomson Learning, 2002), 538–539. Extensive public material on the Social Security Act, its formation, amendments, and administration is available at the Social Security History page: www.ssa.gov/history/history.html

2. The Social Security Amendments of 1939 moved the start of payments of old age benefits to 1940. Public Law 76-379, 76th Cong., 1st sess., sec. 202.

3. Superseded by the Social Security Administration in 1946.

4. Public Law 74-271 (49 Stat. 620), 74th Cong., 1st sess., sec. 202, 210.

5. *Historical Statistics of the United States. Colonial Times to the Present* (New York: Basic Books, 1976), 347.

6. This would gradually rise to 3% in 1949. Social Security Act, 14 August 1935, Public Law 74-271 (49 Stat. 620), 74th Cong., 1st sess., sec. 202, 210, 801–804.

7. One-half of 1% of the accumulated amount up to $3,000 was paid, with decreasing percentages for larger amounts. Public Law 74-271 (49 Stat. 620), 74th Cong., 1st sess., sec. 202.

8. Olga S. Halsey, "British Old Age Pensions and Old Age Insurance" (1935); Marianne Sakmann, "Financial History of the Workers' Invalidity, Old Age and Survivors' Insurance of Germany" (1935); both staff reports by the Committee on Economic Security, 1934–1935, vol. 2 (Old Age Security), Social Security Administration Historical Archive, Baltimore; lecture by Frances Perkins, "The Roots of Social Security," 23 October 1962, Social Security Administration Historical Archive; Michael A. Cronin, "Fifty Years of Operations in the Social Security Administration," *Social Security Bulletin*, 48:6 (1985): 14–26, on 15.

9. "History of Division of Accounting Operation," typescript 1950, 6–10–12, 17–18, Social Security Administration Historical Archive.

10. Public Law 74-271 (49 Stat. 620), 74th Cong., 1st sess., sec. 807, 808.

11. Lars Heide, "The Antitrust Case against IBM and Remington Rand in 1932 and U.S. Anti-trust Regulation," unpublished manuscript.

12. Message from E. J. May to the Social Security Board regarding Account Number System, 1 June 1936, and Social Security Board Minutes, 2 June 1936, box 3, PI-183, entry 12, RG-47, National Archives, Washington DC (NA); Cronin, "Fifty Years," 15–16; "History of Division of Accounting Operation," 21–22, 23–26.

13. The Bureau of Internal Revenue had sixty-four collection districts throughout the United States, Alaska, and Hawaii.

14. Social Security Board Minutes, 8 June 1936, box 4, entry 12, PI-183, RG-47, NA; "History of Division of Accounting Operation," 26–28.

15. "History of Division of Accounting Operation," 41–43, 44–47; Cronin, "Fifty Years," 16–17.

16. A. I. Ogus, "Great Britain," in *The Evolution of Social Insurance, 1881–1981*, ed. Peter A. Köhler and Hans F. Zacher (London: Frances Pinter, 1982), 209–214; *Ministry of Labour, Twenty-First Abstract of Labour Statistics of the United Kingdom (1919–1933)* (London: His Majesty's Stationery Office [HMSO], 1934), 170–171.

17. Yves Saint-Jours, "France: Evolution of Social Insurance," 93–95, 117–120, 121–122, in Peter A. Köhler and Hans F. Zacher (eds), *The Evolution of Social Insurance, 1881–1981* (London: Frances Pinter, 1982); Paul Durand, *La politique contemporaine de sécurité sociale* (Paris: Dalloz, 1953), 32.

18. "Social Security Board: General Statement of the Problems Confronting the Records Division of the Bureau of Federal Old-Age Benefits," 15 June 1936, box 3, PI-183, entry 12, RG-47, NA; "History of Division of Accounting Operation," 9–10, 28–32.

19. Social Security Board Minutes, 16 September 1936, box 4, PI-183, entry 12, RG-47, NA; "History of Division of Accounting Operation," 32–40.

20. "History of Division of Accounting Operation," 62–63.

21. The wages were reported semiannually in 1937.

22. Internal IBM correspondence on the origin of the collator, 1946–1947; *Historical Record of Card and Machine Development*, 1955, 20; both in box A-25-3, International Business Machines (IBM) Archives, New York; *IBM Electric Punched Card Machines Principles of Operation: Collator Type 077* (New York: IBM, 1945); Emerson W. Pugh's interviews with Stephen W. Dunwell, 31 July 1991, no. 1, 31–36 and no. 2, 13–14, interview C176, Technical History Project, IBM Archives.

23. *Historical Record of Card and Machine Development*, 19, 21; "History of Division of Accounting Operation," 80–82.

24. "Social Security Board regulation no. 1," adopted by the Social Security Board, 15 June 1937, Social Security Administration Historical Archive; Arthur Joseph Altmeyer, *The Formative Years of Social Security* (Madison: University of Wisconsin Press, 1966), 58–59.

25. No substantiation was found for the possible influence of the U.S. Social Security register on the development of the French national register.

26. Euginia C. Kiesling, *Arming Against Hitler: France and the Limits of Military Planning* (Lawrence: University Press of Kansas, 1996), 85–91; Henry Dutailly, "L'effondrement," in *Histoire militaire de la France*, ed. Guy Pedroncini, Paris: PUF, 1992), vol. 3, 382.

27. Note, Affaire recrutement-mobilisation, 8 August 1935, folder 1, box 3.11, Archives Bull, Paris; Dutailly, "L'effondrement," 349–350.

28. Tabulating Machine Company, Perfectionnements aux enregistreuses pour machines tabulatrices, [French] Patent 677,427, filed and issued 1929, the equivalent of Clair D. Lake, Record sheet for tabulating machines," [U.S.] Patent 1,772,492, filed 1928 and issued 1939.

29. René Carmille, "Note no. 306"; René Carmille, "Rapport Particulier No. 105: Concernant le position de la Compagnie Bull vis-à-vis du Département de la Guerre," 11 March 1938, 13; René Carmille, "Rapport Particulier No. 68: Études de l'urtication de procédés mécaniques mécanographiques par les Centre de Mobilisation," 5 February 1935, folder 6, box 8 N77, series N, Archives du Service historique de l'Armée de terre, Château de Vincennes, Paris.

30. René Carmille, "Rapport Particulier No. 105," 13–14; René Carmille,

"Rapport Particulier No. 108 au sujet des machines à cartes perforées et de machines à clichés service du Recrutement," 30 June 1938, 3–4, folder 6, box 8 N77, series N, Archives Bull.

31. René Carmille, "Rapport Particulier No. 104: Essai de construction d'un Bureau régional mécanisé de Recrutement dans la 3ème Région Militaire," 16 February 1938; "Annexes de Rapport Particulier No. 104," both in folder 4, box 8 N77, series N, Archives Bull.

32. René Carmille, "Rapport Particulier No. 112 concernant la fourniture de machines à clichés nécessaires à l'Atelier de Mécanographie du Bureau de Recrutement de Rouen," 21 March 1939, folder 6, box 8 N77, series N, Archives Bull.

33. Peter Jackson, *France and the Nazi Menace: Intelligence and Policy Making 1933–1939* (Oxford: Oxford University Press, 2000); Kiesling, *Arming against Hitler*, 173–184; Marc Olivier Baruch, *Servir l'État français* (Paris: Fayard, 1997). It is noteworthy that the efforts to mechanize the French conscript administration started in 1933 when information was collected from the army in Germany. René Carmille, "Note No. 306: Concernant l'historique de l'origine des études sur la mécanisation du Service de Recrutement," 13 June 1938, unnumbered folder, box 8 N77, series N, Archives Bull.

34. The French Armistice Army was not specified in the Armistice Agreement, but it was subsequently specified. *La Délégation Française auprès de la Commission Allemande d'Armistice,* vol. 1, article 4, 2 (Paris: A. Costes, 1947); André Martel, ed., *Histoire militaire de la France,* (Paris: PUF, 1994), vol. 4, 31.

35. Baruch, *Servir l'État*, 97–170.

36. Note relative au Service National des Statistiques (Services des Enquêtes Démographiques), 1944, box B 55,364, Centre des archives économiques et financières, Savigny-le-Temple; Gaston Marie Gaston Marie, "René Carmille. Son oeuvre" [obituary], *Journal de la Société de statistique de Paris,* 86 (1945): 145–148; Alfred Sauvy, "Statistique Générale et Service National de Statistique de 1919 à 1944," *Journal de la Société de Statistique de Paris*, 1975, 40–41; Michel Volle, "Naissance de la statistique industrielle en France, 1930–1960," in *Pour une histoire de la statistique* (Paris: Institut National de la Statistique et des Études Économiques [INSEE], 1977), 347–352; Henri Amouroux, *Quarante millions de Pétainistes, Juin 1940–Juin 1941* (Paris: R. Laffont, 1977), 352–353.

37. Alfred Sauvy, *De Paul Raynaud à Charles de Gaulle. Scènes, tableaux et souvenirs* (Paris: Casterman, 1972), 135.

38. Mireille Moutardier et al., eds., *Cinquante ans d'INSEE* (Paris: INSEE, 1996), 26–45; letter with attachment, General Allard, to Minister of War, "Situation Militaire du Service du Recrutement et des Statistiques," 4 October 4, 1944, folder 4, box 7–P82, series P, Archives Bull.

39. "Étude de la chaîne de travail pendant la période de mechanisation, Ser-

vice technique," n.d. [before 12 June 1941], box B 55,359, Centre des Archives économiques.

40. Compatibility between IBM and Bull alphanumeric punched card standards was only enabled by the introduction of computers after the Second World War.

41. "Recensement des activités professionnels. La chaîne de travail pendant la période de mécanisation, Direction de la demographie," 22 July 1941, box B 55,359, Centre des Archives économiques.

42. Project de carnet individual, 1 February 1941, box B 55,358, Centre des Archives économique; "Étude de la chaîne de travail"; René Carmille, *La mécanographie dans les administrations*, 2nd ed. (Paris: Recueil Sirey, 1942), 122–130.

43. Régences d'Avances, État français, 2 August 1943, box B 55,349, Centre des Archives économiques; Jean-Pierre Azéma, Raymond Lévy-Bruhl, and Béatrice Touchelay, *Mission d'analyse historique sur le système de statistique français de 1940 à 1945* (Paris: INSEE, 1998), 15, 49.

44. Le Ministre, Secrétaire d'État à l'Economie Nationale et aux Finances, to the Interior Minister, 7 April 1941; Le Directeur Régional de la Demographie à Marseille, to Service de la Demographie, Lyon, 12 November 1941, 28 November 1941, and 3 March 1943; Direction de l'Administration de la Police, to Service National des Statistiques (SNS), 11 August 1943; all five documents in box B 55,358, Centre des Archives économiques.

45. Report, SNS Toulouse, to SNS Lyon, 7 March 1944, box 55,359, Centre des Archives économiques.

46. Note relative au Service National des Statistiques (Services des Enquêtes Démographiques), 1944, box B 55,364, Centre des Archives économiques; A. Caffot, "A propos d'un anniversaire ou de la statistique au camouflage," *Bulletin de l'AIS* 25 (1955): 34–50; Azéma, Lévy-Bruhl, and Touchelay, *Mission d'analyse*, 44–45; Robert Carmille, *Services statistiques français pendant l'occupation* (Saint-Cloud: Le Cornec, 2000), 23–27. The endeavor, in 1941–1942, to reorganize the police in the "occupied zone" is a similar example; see Baruch, *Servir l'État*, 384–388.

47. A register of 120,000 people of special military importance was kept until the liberation of France in 1944, a number by and large within the limit of the armistice agreement; Carmille, *Services statistiques français*, 28.

48. Azéma, Lévy-Bruhl, and Touchelay, *Mission d'analyse*, 26, 29.

49. Azéma, Lévy-Bruhl, and Touchelay, *Mission d'analyse*, 45–46; Carmille, *Services statistiques français*, 61–62.

50. Carmille, *Services statistiques français*, 196; Jacques Adler, *The Jews of Paris and the Final Solution*, (New York: Oxford University Press, 1987), 4.

51. Donna F. Ryan, *The Holocaust and the Jews of Marseille* (Urbana: University of Illinois Press, 1996), 24–25; Baruch, *Servir l'État*, 128–133.

52. Azéma, Lévy-Bruhl, and Touchelay, *Mission d'analyse*, 53.

53. Joseph Billig, *Le commissariat général,* 195–210; William Seltzer, "Population Statistics, the Holocaust, and the Nuremberg Trials," *Population and Development Review* 24 (1998): 511–552, on 520–522; Azéma, Lévy-Bruhl, and Touchelay, *Mission d'analyse*, 51–55.

54. "Êtes-vous de race juive?" Bulletin individuel de recensement des activités professionnel, 1941, box 55,359, Centre des Archives économiques.

55. Azéma, Lévy-Bruhl, and Touchelay, *Mission d'analyse*, 19–20. Edwin Black, in *IBM and the Holocaust: The Strategic Alliance between Nazi Germany and America's Most Powerful Corporation* (New York: Crown, 2001), 580, claims that the information about people being Jewish was never punched, but he provides no valid substantiation. That information on whether people were Jewish was punched before December 1942 has been substantiated in Ministre de l'Economie Nationale et des Finances, "Note sur l'établissement et la mise à jour du bulletin individuel Activités Processionnelles," 17 December 1942, box 55,359, Centre des Archives économiques.

56. Adler, *The Jews of Paris*, 7; John F. Sweets, *Choices in Vichy France: The French under Nazi Occupation* (New York: Oxford University Press, 1986), 119.

57. René Carmille, "Sur le germanisme," *Revue politique et parlementaire*, September–October 1939, 31–52.

58. Billig, *Le commissariat général,* 1957, 210–211; Azéma, Lévy-Bruhl, and Touchelay, *Mission d'analyse*, 53.

59. Azéma, Lévy-Bruhl, and Touchelay, *Mission d'analyse*, 54.

60. SNS Toulouse to SNS Lyon, 7 March 1944, box 55,359, Centre des Archives économiques.

61. The several distinct censuses and Jewish refugees in France complicate this estimate. Adler, *The Jews of Paris*, 6.

62. Michel Lévy, "Le numéro INSEE: De la mobilisation clandestine (1940) au projet Safari (1974)," *Dossier et Recherches* 86 (2000): 23–34, on 28; Seltzer "Population Statistics," 521–523; Azéma, Lévy-Bruhl, and Touchelay, *Mission d'analyse*, 56; Adler, *The Jews of Paris*, 6.

63. Azéma, Lévy-Bruhl, and Touchelay, *Mission d'analyse*, 29, 34, 39, 50–51.

64. "Grundlage für eine Geschichte der deutschen Wehr- und Rüstungswirtschaft" (c. 1942), 19–20, frames 717,725–717,726, roll 5, T-77, National Archives (NA); Gregor Janssen, *Das Ministerium Speer: Deutschlands Rüstung im Krieg* (Frankfurt am Main: Verlag Ullstein, 1968), 13–15.

65. Minutes of the Army Lochkarten-Ausschuss meetings between 10 October and 31 December 1937; Dr. Rudelsdorff, "Übersicht über die Vorarbeiten

zur Einführung des Lochkartenverfahrens für de Mob-Planung und Friedensrohstoffbewirtschaftung," 1 October 1937; all in frames 176,421(176,486, roll 339, T-77, NA.

66. Walther Lauersen, "Organisation und Aufgaben des Maschinellen Berichtswesens des Reichsminister für Rüstung und Kriegsproduktion," typescript, Hamburg, 5 December 1945, vol. 3, R-3/17.a, Bundesarchiv, Berlin-Lichterfelde; Kurt Passow, "Das 'Maschinelle Berichtswesen' als Grundlage für die Führung im II. Weltkrieg," *Wehrtechnische Monatshefte*, 62 (1965), 3–6; Hartmut Petzold, *Rechnende Maschinen. Eine historische Untersuchung ihrer Herstellung und Anwendung vom Keiserreich bis zur Bundesrepublik* (Düsseldorf, Germany: VDI Verlag, 1985), 246–257.

67. Meckenstock and Schüssler, "Das Lochkartenverfahren beim 'Hebedienst für Elektrizität, Gas und Wasser' in Frankfurt a. M.," *Hollerith Nachrichten*, 61 (1936): 831–848; Fritz Abendroth, "Die Anwendung des Hollerith-Lochkartenverfahrens im Einziehungswesen Der Berliner Städtischen Wasserwerk A.G.," *Hollerith Nachrichten*, 66 (1936): 918–920; Pankraz Görl, "Technisierung der Administration. Maschinelle Datenverarbeitung und die Rationalisierung der Verwaltung in Deutschland 1924–1945" (unpublished MA thesis, Ludwig-Maximilians-Universität, Munich, 1993), 87–90.

68. "Der Hollerith-Rechenlocher," *Hollerith Nachrichten*, 80 (1938): 1137–1138; Friedrich W. Kistermann "The Way to the First Automatic Sequence-Controlled Printing Calculator: The 1935 DEHOMAG D11 Tabulator," *IEEE Annals of the History of Computing* 17:2 (1995): 33–49, on 46.

69. H. Rottke to W. Heidinger, Neue Modelle/Alphabet-Tabelliermaschinen, 26 June 1940, and H. Rottke to W. Heidinger, D 11 A-Tabelliermaschine, 3 September 1940, both in B-95/71, Wirtschaftsarchiv Baden-Württemberg, Stutgart-Hohenheim; H. Rottke, Rundschreiben Nr. 42/6: Betrifft: Tabelliermaschine Type D 11, 26 April 1942, B-95/85, Wirtschaftsarchiv Baden-Würtemberg; Fellinger, Niederschrift über die Entwicklung der Beziehungen zwischen der Dienststelle M.B . . . . und der Dehomag, 8 October 1943, 8; Willy Heidinger, Aktennotiz betr. MB-Dienststelle . . . , 2 March 1944; both in R-87/6249, Bundesarchiv, Berlin-Lichterfelde; Fellinger, International Business Machines Corporation, Teilbericht no. 6, 10 August, R-87/6250, Bundesarchiv, Berlin-Lichterfelde.

70. Note, Von der Wehrmacht übernommene Maschine amerikanischen Ursprungs ohne bisher erteilte Devisengenehmigung (Equiment of U.S. origin assumed by German military 11 October 1939, B-95/120, Wirtschaftsarchiv Baden-Württemberg.

71. Richard J. Overy, *War and Economy in the Third Reich* (Oxford: Clarendon Press, 1994), 242.

72. Lauersen, "Organisation und Aufgaben," 45–51; Passow, "Das 'Maschinelle Berichtswesen,'" 136.

73. Rüstungsinspektion des Wehrkreises XVII (German armament inspection in Vienna), to Oberkommando der WehrmachtUnterteilung der Personalmeldung nach Gerätegruppe, 12 August 1941, frame 1,633,318, roll 470, T-77, NA; Lauersen, "Organisation und Aufgaben," 2, 20–21; Passow, "Das 'Maschinelle Berichtswesen,'" 136–138.

74. Janssen, *Das Ministerium Speer*, 33, 38–39.

75. Albert Speer, *Erinnerungen* (Berlin: Propyläen Verlag, 1969), 233; Overy, *War and Economy*, 354–355.

76. Overy, *War and Economy*, 355–356; Joachim Fest, *Speer. Eine Biographie* (Frankfurt am Main: Fischer Tachenbuch Verlag, 2001), 182–189.

77. Speer, *Erinnerungen*, 223–224; Janssen, *Das Ministerium Speer*, 42–48, 56–57.

78. Passow, "Das 'Maschinelle Berichtswesen,'" 10; Janssen, *Das Ministerium Speer*, 52.

79. Tables, Zahl vorhandenen Lochkartenmaschinen, October 1944 and Übersicht des Fachpersonals, October 1944, R-3/1294, Bundesarchiv, Berlin-Lichterfelde; Speer, *Erinnerungen*, 224.

80. Rolf Wagenführ, *Die deutsche Industrie im Kriege 1939–1945* (Berlin: Duncker & Humblot, 1954), 139.

81. Formular AG 310 Beschäftigtenmeldung, 20 August 1942; Formular AG 311: Industriebericht/Umsatz; numeric wiring scheme (D-11) for processing AG 310, no date; Bearbeitungsplan für Beschäftigungsmeldung und Industriebericht, 10 November 1942; wiring scheme for an alphanumeric IBM 405 tabulator; proposal for report from an unidentified company, no date, 39; all five in R-3/25, Bundesarchiv, Berlin-Lichterfelde; the text of a lecture by an unidentified individual; Planung der Beschäftigten Meldung in den Betrieben der Rüstungswirtschaft, 29 September 1942; vol. 1, R-3/1086, Bundesarchiv, Berlin-Lichterfelde. Circular, Reichsminister für Bewaffnung und Munition für die deutschen Betriebsführer, Einführung zum Beschäftigtenmeldebogen, 28 July 1942; Notiz für die Presse über die am 14. August 1942 neu durchführende Beschäftigtenmeldung mit Industriebericht, 27 July 1942; both in frames 1,633,473–1,633,487, roll 470, T-77, NA.

82. Wagenführ, *Die deutsche Industrie*, 66–67.

83. Lauersen, "Organisation und Aufgaben," 143.

84. The construction of an alphanumeric punched card register of ill soldiers appears to date back to 1941. Lauersen, "Organisation und Aufgaben," 40; Passow, "Das 'Maschinelle Berichtswesen,'" 107–109; Götz Aly and Karl Heinz Roth, *Die restlose Erfassung. Volkszählen, Identifizieren, Aussondern im Nationalsozi-*

alismus (Berlin: Rotbuch Verlag, 1984), 132. Their allusion to earlier use of the German personal number is not substantiated.

85. Passow, "Das 'Maschinelle Berichtswesen,'" 98–101.

86. Note, W. Heidinger, Betr.: MB-Dienststelle bzw. Beirat, 2 March 1944; report, Fellinger to Reichskommissar (Fellinger was Dehomag's custodian, i.e., a German-government-appointed president), Betrifft: Deutsche Hollerith Maschinen Gesellschaft m.b.H., 9 August 1944; both in vol. 2, R-87/6249; Lauersen, "Organisation und Aufgaben," 44–45.

87. Aktenvermerk Betr. Durchführung der Personaleinzelerfassung bei den Gühring-Werken, Waldeburg, 29 November 1944, vol. 2, R-3/1293, Bundesarchiv, Berlin-Lichterfelde; Lauersen, "Organisation und Aufgaben," 32–34; Aly and Roth, *Die restlose Erfassung*, 131–132.

88. Report, "Die Reichspersonalkartei, MB," 1 August 1944; report, Herbst, "Die Personaleinzelerfassung, Ansbach," 29 September 1944; "Bericht über die Personaleinzelerfassung in Ansbach," 2 November 1944; "Aktenvermerk betr. Personaleizelerfassung," 16 November 1944; V. Nullau, "Mitteilung betrifft: Einführung der Personal-Hollerith-Kartei," 16 January 1945; all in vol. 2, R-3/1293, Bundesarchiv, Berlin-Lichterfelde; Lauersen, "Organization und Aufgaben," 33–36; Passow, "Das 'Maschinelle Berichtswesen,'" 143–144; Aly and Roth, *Die restlose Erfassung*, 133–140.

89. T. G. Belden and M. R Belden, *The Lengthening Shadow: The Life of Thomas J. Watson* (Boston: Little, Brown and Co., 1962), 194–195; Black, *IBM and the Holocaust*, 132–134.

90. Liste der z.Zt. in Deutschland befindlichen Maschinen der Compagnie Electro-Comptable, 15 January 1944, vol.1, R-3/1154, Bundesarchiv, Berlin-Lichterfelde.

91. Group of leasing contracts of the German trustees in Brussels and Amsterdam of the International Business Machines Corporation, New York, with the Maschinelles Berichtwesen concerning various types of machines, 1942, frames 1,053,937–1,054,002, roll 8, T-73, National Archives; Vertrag vom 23.6.1944 zwischen M.B. [Maschinelles Berichtwesen] Berlin und Compagnie Electro Comptable, vol. 1, R-3/1154, Bundesarchiv, Berlin-Lichterfelde; note, An Firmen . . . an geliehene bzw. abgegebene CEC [Compagnie Electro Comptable]). Maschinen. 26.6.1944, vol.2, R-3/1274, Bundesarchiv, Berlin-Lichterfelde.

Historians Götz Aly and Karl Heinz Roth accuse IBM of having leased punched-card machines to German customers after the United States entered into the war. The contract they cite is one of the contracts mentioned above. Aly and Roth, *Die restlose Erfassung*, 123–124. The contract in frames 1,003,937–1,003,938, roll 8, T-77, National Archives. It is not signed by IBM, but by a German custodian.

92. Belden and Belden, *The Lengthening Shadow*, 195–196, 207.

93. W. Heidinger, Aktennotiz Betr. Watson, 23 August 1940; T. J. Watson, to W. Heidinger, 2 October 1941 (transmitted by the Foreign Service of the U.S.); Rottke, Bericht über die Besprechung zwischen Herrn Major Passow . . . Betrifft: Watson, 14 December 1940; note, W. Heidinger, Betrifft. Watson, 4 February 1941; Westerholt, Bericht über eine Besprechung beim OKH, Major Passow, 4.3.41 . . . ; W. Heidinger, Bericht des Aufsichtsrat, Betrifft Verdeutschung der Dehomag, 12 November 1941; all in B-95/39, Wirtschaftsarchiv Baden-Württemberg; Black, *IBM and the Holocaust*, 218–258, 270–273, 277–291.

94. Hartmut Petzold, *Moderne Rechenkünstler. Die Industrialisierung der Rechentechnik in Deutschland* (Munich: Verlag C. H. Beck, 1992), 92–93, 96–100.

95. Wanderer-Werke vorm. Winklhofer & Jaenicke A.-G. in Schönau b. Chemnitz, Verbindung einer Rechenmaschine mit einer Kartenlochmaschine, [German] Patent 576,616, filed 1929 and issued 1933; Michael Schneider, *Unternehmensstrategien zwischen Weltwirtschaftskrise und Kriegswirtschaft: Chemnitzer Maschinenbauunternehmen während der NS-Zeit 1933–1945* (Essen, Germany: Klatext, 2005), 288–292; Michael Schneider, "Business Decision Making in National Socialist Germany: Machine Tools, Business Machines, and Punch Cards at the Wanderer-Werke AG," *Enterprise & Society* 3 (2002): 396–428, on 414–417.

96. Voigt, Aktennotiz, 21 May 1940; OKH to Siemens, Betr.: Patentanmeldungen des Herrn Obering. Voigt, 7 December 1940; letter, OKH to Siemens, 12 June 1941; all in 21/Lg889, Siemens Archiv.

97. Organizational diagram dated 16 November 1940, R3/541, Bundesarchiv, Berlin-Lichterfelde.

98. Schneider, *Unternehmensstrategien*, 293–299; Schneider, "Business Decision Making," 417–420. A similar contract was concluded on the Dutch Kamatec punched card machines to no avail. Schneider, "Business Decision Making," 273–277, 370–371.

99. Letter, Fellinger, to Reichskommissar, Betrifft deutsche Hollerith Maschinen Gesellschaft m.b.H, 10 October 1943; report, Fellinger, International Business Machines Corporation. Anlage . . . betreffen Dienststelle Maschinelles Berichtwesen und Wanderer-Werke A.G., 25 July 1945; both in vol. 2, R-87/6250, Bundesarchiv, Berlin-Lichterfelde; Schneider, *Unternehmensstrategien*, 305–308; Schneider, "Business Decision Making," 420–421.

100. Schneider, *Unternehmensstrategien*, 309–316, 403–405.

101. Note, Voigt to P. Storch, Betr.: Entwicklungsarbeit auf dem Gebiete der Lochkarten-Maschinen, 9 September 1941, 21/Lg889, Siemens Archiv; Report, Fellinger, International Business Machines Corporation. Anlage . . . betreffend Dienststelle Maschinelles Berichtwesen und Wanderer-Werke A.G., 25 July 1945,

vol. 2, R-87/6250, Bundesarchiv, Berlin-Lichterfelde; Schneider, *Unternehmensstrategien*, 312–313. Schneider does not provide reasons for his assessment of divergent strategies in the two contracts.

102. Schneider, *Unternehmensstrategien*, 404–409, 424–426.

103. Note, W. Heidinger, Konkurrenz Wanderer, Bull resp. MBD und JBM-Vertrag, 14 June 1944, and note, W. Heidinger, Einführer von Konkurrenzmaschinen, 14 June 1944, both in B-95/102, Wirtschaftsarchiv Baden-Württemberg; report, Fellinger, International Business Machines Corporation, Anlage . . . betreffend Dienststelle Maschinelles Berichtwesen und Wanderer-Werke AG., 25 July 1945; both in vol. 2, R-87/6250, Bundesarchiv, Berlin-Lichterfelde; Schneider, *Unternehmensstrategien*, 415–424.

104. The incomplete Maschinelle Berichtwesen archives contains the article "'Wocci' Knows All about the Army," *Daily Herald*, 23 November 1944, on a British draft service punched card register, vol. 2, Akten Nr.921/45, Bundesarchiv, Berlin-Lichterfelde.

105. Passow, "Das 'Maschinelle Berichtswesen,'" 65, 103, 105, 106.

106. W. Heidinger, Betrifft Beirat, 18 June 1943; Fellinger, Niederschrift über die Entwicklung der Beziehungen zwischen der Dienststelle M.B. des Reichsministers für Rüstung und Kriegsproduktion und der Dehomag, 8 October 1943; W. Heidinger, Aktennotiz betr.: MB-Dienststelle bzw. Beirat, 2 March 1944; report, Fellinger, International Business Machines Corporation. Anlage . . . betreffend Dienststelle Maschinelles Berichtwesen und Wanderer-Werke A.G., 25 July 1945; all in vol. 2, Akten Nr. 921/45, R-87/6249, Bundesarchiv, Berlin-Lichterfelde.

107. Report, Fellinger to Reischskomissar, Betrifft: Deutsche Hollerith Maschinen Gesellschaft m.b.H., 24 November 1943; report, Fellinger, International Business Machines Corporation, Teilbericht Nr. 6: Deutsche Hollerith Maschinen Gesellschaft m.b.H., 10 August 1945; both in in vol.2, Akten Nr. 921/45, R-87/6249, Bundesarchiv, Berlin-Lichterfelde.

108. Schneider, *Unternehmensstrategien*, 409–415; Schneider, "Business Decision Making," 423–424.

109. Reports, Fellinger, to Reischskomissar, Betrifft: Deutsche Hollerith Maschinen Gesellschaft m.b.H., 11 October 1944 and 24 November 1943; Fellinger, International Business machines Corporation, Teilbericht 5: Compagnie Electro-Comptable, Paris, 30 July 1945; Teilbericht Nr. 6: Deutsche Hollerith Maschinen Gesellschaft m.b.H., 10 August 1945, Anlage I zum teilbericht 6 betreffend Dienststelle Maschinelles Berichtwesen and Wandere-Werke A.G., 25 July 1945; all in vol.2, Akten Nr. 921/45, R-87/6249, Bundesarchiv, Berlin-Lichterfelde.

110. Lauersen, "Organization und Aufgaben," 36.

111. Speer, *Erinnerungen*, 325–328, 340–356; Fest, *Speer*, 246–251, 263–298.

112. Aly and Roth, *Die restlose Erfassung* and Edwin, *IBM and the Holocaust*, did not provide any new insight on the alleged Dehomag complicity in the Holocaust.

113. Ludwig Hümmer, "Die Aufbereitung der Volks- und Betriebszählung 1933 im Hollerith-Lochkartenverfahren," *Hollerith Nachrichten*, 28 (1933): 343–355; Friedrich Burgdörfer, "Die Volks-, Berufs- und Betriebszählung 1933," *Allgemeine statistische Archiv*, 23 (1933/34): 145–171; [Statistische Reichsamt] "Erhebung- und Bearbeitungsplan der Volks, Berufs- und Betriebszählung 1933," *Statistik des Deutschen Reichs* 467 (1939): 5–11; Aly and Roth, *Die restlose Erfassung*, 20–21; Friedrich W. Kistermann, "Locating the Victims: The Nonrole of Punched Card Technology and Census Work," *IEEE Annals of the History of Computing* 19 (1997), 31–45, on 38.

114. Roderich Plate, "Die erste Großdeutsche Volks-, Berufs- und Betriebszählung 1939," *Allgemeines Statistische Archiv*, 28 (1938/39): 421–436; Bieler, "Lochkartenmaschinen im Dienste der Reichsstatistik," *Allgemeines Statistische Archiv*, 28 (1939), 94–95; "Die Juden und jüdische Mischlinge im Deutschen Reich," *Wirtschaft und Statistik*, 20:5/6 (1940), 84–87; "Volks- Berufs und Betriebszählung vom 17. Mai 1939," *Statistik des Deutschen Reiches*, 552:1 (1943); David Martin Luebke and Sybil Milton, "Locating the Victim: An Overview of Census-Taking, Tabulation Technology, and Persecution in Nazi Germany," *IEEE Annals of the History of Computing*, 16:3 (1994): 25–39, on 28, 30–31; Kistermann, "Locating the Victims," 38–40.

No punched card from the processing of this census was actually located, but there was also no indication of a shift from punched cards as a tool for processing statistics to a register tool before 1942.

115. Luebke and Milton, "Locating the Victim," 25–39.

116. Martin Brozat, *Der Staat Hitlers* (Munich: Deutscher Taschenbuch Verlag, 1969), 155, 425; Luebke and Milton, "Locating the Victim," 26, 32.

117. Erich Liebermann von Sonnenberg and Arthur Kääb, eds., *Die Reichsmeldeordnung: Handausgabe mit Erläuterungen* (Munich: J. Jehle, 1942), 1–19, 104–130, first edition in 1938; Aly and Roth, *Die restlose Erfassung*, 39–43; Luebke and Milton, "Locating the Victim," 28–29.

118. Erich Liebermann von Sonnenberg and Arthur Kääb, *Die Volkskartei: Ein Handbuch* (Munich: J. Jehle, 1940), 136–141.

119. Luebke and Milton, "Locating the Victim," 29.

120. Germany reintroduced conscription in 1935.

121. "Verordnung über die Errichtung einer Volkskartei vom 21. April 1939," *Reichsgesetzblatt*, 1939, vol. 1, 823; von Sonnenberg and Kääb, *Die*

Volkskartei; Klaus Heinicken, "Die Volkskartei," *Allgemeines Statistisches Archiv* 31 (1941–42): 39–44; Aly and Roth, *Die restlose Erfassung*, 44–50; Luebke and Milton, "Locating the Victim," 31.

122. Luebke and Milton, "Locating the Victim," 31.

123. Luebke and Milton, "Locating the Victim," 32–34.

124. *Historical Record of Card and Machine Development* (New York: IBM, 1955), IBM Archives.

125. J. F. Brennan, *The IBM Watson Laboratory of Columbia University: A History* (New York: IBM, 1971), 3–5; Wallace John Eckert, *Punched Card Methods in Scientific Computation* (New York: Columbia University, 1940), 23–24.

126. George Walter Baehne, ed., *Practical Applications of the Punched Card Method in Colleges and Universities* (New York: Columbia University Press, 1935), 177–436; Eckert, *Punched Card Methods in Scientific Computation*; Leslie John Comrie, "The Applications of Commercial Calculating Machines to Scientific Computation," *Mathematical Tables and Other Aids to Computation* 2 (1946): 157–158.

127. Herman H. Goldstine, *The Computer from Pascal to von Neumann* (Princeton, NJ: Princeton University Press, 1972); Petzold, *Moderne Rechenkünstler*; Martin Campbell-Kelly and William Asprey, *Computer: A History of the Information Machine* (New York: Basic Books, 1996).

128. Charles J. Bashe, Lyle R. Johnson, John H. Palmer, and Emerson W. Pugh, *IBM's Early Computers* (Cambridge, MA: MIT Press, 1986); Martin Campbell-Kelly, *ICL: A Business and Technical History* (Oxford: Oxford University Press, 1989); JoAnne Yates, "Co-evolution of Information-Processing Technology and Use: Interaction Between the Life Insurance and Tabulating Industries," *Business History Review* 67 (1993): 1–51; Lars Heide, *Hulkort- og edb i Danmark, 1911–1970* (Århus, Denmark: Systime 1996).

129. Leslie John Comrie, "The Application of the Hollerith Tabulating Machine to Brown's Tables of the Moon," *Monthly Notices of the Royal Astronomical Society* 92 (1933): 694–707; Leslie John Comrie, "Computing of the Nautical Almanac," *The Nautical Magazine* 130 (1933): 44–48.

130. Bashe, Johnson, Palmer, and Pugh, *IBM's Early Computers*, 459–460; Campbell-Kelly, *ICL*, 193–194.

Conclusion

1. Wiebe E. Bijker, *Of Bicycles, Bakelites, and Bulbs* (Cambridge: MA: MIT Press, 1995), 122–126, 143, 190–196, 260–267.

2. For example, Thomas P. Hughes, *Networks of Power: Electrification in*

Western Society (Baltimore: Johns Hopkins University Press, 1983); Hans-Liudger Dienel, *Ingenieure zwischen Hochschule und Industrie: Kältetechnik in Deutschland und Amerika, 1870–1930,* (Gottingen, Germany: Hoechks & Rupprecht, 1995); Eda Kranakis, *Constructing a Bridge: An Exploration of Engineering Culture, Design, and Research in Nineteenth-Century France and America* (Cambridge, MA: MIT Press, 1997).

3. Mikael Hård and Andreas Knie, "The Grammar of Technology: German and French Diesel Engineering, 1920–1949," *Technology and Culture* 40 (1999), 26–46. Similarly, historian Matthias Heymann only managed to uncover elements of national style in his article "Signs of Hubris: The Shaping of Wind Technology Styles in Germany, Denmark, and the United States, 1940–1990," *Technology and Culture* 39 (1998), 641–670.

4. Hughes, *Networks of Power.*

5. Russia had its first general census in 1897. The next census was in 1926.

6. David E. Nye, *American Technological Sublime* (Cambridge, MA: MIT Press, 1994).

7. Alfred D. Chandler, *Scale and Scope: The Dynamics of Industrial Capitalism* (Cambridge, MA: Harvard University Press, 1990).

8. Maurice Lévy-Leboyer, "The Large Corporation in Modern France," *Managerial Hierarchies: Comparative Perspectives on the Rise of the Modern Industrial Enterprise,* ed. Alfred J. Chandler and Herman Daems (Cambridge, MA: Harvard University Press, 1980), 117–160, on 118–119.

9. For example, Chandler, *Scale and Scope*; Geoffrey Jones, "Great Britain: Big Business, Management, and Competitiveness in Twentieth-Century Britain," *Big Business and the Wealth of Nations,* ed., A. D. Chandler et al. (Cambridge: Cambridge University Press, 1997), 102–138.

Essay on Sources

The shaping of punched-card technology is the focal point of this book. Individuals shaped the technology and, in most cases, they worked in groups of engineers and craftsmen. Their options and choices were based on their training and professional experience as well as their interaction with people in their work, including end users and customers, real or imagined.

The preserved material limits the analysis of the shaping of technology by the various punched-card producers in two ways: little material is preserved from the shaping processes of the early producers and only larger producers have well-preserved archives. First, the few written records that have been preserved from the development process include, for example, drawings, letters, and reports. Further, the shaping process was based on nonrecorded knowledge. To cope with these limitations on the preserved material, it was necessary, to a large extent, to refer to reflections of this process on different company-based material, which include papers and minutes from different boards of directors and management boards, memoirs, company histories based on reminiscences, and records that have vanished since the company history was written. Much of this material was created on a company level, which limits the amount of details on how the technology evolved, and yet contributes to a reasonably detailed representation of the industry's shaping of the technology.

Second, the preserved material in the archives of the various punched-card producers is limited by its unevenness. While IBM and its predecessors and the French Compagnie des machines Bull have extensive archives from the period up to the Second World War, much less material exists from the smaller producers and very little material exists from foreign agencies and subsidiaries with little or no innovation. Therefore, the materials handed down from the bigger companies tend to dominate the representation.

To counteract this imbalance, three additional groups of material have been used: public antitrust records, public patent records, and records on punched-card users. This material facilitated locating information about contributions to the shaping process by individuals or companies without well-preserved and accessible archives. Also, this material complements the preserved and accessible company material. For example, it facilitates the study of details in the complex processes of the shaping of punched cards for bookkeeping at the Tabulating Machine Company and IBM between 1911 and 1933 and at the Bull company in the 1930s.

The public records from the antitrust suit against IBM and Remington Rand between 1932 and 1936 provide an overview of the punched-card industry in the United States. Also, this material identifies the essential technical features in the development of this technology and the importance of the patents on automatic group control.

A patent contains a description of the patented device, and in many cases it outlines the envisaged applications. First, patents have contributed considerably to identifying significant new technical features that were discarded subsequently, and they facilitate understanding the potential and implications of essential features. Second, patents provide the same information about all members of the industry. Particularly, this has contributed to understanding the development of members of an industry in which few company records on technological development have survived.

Patents also provide access to fruitful related information on the processing of the original patent application, the litigation on patents, and the variations between the patent laws in different countries, which had considerable implications for the industry. However, the archival material on the processing of patent applications and patent litigation is huge and the archives on patent litigation are dispersed throughout the various countries, as patent cases were heard in the lower courts first (with the exception of Herman Hollerith's patent infringement suit against his government in 1910–1912). Therefore, this material is only used in this book in situations concerning the processing of patents and for court cases indicated elsewhere in the material that are considered to be important for the analysis.

Actual and envisioned users were important for how the technology was shaped in that inventors selected features with them in mind. Furthermore, customers were essential for the success or failure of the produced technology and for the shaping of a company's organization, particularly its sales departments. Sales meant the collection of information about experience from the actual use of the technology and user wishes and ideas concerning additional features. As most interactions with users were simple, secondary sources were sufficient documentation. However, in several cases the interaction with a user was essential for the shaping process because a new application or new technical feature was suggested. In this book, a detailed analysis is made of producer relations with several of the users seen as essential in the shaping process, including census offices, the Bureau of the Census, and the Social Security administration in the United States; the Corps du contrôle de l'administration de l'Armée de terre and the Service de la démographie in France; and the Maschinelles Brichtswesen in Germany. A detailed analysis of additional users in the four countries is not included, but would have produced different details, though their addition apparently would not have resulted in a significantly different story.

Using diverse groups of material established a robust and detailed net of information. This net provides a robust representation of the main features of the development of the technology and facilitates distinguishing between the technology's varying characteristics in the four countries.

Archives on Producers

IBM was the prime mover of punched cards in the United States, dating back to Herman Hollerith's work. The IBM Archives, which are in New York, contain mainly company level material and the minutes of board meetings and of various executive committees for the period up to the Second World War. In addition, the archives hold extensive material from the company's own technical history project, which ran from the 1960s up to 1993. The researchers on these projects conducted oral interviews with key people in the technical development process and collected archives established by engineers during their work at the company that would not otherwise have been preserved. (Copies of these interviews are in the archives of Charles Babbage Institute at the University of Minnesota, Minneapolis, Minnesota.) Also, the IBM Archives hold material on several competing inventors and producers, notably inventors John Royden Pierce and Gustav Tauschek, who IBM hired and whose patents they acquired.

The IBM Archives have a collection of material related to Herman Hollerith and his activities, but Hollerith's private archive is kept in the Manuscript Division of the Library of Congress, Washington, D.C. The correspondence between Hollerith and George F. Swain is in the Baker Library at Harvard Business School in Boston, Massachusetts, and the Pusey Library at Harvard University in Cambridge, Massachusetts.

Two sets of material from public archives were used to complement the handed-down material in the IBM Archives. First, there were the records of Hollerith's patent infringement suit against the Bureau of the Census, 1910–1912, Equity Case No. 29,065 of the Supreme Court of the District of Columbia. The records are in the Washington National Records Center in Suitland, Maryland. Second, there are the court files of the government's antitrust suit against IBM and Remington Rand, 1932–1935, first read as Equity Case No. 66-215 in the Southern District of New York. Its records are in the National Archives, Northeast Region, in New York. The appeal to the Supreme Court of the United States, File No. 758 of 1935 is in the Washington National Records Center. The file on IBM in the Antitrust Division of the Department of Justice is in the records of the Department of Justice (RG-60) in the National Archives in College Park, Maryland.

The machine workshop in the Bureau of the Census was the first challenger to Hollerith's punched-card monopoly. The archives of the Bureau of the Census (RG-29), which are located in the National Archives in College Park, Maryland, hold some material on the activities in the machine shop.

Powers Accounting Machine Company was the first commercial challenger to Hollerith's company. The preserved records from this company are in the archives of Remington Rand Corporation (Accessions 1825 and 1910) at the Hagley Museum and Library in Wilmington, Delaware. However, very little is preserved from the Powers company. Further, a personnel file on James Powers from his time at the Census Bureau machine shop is in the National Personnel Records Center in St. Louis, Missouri. In addition, a small collection from Powers' private archives is in the National Museum of American History in Washington, D.C. (No.1991.3180.09).

In Great Britain, the British Tabulating Machine Company and the British Powers Machine Company merged in 1959 and, subsequently, the new company was named International Computers Limited (ICL). The preserved parts of the archives from the British Tabulating Machine Company and the British Powers Machine Company, however, are in three locations. First, board minutes and papers are kept by ICL, their exact location, though perhaps in London, is unknown to the author, who tried in vain to get access to them. Martin Campbell-Kelly, however, had access to this material when he wrote the book, *ICL: A Business and Technical History* (Oxford: Oxford University Press, 1989) and was so kind as to lend his notes and copies to the author. Second, their trade journals sit in the Library of Science Museum in South Kensington. Third, diverse material on technical development and sales activities are in the National Archive for the History of Computing in Manchester, England.

In Germany, the archives of Deutsche Hollerith Maschinen Gesellschaft mit beschränkter Haftung (Dehomag) (B-95) are in Wirtschaftsarchiv Baden-Württemberg, Stuttgart-Hohenheim. The archives of the company custodian during the First World War, the Zwangsverwaltung der feindlischen Beteiligung, are in the Geheimes Staatsarchiv, Preussischer Kulturbesitz, in Berlin. The archives of the company custodian during the Second World War, the Reichskommissariat für die Behandlung feindlichen Vermögens (R-87), are in the Bundesarchiv in Berlin-Lichterfelde.

No archives from the German Powers companies were located. The archives of the company custodian during the First World War, the Zwangsverwaltung der feindlischen Beteiligung, are in the Geheimes Staatsarchiv, Preussischer Kulturbesitz, in Berlin. The archives of the company custodian during the Second World War, the Reichskommissariat für die Behandlung feindlichen Vermögens (R-87), are in the Bundesarchiv in Berlin-Lichterfelde. Siemens' design and production of punched-card equipment was related to the second German Powers company. Papers related to this activity are in the Siemens' Archive in Siemens Forum, Munich.

In France, the archives of the Compagnie des machines Bull are in St. Denis, and the author obtained access through the Fédération des Équipes Bull in France. Supplementary material was obtained form the archives of Bull A/S in Oslo, Norway. In addition, the author was granted access to the diaries of Henrik Hartzner (in Danish) by his family. Files on the patent infringement suit by IBM against the Bull company are in the archives of the Cour d'Appel de Paris, Greffe Civil, Nos. H-1440 (1947) and H-1441 (1947) in the Archives de Paris.

Patents

Most of the patents since the 1880s are in national publications with yearly indexes. The patent publications are available at the national patent authorities, which have publications of foreign patents as well. For example, the Danish national patent authority has the printed patents from France, Germany, Great Britain, and the United States. In Austria and France, printing granted patents only started just before 1900. Lists of earlier patents exist at the national patent authority, and these patents are available in archives. In Austria they are located in the Kaiserliche und königliche Privilege Archiv in the Österreichisches Patentamt in Vienna, while in France, they are located at the Archive de brevet of the Institut national de la propriété industrielle in Paris.

All patents granted in the United States are available on the Internet (www.uspto.gov/patft/), and although they are accessible via their patent number, search tools for locating patents by, for example, inventor and subject are not yet available for the period before the interwar years. As a result, it is still necessary to use printed annual indexes to locate patents. The various European patent authorities are also engaged in publishing their patents on the Internet (www.espacenet.com). However, most European countries have fewer complete Internet publications for the start of the twentieth century than the United States does, which refers researchers to the printed national publications of patents.

A special service in the United States is the national patent authority's recording of the transfer of patent rights, which is a valuable entry for studying the business context of patents. The digest of the assignment of patents are in the records of the Patent and Trademark Office (RG-241) in the National Archives at College Park, Maryland.

Archives on Users

In the United States, the main organizational users in this book were those processing census returns in the United States between 1880 and 1940 and the Social Security administration. The records of the census offices in 1880, 1890, and 1900 and the permanent Bureau of the Census are in the Records of the Bureau of the Census (RG-29), which is in the National Archives in College Park, Maryland. The Records of the Social Security Administration (RG-47) are in the same location. Additional material lies in the Social Security Administration History Archives in Baltimore, Maryland. Its website holds extensive public material on Social Security (www.ssa.gov/history/history.html).

For Great Britain, Martin Campbell-Kelly published studies of data processing in five organizations in the nineteenth and early twentieth centuries: "Data Processing and Technological Change: The Post Office Savings Bank, 1861–1930," *Technology and Culture*, 39(1998): 1–32, "Information Technology and Organizational Change in the British Census, 1801–1911," *Information Systems Research*, 7(1996), 22–36, "Large Scale Data Processing in Prudential, 1850–1930" [in England], *Accounting, Business and Financial History*, 2(1992): 117–139, "Punched-Card Machinery," in *Computing before Computers*, ed. William Asprey (Ames: Iowa State University Press, 1990), 122–155, and "The Railway Clearing House and Victorian Data Processing," in *Information Acumen: The Understanding and Use of Knowledge in Modern Business*, ed. Lisa Bud-Frierman (London: Routledge, 1994), 51–74. Additional information is in the archives of the Records of the General Register Office (RG) and the Stationery Office (STAT) at the Public Record Office in Kew, London.

For Germany, an overview of data processing in private and public administration is in Pankraz Görl, *Technisierung der Administration: Maschinelle Datenverarbeitung und die Rationalisierung der Verwaltung in Deutschland 1924–1945* (unpublished M.A. thesis, Ludwig-Maximilians-Universität, Munich, 1993). Most of the archives of the Maschinelles Brichtswesen were destroyed during and just after the Second World War. The remaining parts are in the archives of the Reichsministerium für Rüstung und Kriegsproduktion (R-3) in the Bundesarchiv in Berlin-Lichterfelde.

In France, the archive of the Corps du contrôle de l'administration de l'Armée de terre is in the Archives du Service historique de l'Armée de terre in Château de Vincennes, Paris. The archives of the Service de la démographie are in the Centre des archives économiques et financières at Savigny-le-Temple near Paris.

Studies of Office Technologies

First, there is an extensive bibliographic guide by James W. Cortada, *A Bibliographic Guide to the History of Computing, Computers, and the Information Processing Industry* (New York: Greenwood Press, 1990). It is particularly useful for locating diverse publications on the various uses of information technologies. Second, the studies of office mechanization can be divided up into three groups: Technical studies, business histories, and recent studies focusing on the interaction between the technology and the related social structures.

Technical history studies on office technologies focus on the artifacts, rather than on how the artifacts relate to the cultural, economic, political or social context. The main emphasis is on telling the story of the emergence of the electronic computer and its subsequent history, with only a few studies appearing on precomputer technologies; see Hartmut Petzold, *Rechnende Maschinen; Eine historische Untersuchung ihrer Herstellung und Anwendung vom Kaiserreich bis zur Bundesrepublik* (Düsseldorf: VDI Verlag, 1985) and Hartmut Petzold, *Moderne Rechenkünstler; Die Industrialisierung der Rechentechnik in Deutschland* (Munich: C. H. Beck, 1992); Robert Ligonnière, *Préhistoire et histoire des ordinateurs* (Paris: Robert Laffont, 1987); Werner Lange, *Buchungsmachinen. Meisterwerke feinmechanischer Datenverarbeitung 1910 bis 1960* (Munich: R. Oldenbourg, 1986); Friedrich W. Kistermann, "The Invention and Development of the Hollerith Punched Card: In Commemoration of the 130th Anniversary of the Birth of Herman Hollerith and for the 100th Anniversary of Large Scale Data Processing," *IEEE Annals of the History of Computing*, 13(1991): 245–259, Lars Heide, "From Invention to Production: The Development of Punched-Card Machines by F. R. Bull and K. A. Knutsen, 1918–1930," *IEEE Annals of the History of Computing*, 13(1991): 261–272; Friedrich W. Kistermann, "The Way to the First Automatic Sequence-Controlled Printing Calculator: The 1935 DEHOMAG D11 Tabulator," *IEEE Annals of the History of Computing*, 17:2(1995): 33–49; Peggy Aldrich Kidwell, "The Adding Machine Fraternity at St. Louis: Creating a Center of Invention, 1880–1920," *IEEE Annals of the History of Computing*, 22:2(2000): 4–21; Peggy Aldrich Kidwell, "'Yours for Improvement': The Adding Machines of Chicago, 1884–1930," *IEEE Annals of the History of Computing*, 23:3(2001): 3–21.

These historians documented the technical development, but their narrow approach curtailed their contribution to understanding how this technology was shaped and its impact on society. Notable exceptions are Paul N. Edward's study of the role of computers in Cold War discourse in the United States in *The Closed World: Computers and the Politics of Discourse in Cold War America* (Cambridge, MA: MIT Press, 1997) and Jon Agar's history of the British civil service in *The Government Machine: A Revolutionary History of the Computer* (Cambridge, MA: MIT Press, 2003).

Another factor to consider is that the business history studies of the office machine industry have a narrow approach. They focus on the business context of the technology, providing important insights into the history of office machine producers. Their information about the importance of establishing sales organizations is useful, as it draws attention to the role of the customer. However, the relations between demand and the shaping of the products is not analyzed; see Saul Engelbourg, *International Business Machines: A Business History* (New York: Arno Press 1976); Gunnar Nerheim and Helge W. Nordvik, *'Ikke bare maskiner'. Historien om IBM i Norge 1935–1985* (Oslo: Universitetsforlaget, 1986); Geoffrey Austrian, *Herman Hollerith: Forgotten Giant of Information Processing* (New York: Columbia University Press, 1982); Martin Campbell-Kelly, *ICL: A Business and Technical History* (Oxford: Oxford University Press, 1989); Martin Campbell-Kelly, "Punched-Card Machinery," in *Computing before Computers*, ed. William Asprey (Ames: University of Iowa Press, 1990); Pierre-Eric Mounier-Kuhn, "Bull: A World-Wide Company Born in Europe," *Annals of the History of Computing*, vol.11 (1989), 279–29; James W. Cortada, *Before the Computer: IBM, NCR, Burroughs, and Remington Rand and the Industry They Created, 1865–1956* (Princeton, NJ: Princeton University Press, 1993).

In *Structuring the Information Age: Life Insurance and Technology in the Twentieth Century* (Baltimore: Johns Hopkins University Press, 2005), JoAnne Yates analyzed the social shaping of office technology from the perspective of the American life insurance industry and office technology card producers. Her study convincingly demonstrates the potentials of analyzing the interaction between producers and users.

In contrast to the top-down perspectives of producers and managers, studies in the history of office workers provide a bottom-up perspective. As society entrusted more and bigger assignments to offices, these offices changed in several ways. In addition to the new machines, offices gained a new social and educational composition, a different organizational structure, and a new look. The crucial story of how female secretaries and typists outdid the better-educated male clerks was the subject of several studies. Analysis threw light on the critical cultural, social, and economic changes within the office, which facilitated the introduction of office machines; see *The White Blouse Revolution: Female Office Workers since 1870*, ed., Gregory Anderson (Manchester: Manchester University Press, 1987); Margery W. Davies, *Woman's Place Is at the Typewriter: Office Work and Office Workers 1870–1930* (Philadelphia: Temple University Press, 1982); Samuel Cohn, *The Process of Occupational Sex-Typing: The Feminization of Clerical Labor in Great Britain* (Philadelphia: Temple University Press, 1985); Delphine Gardey, *Un mode en mutation: Les employés de bureau en France: Féminisation, mécanisation, rationalisation* (doctoral dissertation, Université Paris 7, 1995).

Index

Accounting and Tabulating Machine Company of Great Britain. *See* Powers company, British
adding machine, keyboard, 31, 32, 39, 40, 47, 51, 70, 71, 81, 86, 89, 98, 108, 141, 154, 198, 199, 207, 217, 358
adding tabulator. *See* tabulator
address plate, 117, 168–70, 173, 224, 225
agency. *See* International Business Machines; Powers Accounting Machine Company; subsidiary/subsidiaries; Tabulating Machine Company
American Arithmometer Company. *See* Burroughs Adding Machine Company
antitrust, 10, 34, 74, 82, 90, 111, 116, 214, 257, 352, 354
army punched-card service, German. *See* Maschinelles Berichtswesen
audit(s), 145, 148; railroad, 45, 46, 67
Aussedat. *See* Société des Papeteries Aussedat
Austria, 18, 41, 129–31, 133, 135, 136, 144, 147, 152, 257, 262, 356
Austrian, Geoffrey D., 54, 359
automatic group control. *See* tabulator

Baltimore (Maryland), 1, 18, 33, 215, 217, 356, 359
banks, 70, 97, 123; British, 356; French, 192–94; German, 122, 180, 183–85; U.S., 89
Bayer and Co. *See* Farbenfabriken vorm. Friedrich Bayer and Co.

Bell, Alexander Graham, 29
Bell Telephone Company, 34, 61
Bijker, Wiebe E., 7, 8, 257, 258
Billings, John S., 22, 23, 29, 32, 253, 258
Boston, 49–51, 67, 354
British Tabulating Machine Company (BTM), 82, 161, 167, 170, 210; establishment of, 139–42; and First World War, 145; inventions of, 142, 176, 177–80; and patents, 143–44; relationship with TMC/IBM, 140, 167, 175–78; sales, 141, 143, 177, 251, 263, 264. *See also* Tabulating Machine Company: agencies of
brush reading, 60, 75, 108
Bryce, James Wares, 113, 124
BTM. *See* British Tabulating Machine Company
Bull, Compagnie des Machines. *See* Compagnie des Machines Bull; Egli-Bull
Bull, Fredrik Rosing, 198, 358
Bundy Manufacturing Company, 106
Bureau of Internal Revenue (now Internal Revenue Service), 213, 215, 218
Bureau of the Census (U.S.). *See* census office, U.S.
Burroughs, William Seward, 70
Burroughs Adding Machine Company, 32, 52, 53, 70, 71, 78, 81, 98, 99, 108, 217, 359
Burroughs Corporation. *See* Burroughs Adding Machine Company

361

Cailles, Jacques, 201, 203
calculating machine(s): key office machine, 40, 70, 71, 123, 151, 191, 217; punched-card tabulator, 5, 87, 170, 174, 183, 184, 196, 235
Carmille, René, 196–98, 203, 222, 226, 227, 229, 230, 232, 233
Carroll, Fred Merchant, 112, 113
CEC. *See* International Business Machines, French company of
Census Bureau machine shop, 72–82, 85, 99, 103, 126, 354
census office, non-U.S.: Austria 18, 131, 152; France, 18, 35, 133–36, 152–54, 226, 229, 231–33; Germany, 20, 35, 135; Great Britain, 18, 20, 35, 135, 139, 141, 142, 155, 357; Norway, 132; Prussia, 18, 135; Russia, 133
census office, U.S.: discussion of permanent office, 3, 14, 15, 17–20, 35, 37; U.S. Census Bureau (Bureau of the Census), 13, 18, 35, 54–59, 65, 66, 74–82, 97, 103, 104, 108, 126, 142, 155, 253, 254, 353, 354, 356; U.S. Census Office of 1870, 20, 21; U.S. Census Office of 1880, 20–22, 28, 29, 31, 32, 253, 258, 356; U.S. Census Office of 1890, 3, 5, 11, 15, 19, 20, 22–28, 33–36, 38–42, 43, 67, 86, 139, 259, 262, 356; U.S. Census Office of 1900, 27, 51–54, 74, 356
census, population, 3, 15, 33, 35, 40, 47, 50, 52, 79, 81, 128, 153, 258. *See also* census office, non-U.S.; census office, U.S.
Chandler, Alfred D., 9, 10, 38, 102, 116, 161, 262
Chicago, 67, 110, 132, 133, 262, 358
Civil War (U.S.), 17, 38, 63, 68, 226
classi-compteur, 134, 153–55
CMC. *See* Compagnie des Machines Bull
Coase, Ronald H., 10
collator. *See* punched-card machines
Columbia College (now Columbia University), 18, 28, 96, 249, 250, 354, 359
Columbia University. *See* Columbia College

Compagnie des Machines Bull (CMB; Bull Machine Company), 194, 227, 255–57, 260, 351, 352, 355, 359; establishment of, 198–203; and French army, 196–98, 223–24; and German takeover of France, 242–43; inventions of, 204–8; sales, 263
Compagnie électro-comptable. *See* International Business Machines, French company of
Computing Scale Company, 106
Computing Tabulating Recording Company (CTR), 89, 105–7, 113, 115, 158, 159; sales, 111, 112. *See also* International Business Machines; Tabulating Machine Company
Congress, U.S., 21, 37, 56, 213, 253, 353
conscription. *See under* First World War; Second World War
Constitution: French, 193, 225; U.S., 5, 17, 19
Corps du contrôle de l'administration de l'Armée de terre, 196, 353, 357
CTR. *See* Computing Tabulating Recording Company
current account. See *Kontokorrent*

D11 tabulator. *See* Deutsche Hollerith-Maschinen Gesellschaft
Dalton Adding Machine Company, 86, 89
DeCamp and Sloan Manufacturing Company, 100. *See also* Sloan and Chance Manufacturing Company
Dehomag. *See* Deutsche Hollerith-Maschinen Gesellschaft
Demographic Service. *See* Service de la démographie
Department of Agriculture, U.S., 18
Department of Justice, U.S., 90, 116, 354
Deutschen Gesellschaft für Addier- und Sortiermaschinen. *See* Powers company, German
Deutsche Hollerith-Maschinen Gesellschaft (Dehomag): as agency of Tabulating Machine Company, 145, 150–51, 161, 162; D11 tabulator, 185, 187,

209, 235, 244, 245, 358; and German army, 234–36; machine building and production, 124, 150, 162, 163, 177, 183–85, 187, 264; and Ministry for Armaments and Munitions, 239–42, 244; operations in Germany, 182–85, 187–89, 209, 235, 236, 257, 263; operations outside of Germany, 132, 142; patent suits, 148, 149, 186, 187; sales, 147, 161, 189, 234, 263; as subsidiary, 151, 152
Doury, Elie, 194, 200, 203

Eastman Kodak, 34, 67, 159
École polytechnique (Paris), 153, 194, 196
Edison, Thomas Alva, 29, 67
Egli adding machine company (Zürich), 198–200, 204, 207, 209
Egli-Bull, 198–200. *See also* Compagnie des Machines Bull
electromechanical technology, 29–31, 51, 85, 104, 117, 153, 154, 204, 249, 250
Equitable Life Assurance Society, 218. *See also* insurance companies: U.S.

Farbenfabriken vorm. Friedrich Bayer and Co., 146, 147, 162
feed. *See* punched-card machine(s)
Felt, Dorr E., 70
Felt and Tarrant, 32, 52, 53, 70, 78, 83, 85, 86, 251
First World War, 62–65, 265; conscription, 33, 144–45; and Powers company, British, 166; and Powers company, French, 156–57; and Powers company, German, 149–52
Ford, Eugene Amzi, 47, 48, 60, 61, 66, 107, 112
France. *See* census office, non-U.S.: in France; First World War: and Powers company, French; International Business Machines, French company of; insurance companies: French; Jew(s); register(s): in France; register(s), national: in France; statistical office: French; Third Republic; Vichy France; *and individual cities, companies, and persons*

Frenchify, 194, 195, 210
Foster, Charles, 166

gang punch. *See* punched-card machine(s)
Germany. *See* census office, non-U.S.: in Germany; Germanness; insurance companies: German; Jew(s); Prussia; register(s), national: in Germany; self-sufficiency, national, Germany; statistical office: German; *and individual cities, companies, and persons*
Germanness, 188
Gestapo, 230, 232, 246, 248
Gore, John K., 43, 44, 48, 52, 53, 55, 75
Great Britain, 18, 35, 135, 136, 139, 142, 167, 355. *See also* London; Scotland, Wales; *and individual cities, companies, and persons*
Greene, Everard, 140, 141

health insurance, 165, 182
Heidinger, Willy, 146–52, 188, 189, 209, 241, 244
His Majesty's Stationery Office (HMSO), 143–45, 165, 354
HMSO. *See* His Majesty's Stationery Office
Hollerith, Herman, 3, 14, 31; business strategies of, 34–35, 41, 50, 58, 61–62, 66–68, 125, 128, 136; and companies outside of U.S., 49, 129–35, 138–40, 145–46; inventions, 39, 40, 51, 54, 59–60, 108–10; and insurance companies, 42–43; patents of, 103, 114, 128, 143–44, 159, 186; and railroads, 45–48; and sales, 41, 50, 58, 66, 67, 136, 137; as student of mechanical engineering, 26–27; and tabulator, 25–26, 29–31, 32–33, 39, 47, 59; and U.S. census office, 22–25, 28, 38, 51–57
Hughes, Thomas E., 8, 9, 12, 125, 259, 262
Hunt, William C., 23, 36, 37

IBM. *See* International Business Machines

INDEX 363

IBM card. *See* patent(s), IBM card; punched card(s): 80-column
IBM company, British. *See* British Tabulating Machine Company
IBM company, French. *See* International Business Machines, French company of
IBM company, German. *See* Deutsche Hollerith-Maschinen Gesellschaft
IBM Type 405, 239
IBM Type 600, 124, 249
IBM Type 601, 124, 249
IBM Type 603, 249
IBM Type 604, 249
Institute of Actuaries of America, 42, 43
insurance companies: British, 165, 166, 168, 169, 179, 242; French, 198, 200, 205, 260; German 147, 186; U.S., 4, 35, 38, 42–45, 52, 58, 75, 91, 99–102, 117, 118, 121, 122, 212, 224, 253, 255, 256, 359
International Business Machines (IBM): agencies of, 82, 132, 139, 140, 143, 145, 146, 148, 151, 152, 162, 176, 180, 209, 264; business strategy, 34, 95, 111, 112, 121, 212, 243, 246, 249, 251, 257, 260, 263; equipment, 81, 93, 102, 121–27, 163, 164, 172, 174, 178, 179, 183–86, 196–98, 203, 206–9, 214, 217, 220, 221, 223, 227, 235, 236, 239–41, 244, 249–51, 255–57; patents, 104; as prime mover, 105, 143; sales, 34, 123, 126, 251; subsidiaries, 152, 158, 159, 183, 187, 189, 201, 204, 210, 234, 240, 241, 244, 255. *See also* Computing Tabulating Recording Company; Tabulating Machine Company; *and individual machines and subsidiaries*
International Business Machines, French company of: Compagnie électro-comptable (CEC), 158, 195; Société française Hollerith, machines comptable et enregistreuse, 195; Société internationale de machines commerciales (SIMC), 158, 159
International Chamber of Commerce, 240, 241

International Time Recording Company, 106, 113, 158
interpreter. *See* punched-card machines

Jew(s): discrimination against in France, 225, 226, 230–32; location of, for deportation in France, 5, 222, 232, 234; location of, for deportation in Germany, 246–48
Johns Hopkins University, 18, 359

Keen, Harold Hall, 29, 55, 145, 178, 179
keyboard adding machine. *See* adding machine
key punch. *See* punch
Kontokorrent, 183, 184, 193
Knutsen, Knut Andreas, 198, 204–7, 358

Lake, Clair Dennison, 112–14, 118–21
Lanston, Tolbert, 31, 32, 39, 40, 47
Lasker, William W., 82, 85–87, 89, 91, 93
Library Bureau (Boston), 49, 50, 134
London, 41, 49, 91, 129, 130, 134, 135, 138, 141, 165, 167, 171, 176, 262, 354, 357

March, Lucian, 134, 153, 154
Maschinelles Berichtswesen, 191, 234–37, 241, 243
Massachusetts, 18, 22, 23, 28, 33, 41, 50, 61, 66, 107, 112, 354
Massachusetts Institute of Technology (MIT): and Herman Hollerith, 28; and teaching of statistics, 18;
master card (punched card), 100, 101, 217, 219
Metropolitan Life Insurance Company, 42, 91, 100–102, 118. *See also* insurance companies: U.S.
Ministry for Armaments and Munitions. *See* Reichsministerium für Bewaffnung und Munition
Ministry for Munitions (Great Britain), 145
Ministry for Trade and Industry, National. *See* Reichswirtschaftsministerium

mobilization. *See* register(s): and mobilization of troops
monopoly/monopolies, 55, 73, 74, 82, 116, 125, 254, 354
Monroe Calculating Machine Company, 217
Morland and Impey (Britain), 159, 169, 171

National Cash Register Company (NCR), 51, 98, 110–12, 116, 359; sales, 110, 111
national registers. *See* register(s)
National Statistics Service. *See* Service national des statistiques
New York Central and Hudson River Railroad, 45–50, 52–54, 58, 59, 65–67
New York City: Columbia College / Columbia University, 18, 28, 249–50; and Herman Hollerith, 27–28, 32–34, 42, 51; and John Royden Peirce, 96, 99–100, 118; and punched-card technology, 67, 91, 143
New York State, 27, 41, 70, 89, 106, 353
North, Douglass C., 10
North, Simon Newton Dexter, 55–57, 75, 81
Norway, 41, 129, 130, 132, 133, 135, 136, 152, 198, 199, 204, 262, 355
number, identifying people with: in Germany, 246; in U.S., 63, 214
numeric printing, and key office machine, 98. *See also* tabulator

office machine. *See individual machines and inventors*
old age pension, 1, 165, 211–13, 216, 251
operational statistics. *See* statistics: operational

paper strip, 29, 32
Paris, 35, 42, 49, 93, 128–30, 134, 135, 146, 148, 153, 156, 159, 194, 196–201, 204, 225, 227, 242, 243
Passow, Kurt, 239, 241–44

patent(s): automatic group control, 90, 93, 103, 104, 109, 114, 116, 117, 125, 143, 144, 191, 352; Hollerith's original punched-card, 12, 31, 56, 77, 80, 88, 103; IBM card, 93, 119, 173, 186, 187, 206, 209; key punch, 45, 75; Lanston adding machine, 32, 40; legislation on, 12, 54, 129, 130, 138, 143, 144, 234, 257, 352; litigation on, 54, 79, 114, 116, 144, 149, 159, 160, 185, 186, 206, 223, 352, 354, 355; sorter, 75, 79, 80, 88
patent infringement. *See* patent(s): litigation on
Patent Office, U.S., 12, 28, 29, 114, 124. *See also under patent(s)*
Patterson, John Henry, 111
Peirce, John Royden, 96; as challenger to Herman Hollerith, 72–73, 108; inventions of, 96–102, 104, 117–19; and patents, 114, 117–19, 121–26. *See also* Taft-Peirce Manufacturing Company
Pennsylvania Railroad Company, 45
Pétain, Philipe, 2, 225
Phillpotts, Ralegh, 139
Pidgin, Charles F., 23, 36, 37, 50, 51, 55, 56, 74
plugboard. *See* tabulator
polytechnicien, 194, 196, 202, 203
Porter, Robert Percival, 139
Powers, James, 76; business strategies of, 89, 142–43, 148; challenger to Herman Hollerith, 73, 79–80, 104; inventions of, 78–79, 82–86, 108, 109; and patents, 88, 93, 104, 109, 114–16, 148, 159, 186
Powers Accounting Machine Company, 84–91, 93, 114–16, 127, 166–72, 192–94, 208, 241, 255, 260; agencies of, 138, 143, 148–49, 160, 164, 186; sales, 88, 115. *See also* Remington Rand; *and individual agents and subsidiaries*
Powers company, British (Powers-Samas), 91, 169–75, 192, 205, 206, 208–10, 255, 264, 354; Accounting and Tabulating Machine Company

Powers company, British *(continued)* of Great Britain, 143; and First World War, 166; inventions of, 91, 169–70, 174–75; relationship with parent company, 159–60, 166–68; sales, 169, 171, 251, 263, 264; and SAMAS, 194–95

Powers company, French (Société anonyme des machines à statistique, or SAMAS), 159, 160, 167, 171–73, 175, 194, 195, 199, 202, 204, 210, 223, 224, 251, 255

Powers company, German (Deutschen Gesellschaft für Addier- und Sortiermaschinen; Powers Gesellschaft), 143, 148–50, 164, 183, 185–87, 189–92, 208–10, 243, 244, 255, 264, 355

Powers company, U.S. *See* Powers Accounting Machine Company; Remington Rand

Powers-Four, 172–74

Powers-One, 174, 175

Powers-Samas. *See* Powers company, British

Prudential Assurance Company, 143, 165–68, 208, 357. *See also* insurance companies: British

Prudential Insurance Company, 43–45, 52, 75, 91, 101, 118, 122, 253. *See also* insurance companies

Prussia, 18, 135, 136, 147, 246

public utilities, punched cards and, 4, 95, 97, 121, 142, 181, 235, 255, 266

punch, alphanumeric, 101; automatic key (Powers), 76, 78–80, 82, 87, 114; conductor's, 33; enforced key (TMC), 114, 115; gang, 112; Gore, 29, 43; Hollerith's original model, 29; key, 5, 27, 47, 48, 53, 87, 96, 97, 110, 112, 114, 115, 173; pantograph, 25, 33, 47, 77–79, 81, 110

punched card(s), organized by columns, 46, 47, 59; 20-column, 79; 53, 77–79; 21-column, 174; 24-column, 24, 47, 53, 58, 76, 79, 81, 179; 26-column, 59, 172, 174; 36-column, 47, 53, 58, 66, 174; 38-column, 179; 40-column, 174; 43-column, 96; 45-column, 47, 59, 63, 66, 81–83, 92–94, 117, 119, 120, 126, 148, 160, 172–74, 191, 192, 205–7, 209, 223, 254–56, 260; 60-column, 172; 65-column, 174, 175, 223; 80-column (IBM card), 47, 81, 93, 119, 120, 172–75, 178–80, 206–7, 209, 223, 224, 227, 256; 86-column, 117, 118, 122; 90-column, 91, 92, 94, 122, 172, 186, 192; 130-column, 223, 224; 135-column, 191

punched card(s), organized by decks: double-deck, 94, 95, 117, 118, 122, 123, 172, 186, 190, 191; triple-deck, 191

punched-card machine(s): collator, 219, 220, 251; gang punch, 112; interpreter, 94, 217; manual feed, 25, 26, 54, 77, 99; mechanized feed, 30, 53, 54, 58, 75, 77, 99, 114; reproducer, 120; speck detector, 108; verifier, 5, 85, 86, 94, 114, 115, 179, 192, 196, 242. *See also* punch; sorter; tabulator

punched-card register. *See* register(s): punched-card

punch strip, 29, 30, 32, 78, 191, 192, 259

Puteaux, France, 197

railroad(s)/railway(s), 32, 38, 44–50, 52–54, 58, 59, 62, 65–67, 117, 141, 142, 147, 159, 160, 181, 196, 200, 253, 254

Rand, James Henry, 89

record management, 4, 255, 256

register(s), 1, 10, 14, 252, 256; in France, 2–3, 123, 198, 203, 222, 224, 226, 227, 229, 232, 233, 260; in Germany, 3, 182, 212, 233, 234, 236, 238–40, 246–48; and mobilization of troops, 2, 3, 123, 222, 226, 227, 229, 230, 233; punched-card, 1, 3, 4, 182, 212, 222, 224, 226, 227, 230, 232–34, 235, 237–39, 245, 246, 248, 256, 266; in U.S., 1, 4, 5, 63, 212, 214–21, 251, 256, 257, 260

register(s), national: in Britain, 136; in France, 3, 5, 164, 222, 226, 230, 232, 233, 240; in Germany, 3, 239, 245, 246, 248

Registrar General (Britain), 135, 141, 142
Reichsministerium für Bewaffnung und Munition, 3, 62, 237
Reichswirtschaftsministerium, 187, 191, 192, 243, 244
Remington Rand (RR), 116, 122, 127, 168, 186, 190, 192, 195, 199–201, 208, 256–57, 352, 354, 359; establishment of, 89; and inventions, 172; Model 2, 94, 186; Model 3, 95, 221; and patents, 90, 116; sales, 90, 92, 171. *See also* Powers Accounting Machine Company
Remington Typewriter Company, 70, 89
Rentenanstalt Zürich (Switzerland), 205
reproducer. *See* punched-card machines
Rimailho, Emile, 202, 203
Roosevelt, Franklin D., 1, 4, 211–13, 221
Roosevelt, Theodore, 55
Royal Statistical Society, London, 41, 139
royalty, 88, 90, 115, 116, 140, 146, 149, 151, 167, 175, 176–79, 242
RR. *See* Remington Rand
Russell, Harold R., 171
Russia, 49, 63, 76, 129, 130, 132, 133, 135, 136
Russian Central Statistical Committee, 133

sales. *See individual companies*
sales analysis, 58, 67, 141, 147, 169
SAMAS. *See* Powers company, French
Sauvy, Alfred, 227
Schaaf, John Thomas, 114, 124
Schäffler, Theodor Otto Hermann, 47, 60, 126, 130, 131, 133, 162, 257
Scotland, 135, 142
Seaton, Charles W., 21, 22, 29, 32, 36
Seaton device, 20–22, 29, 32, 36
Second World War, 14, 65, 164, 212, 221; armaments, 236; conscription and mobilization of troops, 223–30, 238; and German punched-card companies, 244–45; and Jews, 230–33; managing resources during, 233–35.

See also register(s), national; Vichy France; *and individual companies and persons*
Service de la démographie, 227
Service national des statistiques, 227–29, 231, 232
Sevran-Livry, France, 197, 198
self-sufficiency, national, Germany (autarchy), 187, 189, 190, 204, 210, 236, 241, 257, 265
Siemens, 182, 189–92, 242, 355
SIMC. *See* International Business Machines, French company of
Sloan and Chance Manufacturing Company, 77. *See also* DeCamp and Sloan Manufacturing Company
SNS. *See* National Statistics Service
social construction of technology, 6–8, 257
Social Security: administration of, in U.S., 1, 95, 218, 222, 251, 256, 260, 262, 353, 356; number, in France, 228; number in U.S., 214–19, 221; pension program, 65, 212, 213, 215–17
Social Security Act, 1, 211–13, 216
Social Security Administration (U.S.), 4, 356
Social Security Board, 211, 213–17, 220, 221
Société anonyme des machines à statistique. *See* Powers company, French
Société des Papeteries Aussedat, 200–202
Société française Hollerith, machines comptables et enregistreuses. *See* International Business Machines, French company of
Société internationale de machines commerciales (SIMC). *See* International Business Machines, French company of
sorter(s): double-deck horizontal (Powers), 83; Gore, 43, 44, 46, 53, 253; hand sorting, 23, 25, 26, 42, 46, 53, 224, 258; Powers's first model, 78–80, 82; single-deck horizontal (Bull), 204; single-deck horizontal (Hollerith,

sorter(s) *(continued)*
   IBM, TMC), 44, 46, 48, 53, 54, 58, 77, 79, 80, 253; single-deck horizontal (Powers), 76, 78, 79, 87; vertical (Hollerith, Tabulating Machine Company), 61, 83, 112. *See also* sorting box
sorting box: Hollerith, 25–27, 32, 43, 44, 46, 53, 56; Pidgin, 51, 56
Speer, Albert, 237, 239–41, 245, 266
Speicer, Charles E., 76, 80
Stationery Office. *See* His Majesty's Stationery Office
statistical office(s), national: Austrian, 18, 130, 131; Belgian, 155; British, 18, 20; Dutch, 155; German, 20, 147, 180, 234, 247; French, 18, 132, 133, 134, 155–57, 226, 227, 230; Norwegian, 132; Prussian, 18, 135; Russian, 132, 133; U.S., 128, 55. *See also* army punched-card service, German; *and under* statistics
statistics: agricultural, 18, 40, 53, 59, 63, 65; business, 4, 39, 40, 71, 117, 140, 145–47, 157, 158, 159, 162, 180; general 14, 73, 103, 104, 108, 110, 125, 131, 132, 138, 152, 160, 161, 198, 252, 259, 260, 262; insurance (actuarial), 43, 45, 84, 99, 142, 147; operational, 11, 14, 38, 39, 63, 65, 142, 147, 150, 158, 161, 162, 180–82, 197, 222, 226, 249, 254, 262, 263; railroad/railway, 45, 67, 147, 165, 166, 253, 254. *See also* census office, non-U.S.; census office, U.S.; sales analysis
sterling (British currency), 139–41, 162, 167, 176, 177, 179, 195, 210
stop card, 87, 109
subsidiary/subsidiaries, 138; of International Business Machines, 124, 129, 151, 152, 158, 177, 178, 202, 208, 209, 234, 235, 241, 257, 263; of Powers (later Remington Rand), 160, 186, 187, 190, 194, 202
Surgeon General, U.S. Army, 33, 34, 128

Tabulating Machine Company (TMC): acquires patent rights, 102, 108; agencies of, 54, 132, 138, 146, 148, 161, 162, 167, 175, 180; history of, 3, 50, 61, 62, 66, 67, 105, 106, 107, 110, 113, 126, 150–52, 254, 255, 263; Hollerith's role in, 66; license to Powers, 87–88, 115, 116; John Royden Peirce and, 108, 117, 118; as prime mover in field, 103, 126; private business of, 57, 58, 116–18; products of, 86, 90, 91, 93, 99, 101, 109, 113–15, 118, 126, 148, 150, 151, 155, 163, 170; relations with Census Bureau, 54–57, 79–82; sources, 352, 354; subsidiaries of, 158, 161, 177, 263. *See also individual products and subsidiaries*
tabulator: adding, 33, 39, 40, 44–48, 52, 59, 65, 77, 79–81, 96, 142, 184; alphabet printing, 91, 92, 94, 102, 118, 121–23, 164, 166–69, 173, 186, 192, 193, 205, 208, 235, 239, 241–45, 263, 264; alphanumeric printing, 91, 94, 95, 119–21, 123, 173, 179, 196, 205, 208, 209, 221, 223, 235, 236, 238, 239, 241, 243–45, 251, 255, 256, 264; automatic group control, 93, 100, 109, 113, 114, 117, 144, 169, 170, 173; counting, 24–27, 52, 54, 59, 77, 80, 81, 142; multiplying, 124, 185; nonprinting, 25, 70, 112, 117, 123, 145, 151, 198, 204; number printing, 73, 77, 78, 80, 82–86, 90, 93, 103, 104, 108, 112–17, 126, 127, 143, 144, 147–49, 152, 160, 163, 170, 173, 174, 177, 183–85, 187, 186, 198, 204–6, 244, 255; plugboard, 47, 60, 131, 162; subtracting, 119, 120, 170, 177, 178, 184, 185, 193, 197, 207
Tabulator Limited. *See* British Tabulator Machine Company
Taft-Peirce Manufacturing Company, 48, 50, 57, 60, 66, 107
Tarrant, Robert, 70. *See also* Felt and Tarrant
Tauschek, Gustav, 124, 125, 353
Taylor, Frederick Windslow, 158
Third Republic (France), 156, 157, 226, 230

Treasury, Department of (U.S.), 18, 102, 213

verifier. *See* punched-card machine(s)
Veterans Bureau, U.S., 118. *See also* War Risk Insurance, Bureau of, U.S.
Vethe, Anders Eirikson, 205, 207
Vichy France, 2, 3, 5, 194, 222, 225–27, 229–33
Vieillard, Georges, 194, 200, 201, 203
Vienna, 126, 130–4, 162, 257, 356
Villingen (Germany), 150, 183
Vincennes (France), 159, 195, 196, 357

wages, administration of, 58, 71, 182, 186, 190, 198, 220, 224

Wales, 135, 136, 142
Wanderer-Werke, 241–43, 245
Wanderer Works. *See* Wanderer-Werke
War Risk Insurance, Bureau of, U.S., 101, 102. *See also* Veterans Bureau, U.S.
Waterbury, John Isaac, 88, 149
Watson, Thomas John, Sr., 105, 110–13, 115, 116, 119, 126, 151, 152, 201, 241
Western Electric Company, 49, 58, 67, 133
Williams, Robert Neil, 146, 148, 149
World War I. *See* First World War
World War II. *See* Second World War

JUL 10 2009

BAKER COLLEGE LIBRARY

3 3504 00521 3683

HF 5548 .H387 2009
Heide, Lars, 1950-
Punched-card systems and the
 early information

PROPERTY OF
BAKER COLLEGE OF AUBURN HILLS
LIBRARY